W0080475

Enzyme Technology

III. Rotenburg Fermentation Symposium 1982
Schloßhotel »Wilhelmshöhe« Kassel
22nd-24th September 1982

Edited by R. M. Lafferty

with the Assistance of E. Maier

in Co-operation with
B. Braun Melsungen AG, Melsungen

With 182 Figures

Springer-Verlag Berlin Heidelberg GmbH 1983

o. Prof. Dr. rer. nat. Robert MacIntyre Lafferty

Institut für Biotechnologie,
Mikrobiologie und Abfalltechnologie
Technische Universität Graz
A-8010 Graz, Österreich

Dr. Erika Maier

Wissenschaftl. Oberrat am
Institut für Biotechnologie,
Mikrobiologie und Abfalltechnologie
Technische Universität Graz
A-8010 Graz, Österreich

ISBN 978-3-540-12479-5   ISBN 978-3-642-69148-5 (eBook)
DOI 10.1007/978-3-642-69148-5

Library of Congress Cataloging in Publication Data
Rotenburger Symposium (3rd: 1982: Kassel, Germany)
    Enzyme Technology.
1. Enzymes Industrial Applications Congresses.
I. Lafferty, R. M. (Robert MacIntyre) II. Maier, E. III. Title
TP248.E5R67 1982 660.63 83-10245

© Springer-Verlag Berlin Heidelberg 1983
Originally published by Springer-Verlag Berlin Heidelberg

2127/3020-543210

Preface

The main subject of the "III. Rotenburger Fermentation Symposium" is enzyme technology. Enzyme technology could be simply defined as the scientific study of proteinaceous catalysts derived from living organisms and the application of the knowledge to solve specific problems. The scope of the application of enzyme technology ranges from medical to industrial uses and in the future even living organisms as a source of enzymes may be replaced by fully synthetic enzymes — "synzymes".

Although enzyme technology still remains a particular field of biotechnology, the extremely rapid rate of expansion and the enormous increase in the diversification of all aspects of enzyme technology during the immediate past has created a certain tendency to separate biotechnology and enzyme technology from each other. Certainly, those areas of biotechnology characterized by astounding advances are enzyme technology, bioreactor development and genetic manipulation as related to biotechnological processes. However, a glance at many of the common problems of biotechnology and enzyme technology such as diffusion barriers, reactor design, mass transport, substrate or product inhibition phenomena and the effect of physical-chemical parameters on process kinetics reveals that these two fields are inseparable with respect to research and application.

Enzyme technology is a relatively young branch of biotechnology. The discovery by Eduard Buchner of the first cell-free enzyme "zymase" catalyzing the alcoholic fermentation in 1897 opened the era of modern biochemistry and enzyme technology. We have experienced astonishing advances since then with respect to the increased use of enzymes for a diversity of purposes such as for medical diagnosis, in the food and textile industries, for animal husbandry, for washing powders, and in the production of pharmaceutical products to mention only a few.

However, proteinaceous catalysts differ vastly from inorganic catalysts with respect to stability and long term use. Therefore, the development of enzyme technology has been characterized by the constant attempt to improve enzyme stability, using a variety of substances. Chemical modification of enzymes such as the use of cross-linking reactions is still another method to stabilize enzymes.

Immobilization of enzymes in or on a matrix has allowed us to vastly increase the stability of enzymes with respect to temperature, pH, solvents, salts and proteolytic effects. The application of immobilized whole cells is the most recent advance pertaining to the optimization of enzyme activity or stability. Both immobilized enzymes and whole cells create the possibility of utilizing enzyme reactors in continuous processes which offer all the inherent advantages of better process control and improved overall economics, especially for large-scale technical processes.

The spectrum of individual topics pertaining to the field of enzyme technology is itself extremely broad and it has not been an easy task for the organizers to select those topics for presentation that are timely and capable of being adequately treated within a limited period of time. The choice of the symposium topics was therefore based on the attempt to achieve a balance between basic scientific research in the field of enzyme technology and the practical application of the scientific results present. At the same time future possibilities and developments of enzyme technology ought to be sufficiently considered.

I should like to very heartily thank all participants for their efforts in attending this symposium. Special thanks are due to all the authors who presented their contributions to the "Rotenburger Symposium" 1982.

Furthermore, I should like to express my own personal thanks to the B. Braun Melsungen company whose generous financial support and excellent organization has made this symposium possible. Acknowledgement are also especially due to Dr. Erika Maier, Mrs. M. Angersbach and to Mr. G. Kothe and Mr. J. Grebe for their unfailing personal efforts for having made this symposium possible.

Graz/Austria, September 1982.   R. M. Lafferty

Because of the international origin of the manuscripts, more attention was given by the editor to the technical merit of the manuscripts than to the correct use of the English language.

Table of Contents

List of Contributors

O. Andresen
Novo Industri A/S, 2880 Bagsvaerd, Denmark
H. Aurich
Physiologisch-chemisches Institut, Martin-Luther-Universität, Hollystraße 1, PSF 184, 4020 Halle (Saale), German Democratic Republic
R. Bergmann
Physiologisch-chemisches Institut, Martin-Luther-Universität, Hollystraße 1, PSF 184, 4020 Halle (Saale), German Democratic Republic
K. Buchholz
Zucker-Institut, Langer Kamp 5, 3300 Braunschweig, Federal Republic of Germany
D. J. Dalietos
Cetus Corporation, 600 Bancroft Way, Berkeley, CA 94710, USA
B. Danielsson
Pure & Applied Biochemistry, University of Lund, Box 740, 220 07 Lund, Sweden
S. K. DeWitt
Cetus Corporation, 600 Bancroft Way, Berkeley, CA 94710, USA
L. Edebo
Department of Clinical Bacteriology, University of Göteborg, 413 46 Göteborg, Sweden
N. Esaki
Institute for Chemical Research, Kyoto University, Uji, Kyoto-Fu 611, Japan
R. K. Finn
School of Chemical Engineering, Cornell University, Ithaca, N.Y. 14853, USA
E. Fiolitakis
Institut für Biotechnologie der Kernforschungsanlage Jülich, Postfach 19 13, 5170 Jülich, Federal Republic of Germany
E. Flaschel
Swiss Federal Institute of Technology, Institute of Chemical Engineering, EPFL-Ecublens, 1015 Lausanne, Switzerland
W. M. Fogarty
Department of Industrial Microbiology, University College, Dublin 4, Ireland
K. Forss
The Finnish Pulp and Paper Research Institute, P.O.Box 136, 00101 Helsinki, Finland
B. Galunsky
Institute of Organic Chemistry, Bulgarian Academy of Sciences, Sofia, Bulgaria
J. Geigert
Cetus Corporation, 600 Bancroft Way, Berkeley, CA 94710, USA
W. Goebel
Institut für Genetik und Mikrobiologie, Universität Würzburg, Röntgenring 11, 8700 Würzburg, Federal Republic of Germany

W. Hartmeier
Institut für Mikrobiologie, Rheinisch-Westfälische Technische Hochschule, 5100 Aachen, Federal Republic of Germany
U. Haufler
Fachbereich Biologie, NW II, Universität of Bremen, 2800 Bremen 33, Federal Republic of Germany

L. Hepner
L. Hepner and Associates, Tavistock House North, Tavistock Square, London, W.C.1, Great Britain
H.-P. Hohn
Institut für Biotechnologie der Kernforschungsanlage Jülich, Postfach 19 13, 5170 Jülich, Federal Republic of Germany
H. Hustedt
Gesellschaft für Biotechnologische Forschung mbH, Mascheroder Weg 1, 3300 Braunschweig-Stückheim, Federal Republic of Germany
V. Kasche
Fachbereich Biologie, NW II, Universität Bremen, 2800 Bremen 33, Federal Republic of Germany
C. T. Kelly
Department of Industrial Microbiology, University College, Dublin 4, Ireland
J. Klein
Institut für Chemische Technologie, TU Braunschweig, Hans-Sommer-Straße 10, 3300 Braunschweig, Federal Republic of Germany
A. M. Klibanov
Laboratory of Applied Biochemistry, Department of Nutrition and Food Science, Massachusetts Institute of Technology, Cambridge, MA 02139, U.S.A.
J. Konecny
Pharmaceutical Division, CIBA-GEIGY Ltd., 4002 Basel, Switzerland
J. Kreft
Institut für Genetik und Mikrobiologie, Universität Würzburg, Röntgenring 11, 8700 Würzburg, Federal Republic of Germany
K. H. Kroner
Gesellschaft für Biotechnologische Forschung mbH, Mascheroder Weg 1, 3300 Braunschweig-Stöckheim, Federal Republic of Germany
M.-R. Kula
Gesellschaft für Biotechnologische Forschung mbH, Mascheroder Weg 1, 3300 Braunschweig-Stückheim, Federal Republic of Germany
J. Lasch
Physiologisch-chemisches Institut, Martin-Luther-Universität, Hollystraße 1, PSF 184, 4020 Halle (Saale), German Democratic Republic
R. Lehmann
Abteilung für Biotechnologie, Henkel KGaA. Henkelstraße 67, 4000 Düsseldorf, Federal Republic of Germany
C. Male
L. Hepner & Associates, Tavistock House North, Tavistock Square, London, W.C.1, Great Britain
S. L. Neidleman
Cetus Corporation, 600 Bancroft Way, Berkeley, CA 97710, USA
H. Pfeiffer
Abteilung für Biotechnologie, Henkel KGaA, Henkelstraße 67, 4000 Düsseldorf, Federal Republic of Germany
P. B. Poulsen
Novo Industri A/S, 2880 Bagsvaerd, Denmark
E. Raetz
Swiss Federal Institute of Technology, Institute of Chemical Engineering, EPFL-Ecublens, 1015 Lausanne, Switzerland
A. Renken
Swiss Federal Institute of Technology, Institute of Chemical Engineering, EPFL-Ecublens, 1015 Lausanne, Switzerland
H. Sahm
Institut für Biotechnologie der Kernforschungsanlage Jülich, Postfach 19 13, 5170 Jülich, Federal Republic of Germany
S. Sawada
Kyoto University of Education, Fushimi-Ku, Kyoto 612, Japan

J. Schindler
Abteilung für Biotechnologie, Henkel KGaA, Henkelstraße 67, 4000 Düsseldorf, Federal Republic of Germany

R. Schmid
Abteilung für Biotechnologie, Henkel KGaA, Henkelstraße 67, 4000 Düsseldorf 1, Federal Republic of Germany

P. F. Schubert
Department of Chemical Engineering, Polytechnic Institute of New York, Brooklyn, NY 11201, USA

H. Schütte
Gesellschaft für Biotechnologische Forschung mbH, Mascheroder Weg 1, 3300 Braunschweig-Stückheim, Federal Republic of Germany

K. Soda
Institute for Chemical Research, Kyoto University, Uji, Kyoto-Fu 611, Japan

H. Sorger
Physiologisch-chemisches Institut, Martin-Luther-Universität, Hollystraße 1, PSF 184, 4020 Halle (Saale), German Democratic Republic

H. Tanaka
Institute for Chemical Research, Kyoto University, Uji, Kyoto-Fu 611, Japan

E. J. Vandamme
Laboratory of General and Industrial Microbiology, Faculty of Agricultural Sciences, University of Ghent, Coupure Links, 653, 9000 Ghent, Belgium

K.-D. Vorlop
Institute of Chemical Technology, TU Braunschweig, Hans-Sommer-Straße 10, 3300 Braunschweig, Federal Republic of Germany

C. Wandrey
Institut für Biotechnologie der Kernforschungsanlage Jülich, Postfach 19 13, 5170 Jülich, Federal Republic of Germany

R. Zöllner
Fachbereich Biologie, Universität Bremen, NWII, 2800 Bremen 33, Federal Republic of Germany

Opening Lectures

Enzyme Technology

R. M. Lafferty

Honoured guests, Ladies and Gentlemen,

I should very much like to welcome you to the third "Rotenburger Symposium" taking place this time in Kassel. Furthermore, I should also hope that this will be an opportunity for you to exchange ideas and information on enzyme technology and that you will profit as much as possible from your stay here during this relatively short period of time.

As in the past, I feel that it is necessary to once again emphasize the aim of this series of symposia. The basic idea is to enable a relatively small group of experts who are involved in a particular area of biotechnology to meet and intensively handle or discuss one or few subjects pertaining to the current and future development of biotechnology.

One important aspect of these symposia is the furthering of contacts between scientists from universities and research institutes and those persons engaged in the industrial application of biotechnological innovations. It is most probably superfluous to explain to you at the present time that biotechnology is one particular type of technology with extremely strong connections to both basic research and practical applications on an industrial basis. It is also more than superfluous to tell you that biotechnology is a science characterized by a rate of expansion that is on the one hand a cause for enthusiasm and on the other hand a cause for worry since too many false hopes are being created.

At the moment we are experiencing a "bio-boom" and one reason for this "bio-boom" is the enormously broad spectrum of possibilities to produce all biological compounds — and others as well — using the methods of bio-technology and the capacity of industrial plants engaged in fermentation.

This expansion has received an enormous impetus from the recognition of the fact that the use of recombinant DNA or protoplast fusion for genetic mani-pulations on microorganisms has created a realm of possibilities hitherto un-suspected.

It has been said or prophesized that the "atomic age" may be surpassed in its long-term effects by the forthcoming "biological age".

The reasons for the "bio-boom" are few and all are connected with the crises facing mankind today. These are: the food crisis, the energy crisis, the environ-mental crises, and to a certain extent the raw material crisis. One expects biotech-nology to solve many of these crises which could be an inherent danger for the sensible development and the trust or faith in biotechnology in the immediate future.

The subject of this symposium is enzyme technology which is one of the most rapidly expanding areas of biotechnology. Other equally rapid expanding areas of biotechnology are bioreactor development and of course "genetic engineering". One cannot separate one field completely from the other since each is intimately connected with the others. In the case of enzyme technology, the aim is the economical production of particular substances and market forces will determine whether or not the process survives beyond the laboratory stage.

Enzyme technology, somewhat simply defined, is the study and the application of proteinaceous catalysts to either produce or alter known biological compounds. It could be expanded to encompass uses for organic synthesis, that is, to obtain substances not necessarily the product of a living organism.

Enzyme technology is a part of biotechnology and today the fine difference between "technology", the scientific study, and "technique", the application of scientific results has almost disappeared, thus we use the word "technology" interchangeably. However, enzyme technology is confronted with many typical problems encountered in fermentation technology. Some of these are, to mention only a few, the problem of mass transfer, diffusion barriers, bioreactor design, substrate and product inhibition phenomena, and the pronounced effects of physical-chemical parameters on the overall kinetics of the process. And in the end it is often the kinetics of the process which will determine the economic feasability of the process using either soluble or immobilized enzyme systems.

The inherent advantage of employing isolated enzyme systems is the separation of complicated growth processes from the catalytic effect of a particular protein. Processes to separate, recover and purify the enzyme in question usually involve a significant loss of the activity of the enzyme. Therefore, much of the historical development of enzyme technology is connected with improving isolation methods to avoid loss of catalytic activity during the various steps involved.

Once isolated, methods had to be developed to retain the activity over longer periods of time. This involved the use of low temperatures and in many cases the use of almost exotic substances as additives or protective agents. Many of these are well known to you all, for instance, the use of metallic ions, polymers, sugars, sugar alcohols, organic compounds or even the corresponding substrate or coenzyme itself to avoid activity losses.

Since enzymes have an extremely high specificity in comparison to inorganic catalysts this automatically means that enzymes are in general complex unstable substances.

This is especially true with respect to the negative effects of physical-chemical parameters such as temperature, pH, ionic strengths and substrate concentrations. These negative characteristics are somewhat balanced out by separating the enzyme from the growing cell. The isolated enzyme system is no longer subject to the extreme complexity of intracellular regulatory mechanisms. For instance, the use of the extensive measuring and control systems in a bio-technological process involving whole cells in a fermenter is nothing less than the attempt to circumvent the intracellular regulatory mechanisms by extracellular controls, usually by variation of physical-chemical parameters during growth and excretion of products. The use of isolated enzyme systems should, therefore, simplify the whole situation if only enzyme stability were no problem.

The culmination of the history of attempts to improve enzyme stability is the use of immobilization techniques. The further development of this trend is the use of immobilized whole cells thus avoiding negative effects arising from the disruption of the physical structure of intracellular enzymes and from the purification and fixation steps necessary with isolated enzymes. A still further development is the combination of both immobilized systems, i.e. immobilized enzymes coupled with immobilized cells in a single bioreactor to obtain a particular product. One example of this is the work of Hägerdal and Mosbach from Sweden. They combined, or co-immobilized, β-glucosidase and yeast cells to first degrade cellobiose and to then convert the glucose so obtained to ethanol.

In the past, plants and animal cells were the primary sources of soluble enzymes. It was a considerable advance when the use of surface cultures of fungi was introduced to produce an enzyme. One of the very first enzyme production processes was "Takadiastase" from *Aspergillus niger*. "Takadiastase" is a multi-enzyme complex used for many years as a digestive aid. The process itself — *Aspergillus niger* growing on bran — was patented in 1894 by Takamine in Japan. However, surface cultures are not always easy to control nor easy to use on a large scale. The breakthrough in the industrial production of enzymes on a large scale was the use of submerged cultures of fungi in a fermenter to specifically produce one enzyme under exactly controlled conditions. "Controlled conditions" meant that larger volumina could be used, infections by unwanted organisms eliminated and optimal physical-chemical conditions maintained. The necessary separation of the biomass — or cells — from the bulk of the liquid led to the development of one of the typical "unit operations" of biotechnology — the separation of cells by continuously operating filters or centrifuges.

Following separation of the cells, and if the enzyme was not an extracellular type, it was necessary to disrupt the cells and to recover, isolate and purify the desired enzyme. Again these steps are an intimate part of the complex development of both biochemistry and enzyme technology. The methods to isolate specific enzymes are well known and vary from simple precipitation reactions with ammonium sulfate to newer, more sophisticated methods such as membrane separation processes involving dialysis, ultra-filtration, the use of molecular sieves, electro-dialysis, electrophoresis, pH-focussing and affinity chromatography to mention only a few. The extreme variety of methods to separate a particular enzyme from a mixture of proteins is one of the most extensive chapters of enzyme technology.

Today, the combination of recombinant DNA technologies and the use of improved fermentation processes offering vastly improved possibilities of process controll has opened completely new opportunities for the expansion of enzyme technology.

The products of enzyme technology find widespread use in the following fields: in the food and the pharmaceutical industry, for biomedical purposes, i. e. for medical diagnosis or clinical analysis, and for therapeutic uses. In addition, the use of enzymes is widespread for technical and industrial analysis as well as for technical and industrial purposes.

Neither time nor space allow one to go into detail regarding these well known and possible new applications deriving from enzyme technology. However, if the activity of research groups and that of the "enzyme industry" is a reliable barometer, then it would appear that enzyme technology is at present in a rapid state of expansion, which applies to all areas of biotechnology as well.

Quite recently it was pointed out that the shares of companies involved in biotechnology have fallen less severely on the stock market than the average industrial company share.

One might emphasize this positive tendency by looking at the development of the industrial use of one enzyme, that of immobilized glucose isomerase. Almost unknown at the beginning of the 1970's, the industrial use of glucose isomerase to produce "high-fructose corn syrup" has expanded to the extent that the carbohydrate market in the USA has changed drastically. Three million tons of "high-fructose corn syrup" are being produced in the USA annually and there is little doubt regarding expansion of the market for this specific product. In comparison to this "bulk product", the sales of digestive enzymes in the USA in 1981 still amounted to over 17 million dollars market value.

During this symposium you will have an opportunity to hear a number of contributions extensively dealing with many aspects of enzyme technology which I have only touched on. These involve enzyme production, cell disintegration, enzyme recovery and purification, the application of soluble enzymes, methods of enzyme and cell immobilization, the development of enzyme reactors, process kinetics, and the situation with respect to industrial enzymes at the present time.

It is to be hoped that these presentations will stimulate an active discussion and further the exchange of ideas during personal talks between all participants. I should like to thank you all in advance for your efforts and wish you a most enjoyable and profitable stay here in Kassel during the third "Rotenburger Symposium".

Graz/Austria, September 1982.

Industrial Enzymes — Present Status and Opportunities

Leo Hepner and Celia Male

Industrial enzymes are enzymes used for catalysing large-scale industrial reactions. They are applied at present mainly in the food and related industries.

Speciality enzymes on the other hand are used in small volume in applications such as diagnostic aids, clinical analysis and research work. Whereas industrial enzymes are usually not highly purified and relatively cheap [the cost per kilogram could be $10], speciality enzymes tend to be highly purified and very expensive. There are many hundreds of enzymes which have been identified in plants, animals, fungi and bacteria. Of these, dozens have speciality application. In our detailed survey on industrial enzymes [1], we identified only 16 enzymes [2] with markets of over $ 0.5 m worldwide.

In the early 1970's, it was projected that immobilised enzymes would supersede soluble enzymes as industrial catalysts. However, of the 16 enzymes identified, only one, glucose isomerase is in immobilised form. Two or three other immobilised enzymes at present either have limited application or are captively produced and hence do not have a commercial market. Enzymes which fall into this category include penicillin acylase, amino acid acylases, and lactase.

Historically, the first enzymes were derived from animal tissues, [i.e. pancreatic protease, calf rennet] or from fungal/plant sources [i.e. malting enzymes, papain].

In the 1950's, surface culture of fungi was used to produce industrial enzymes on a large scale. By the 1960's, submerged fermentation methods had been developed for the large scale production of enzymes from species such as Bacillus and Aspergillus.

In 1982, three categories of industrial enzymes can be identified, including proteolytic, amylolytic and "other". This last category includes glucose isomerase, pectinase, glucose oxidase, and invertase.

The proteolytic enzymes account for over 50% of the market, amylolytic enzymes for 33% and others for the balance. The sales of the 16 individual industrial enzymes may range from $ 0.5 m for a newly introduced enzyme like lactase to over $ 50 m for α-amylase or bacterial protease.

The main industrial sectors in which enzymes are used are detergents, starch and sweeteners, dairy and cheese, brewing, fruit juice and wine, ethanol [both potable and industrial] confectionery, baking, leather, paper and textiles.

The enzymic step is not as well defined as a chemical reaction and there is still a lot of "art" and "know-how" involved in these industries.

In 1980, the total market for industrial enzymes was in the region of $ 300 m and this is projected to grow to $ 500 m by 1985. The US and W. European will each account for a 40–45% share and the rest of the world for the balance.

The are only 20–30 primary producers of enzymes in US and W. Europe, of which nine major companies control 90% of the market, and 15–20 companies the remaining 10%. Almost two thirds of the companies are situated in W. Europe.

For most companies, enzymes represent a very minor part of their chemical/ pharmaceutical business. The three notable exceptions are situated in W. Europe [Novo, Gist-Brocades, Hansen] and enzymes form at least one third of their total turnover.

Over the next few years, the major growth in the enzyme area will be in the increasing markets for all 16 enzymes either in their existing applications or in modified or new processes.

There will be significant growth in amylolytic enzymes and glucose isomerase.

There will be increasing use of enzymes in fine chemicals, pharmaceuticals, flavours, biotransformations and optical resolutions etc., but many of the enzymes for these specialised applications may be produced by companies for their own captive use.

Future success in the enzyme area will lie with companies who understand the industries in which enzymes are used, their technical/process problems, legislation and especially the latest trends in the food and pharmaceutical industries.

References

1. Industrial Enzymes, Market Opportunities, US & W. Europe, 1980–1985, L. Hepner and Associates.
2. Chemical & Engineering News, Volume 59 [1981], page 37.

Parameters Involved in Heterogeneous Biocatalysis

K. Buchholz

Research in the field of enzyme technology still is growing, and more than a decade after the establishment of immobilized enzymes in industrial processes work on engineering aspects is continuing for further improvement of their performance (Lilly, 1978).

Basic aspects

Insight into the fundamental processes underlying immobilization contribute to optimization approaches. *Adsorption phenomena* have been known for long time to interfere with enzymatic activity and to lead to inactivation (c.f. Miller, 1971). It is advisable to apply appropriate derivatization of carrier surfaces in order to reduce strong adsorption and to restrict multiple interaction with surface functional groups (Buchholz et al. 1979a). Inactivation can thus be minimized and adsorption made reversible (Fig. 1).

These aspects are also most important for chromatography (notably affinity chromatography) of enzymes on solid supports (c.f. Borchert et al. 1982). Ancient adsorption materials have found renewed interest, like bentonite and other minerals. They exhibit very high adsorption capacity (up to 2 g protein per g dry carrier) as well as reversibility without significant inactivation, excellent conditions for concentration and purification of enzymes. Also for basic research these systems might be stimulating since non ideal adsorption isotherms with a maximum in adsorption capacity have been obtained waiting for explanation in terms of specific interaction mechanisms (Armstrong and Chesters, 1964; Rodriguez et al., 1977). Obviously the matrix of SiO_4-tetraeder layers can interact in a flexible way with proteins and adjust to their molecular dimensions by expansion of the layer distance. Adsorption kinetics are governed by pore diffusion of proteins, which in turn depends on adsorption phenomena, exhibiting less hindrance e.g. with derivatized silica carriers and reversible adsorption. Systems with pore sizes in the range of 20 to 100 μm can in general be described with diffusion coefficients near those for free solution when adsorption is fully reversible. The rates of covalent immobilization reactions in many cases are rather quick at room temperature (appropriate conversion in the range of 5–20 minutes for coupling with glutardialdehyde or carbodiimide derivatives) (Borchert et al., 1979). They allow for kinetic control of immobilization and specific catalyst design. The role of carrier activation, density of functional groups and reaction conditions have been correlated with overall catalyst activity and immobilization yield (Schlünsen et al., 1979; Monsan, 1977 and 1978).

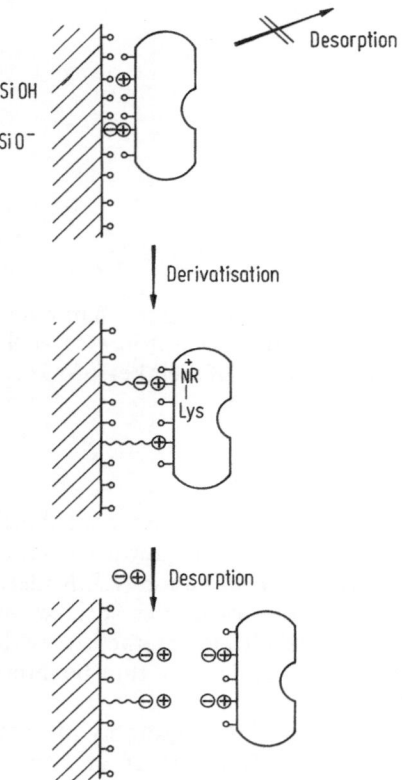

Figure 1. Scheme of restricted enzyme desorption from a silica surface, and of reversible adsorption/ desorption after modification of the surface

Further details of mechanisms in covalent immobilization have been revealed, e.g. the occurance of thermal stabilization by formation of multiple bonds (Koch-Schmidt and Mosbach, 1977).

Advanced knowledge on immobilization allows for optimized *catalyst design*. By means of kinetic control during immobilization enzymes can be immobilized preferentially near the external carrier surface, where they are accessible through short diffusion paths and thus exhibit higher effectiveness under conditions of mass transfer limitation of the catalytic reaction (c.f. discussion of catalyst effectiveness below). Improved effectiveness has been obtained experimentally

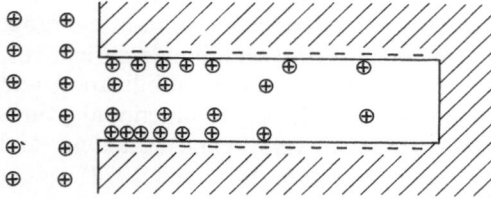

Figure 2a. Schematic illustration of enzyme penetration into a carrier pore, initially resulting in non uniform distribution through the carrier

S · 10^4 (mol/l)   E A · 10^4

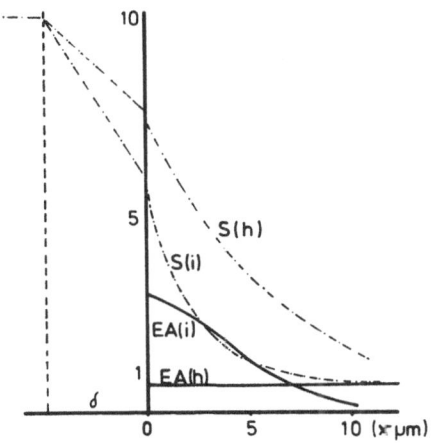

Figure 2b. Calculated profiles of absorbed enzyme density as a function of the distance (x) from the external carrier surface, EA (i) for non uniform, EA (h) for uniform distribution of the same amount of enzyme inside the porous carrier. Calculated profiles for the respective substrate concentrations (S (i) and (h), respectively) indicate a steeper profile, that is a faster overall reaction rate, for non uniform enzyme distribution

(Borchert et al., 1979; Carleysmith et al., 1979 and 1980) and discussed theoretically (Buchholz 1979). Fig. 2 shows profiles of immobilized enzyme for uniform and non uniform distribution and substrate profiles inside the carrier, where the steeper profile of S(i) for non uniform distribution of enzyme signifies a higher reaction rate and effectiveness.

Characterization of immobilized biocatalysts

The transfer of laboratory results to technical development and application is a most difficult step and it depends much on appropriate information concerning the properties, merits and shortcomings of materials and procedures. All data relevant for application of a biocatalyst must be available. The complexity makes it difficult in many cases to find a selection of parameters which meets these requirements. Two approaches have been presented (c.f. Buchholz et al., 1979 b and Poulson et al., 1981), the last giving minimum requirements for characterization of an immobilized biocatalyst. Table 1 presents a rather concise set of parameters considered as most important. They must be determined experimentally and given with all physical and chemical parameters involved. Short comments on them follow.

A *general description* should indicate the enzyme and/or microorganism involved, the carrier type and method of immobilization, the respective reaction scheme and conditions. Organic carriers have a more or less high water content which should be determined under application conditions (pH, temperature). It allows to calculate the catalyst activity inside the actual reaction volume. High values for water uptake (> 5 l/kg dry weight) in general indicate limited mechanical stability in fixed beds, whereas most macroporous polymeres exhibit less water uptake and mechanical rigidity (Manecke et al., 1979; Krämer, 1979).

Table 1. Relevant parameters affecting the performance of heterogeneous immobilized biocatalysts

	Physical and chemical parameters involved[a]
Swelling behaviour of the carrier	pH, I
Mean wet particle diameter (d_p) (distribution, shape)	pH, I
Particle compression behaviour,	Δp
flow resistance in fixed beds,	Δp, h, u, v, d_p
abrasion in stirred vessels	d_i, n
fluidization velocity	d_p, u, $\Delta \varrho$, v
Maximum activity, or initial reaction rates (V_{max}, v)	S, P, t, T, pH, I and buffer conc., d_p, u or d_i, n
Effectiveness η as a function of external mass transfer,	d_p, v, u, or d_i, n
pore diffusion, partition effects, degree of conversion	d_p, V_{max}, K_M, T, D_e, S, P
Operational stability, depending on abrasion, enzyme inactivation, fouling, irreversible adsorption, occlusion etc.	t, S, P, T, pH, conc. of other compounds

[a] Symbols are given at the end of the article

The *mean wet particle diameter* d_p and its distribution is most important, since it determines the pressure drop in fixed beds, the sedimentation velocity in fluidized beds and since it has a significant influence on the catalyst effectiveness. Equation (1) shows the correlation for the *pressure drop* Δp in a fixed bed under laminar flow conditions and when the particles are not compressible.

$$\Delta p = 150 \frac{(1 - \varepsilon)^2}{\varepsilon^3} \cdot \eta \frac{u \cdot h}{d_p^2} \tag{1}$$

The influence of the void volume ε, which depends on the particle size distribution, is obvious from this correlation. If the particles are not mechanically stable, the pressure drop is even more sensitive to d_p, as can be seen from Fig. 3, and occlusion may be a consequence (Buchholz et al., 1978).

Experimental procedures and results giving e.g. the influence of particle diameter, matrix material, cell loading (Klein et al., 1979a) and time of operation (Norsker et al., 1979) have been published. For catalyst application in fluidized beds and stirred tanks it is important to investigate *abrasion*. Minimum dimensions for stirrer and stirring rates (e.g. $d_i = 5$ cm, n = 500 min^{-1}) should be considered (c.f. Fig. 4). Cell loading, shear rate and volume fraction exhibit distinct influence on abrasion (Klein et al., 1979b).

It is also advisable to test the chemical carrier stability under operation conditions. Thus glass or silica matrices exhibit distinct hydrolysis rates at pH > 8. They can be stabilized by surface modification (Krämer et al., 1979).

The *activity* of a biocatalyst represents a key parameter, it should be determined from measurements of initial reaction rates under optimal and/or operational

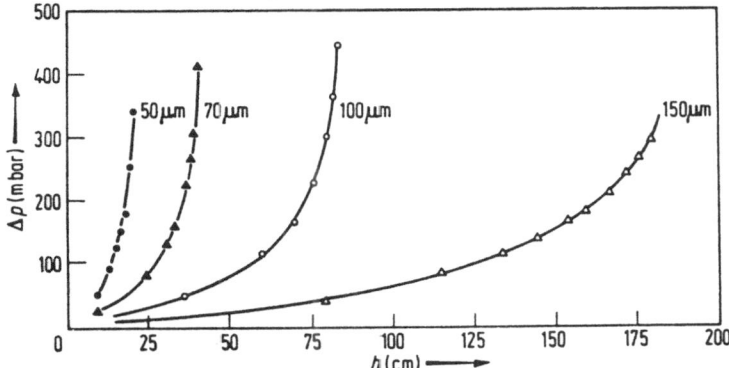

Figure 3. Pressure drop as a function of bed height in a fixed bed reactor with compressible or deformable particles of different diameters; model calculation

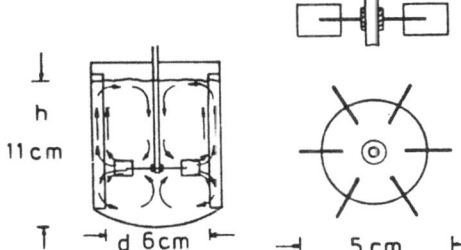

Figure 4. Test vessel and stirrer (internal vessel diameter: 7 cm (baffles each 0.5 cm))

conditions and given together with all relevant parameters (c.f. table 1). It should be based on dry weight of catalyst. The water uptake of the carrier should be stated so that activity per volume can be calculated. The activity allows for the determination of the immobilization yield which is important with respect to the economics of a process. In this respect also measurements of overall conversion (up to 90% of equilibrium) are relevant since they show the catalyst efficiency with decreasing substrate and increasing product concentrations.

In many cases the rate of the catalytic reaction in heterogenous systems is limited by mass transfer of substrates and products. The substrate first must be transported by convection and diffusion through a stagnant liquid layer on the external carrier surface (δ in Figure 2). The rate of this step depends on the flow rate in tubular reactors or on the stirrer dimensions and stirring rate in the respective reactors. Quantitative correlations for the mass transfer coefficient or Sherwood number which describe these effects, are available from the chemical engineering literature (see also Buchholz 1982a). The influence of external mass transfer may be significant with low substrate concentrations or those substrates which exhibit low solubility (e.g. oxygen).

Diffusion inside a porous catalyst is a more important factor in most cases, notably with
— high catalyst activity,
— low substrate concentrations,
— inhibiting products, like protons in hydrolytic reactions, and
— high molecular weight substrates.
Buffers facilitate proton transport and thus may have a significant influence on the overall reaction rate. Its temperature dependence can be an indication for diffusion limitation when it is below that for the native enzyme since diffusion rates depend much less on temperature as compared to (bio)chemical reactions (c.f. Buchholz 1982a).

The *effectiveness* of a catalyst relates the overall reaction rate for a given immobilized enzyme or cell system to that of the same amount of native bio-catalyst under otherwise identical conditions (concentrations etc., see Table 1):

$$\eta = V_{(imm)}/V_{(native)} = f(Sh, \varphi, S_i) \, . \tag{2}$$

It depends most significantly on the Sherwood number (Sh) which characterizes the external mass transfer, and the Thiele modulus (φ) and substrate concentrations (S_i) which characterize the coupled pore diffusion and reaction kinetics. It can be determined in a simple way if the amount of active immobilized enzyme inside the carrier is known. Notably values for η at different substrate concentrations are of interest when referring to application conditions. That substrate concentration which corresponds to half the maximum reaction rate has found special consideration ($K_{M \, (app)}$). The evaluation of $K_{M(app)}$ as a result of linearized plots based on Michaelis kinetics (Lineweaver-Burk or Eadie-Hofstee) are misleading in general, since any interpretation of heterogeneous systems by homogeneous kinetics may lead to erroneous results. Much more useful are plots of conversion versus time, when the experimental details are given completely.

The Thiele modulus summarizes those catalyst parameters which determine the influence of diffusion in the overall reaction:

$$\varphi = R(k_1/D_e)^{1/2} \, , \qquad \text{(first order kinetics)}$$
$$\varphi = R(V_{max}/K_M D_e)^{1/2} \qquad \text{(Michaelis kinetics)} \, .$$

A high Thiele modulus means in most cases severe diffusion limitation and low catalyst effectiveness. Thus for first order kinetics and a highly active catalyst a limiting and approximate solution of the mass balances involved gives:

$$\eta \simeq 1/\varphi \simeq (1/R) \cdot (k_1/D_e)^{-1/2} \, . \tag{3}$$

The catalyst performance in such a case decreases in linear correlation with the inverse particle radius R.

For Michaelis kinetics correlations for η, φ and S have been published manyfold in form of graphics or tables (Goldstein, 1976; Kasche, 1979). The calculation or estimation of φ allows for a rapid information on the catalyst effectiveness η which is to be expected under certain operational conditions, notably substrate concentrations. For practical considerations the substrate conversion range should also be taken into account, since its concentration decreases with reaction time (in batch operation) or with residence time (in a tubular reactor). For such cases an operational effectiveness factor has been calculated (Fig. 5) (Kasche, 1979).

Bioconversion systems may be more complex. Thus two substrates are required in oxidation reactions, and either of both may be limiting by slow diffusion inside the carrier when available only at low concentrations. Since oxygen has a very slow

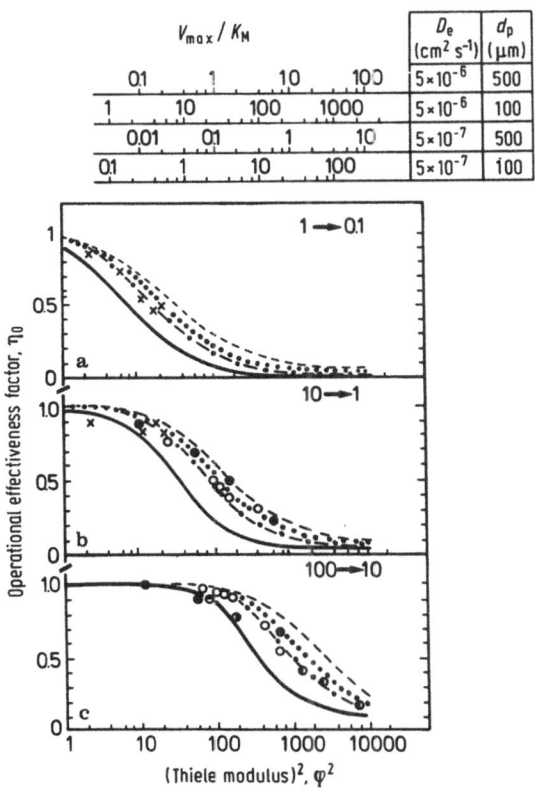

Figure 5. Calculated and experimental data for the operation effectiveness factor (η_0) for 90% substrate conversion, with a one substrate enzyme immobilized in spherical particles, as a function of the square of the Thiele modulus (proportional to the enzyme content), Sherwood number and initial substrate concentration, given in K_M-units, in the upper right hand corner. The upper scales give V_{max}/K_M for different particle dimensions (d_p) and substrate diffusion coefficients (D_e). The corresponding Thiele modulus is given by the lower scale. Curves: calculated data for — — — — Sh′ = 100; .... Sh′ = 16; —·—·—· Sh′ = 8; ———— Sh′ = 2 Experimental data:
  α-chymotrypsin bound to Sepharose 4B;
o trypsin bound to Sepharose 4B;
● trypsin bound to Sepharose CL 2B;
◓ trypsin bound to isothiocyanatostyrol acrylic acid matrix;
◑ trypsin bound to porous glass Servachrom G 550;
◒ trypsin bound to Oxiran 5120 B.
  (Kasche 1979)

solubility, this in many cases will be the limiting parameter. Figure 6 illustrates glucose oxidation by glucose oxidase with different levels of glucose concentration and oxygen saturation of the free solution. Even at moderate glucose concentrations (right), oxygen is consumed readily near the external catalyst surface, which means low catalyst effectiveness ($\eta \approx 0,1$ in this example) (Reuss et al., 1979).

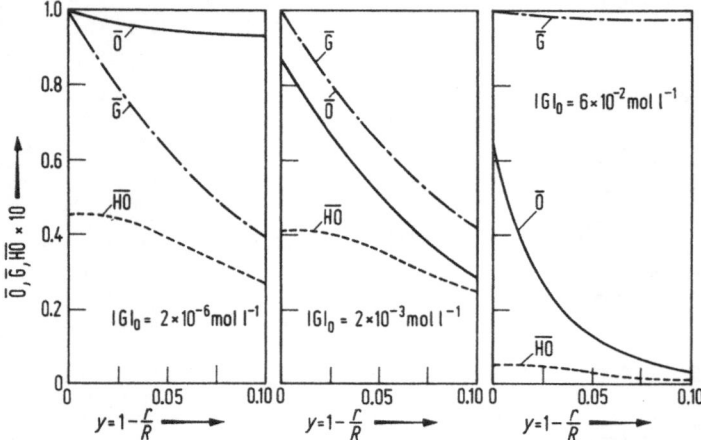

Figure 6. Calculated profiles of substrates for the glucose oxidation inside a catalyst particle, normalized with respect to solution concentrations ($\bar{G}$: Glucose, $\bar{O}$: Oxygen; H_2O_2 as a product is also given: $\overline{HO}$. Abscissa: radial coordinate, up to 1/10 of particle radius (O: particle surface). Three examples for different glucose concentrations in solution at oxygen saturation are shown (Reuss et al. 1979)

Bioconversions in many cases require more than one enzyme and also coenzymes which must be regenerated for reasons of economy. One example is coimmobilized glucose oxidase and catalase. The latter enzyme decomposes H_2O_2 formed in the oxidation reaction. It delivers $1/2$ O_2 back into the reaction cycle and thus improves the efficiency of the system. It also protects, for a limited time, both enzymes from inactivation by the peroxide. In such a system of coupled enzymatic reactions the ratio of the enzyme activities will exhibit an optimum with respect to the efficiency of the overall system (Reuss et al., 1979) and immobilization strategies can be designed with respect to such phenomena (Fernandes, 1975).

Stability is a key parameter for economic application of immobilized biocatalysts. Stability under storage conditions is important when the biocatalyst has to be shipped and/or stored for supplementation in industrial plants located in distant regions which is common in the agrofood sector. Operational stability depends on the many parameters governing the reaction in a technical unit: pH, temperature, chemical compounds (also poisoning or toxic, like traces of metals), enzyme leakage, microbial contamination, fouling, clogging. Since so many diverse factors are involved, only the experimental determination of activity as a function of (operation) time leads to reliable results. Extrapolation does not provide meaningful results. Slow initial inactivation may be due to diffusion limitation of the catalytic reaction and increased rates of inactivation can be found with progressing operation time (Klein et al., 1981).

Productivity quantifies the potential of a biocatalyst for industrial application: the amount of product formed per amount of catalyst during the overall operation time. It is advisable to also state the product concentration, the overall operation time and final catalyst activity with this parameter. For some processes figures are known, e.g. 100–250 kg product per kg catalyst for penicillin acylase, 1000–2000 for glucose isomerase (Poulson, 1981 b) and 650 (milk) or 3500 (whey) for β-galactosidase (Plainer et al., 1981).

Reactors

Stirred tank reactors in batch operation and *fixed beds* in continuous operation have been applied most frequently with industrial processes (Lilly, 1978). Stirred tanks are convenient e.g. when continuous neutralization or intensified mass transfer (e.g. oxygen supply) is necessary. Batch operation provides for better catalyst efficiency, since initially it can operate at high substrate concentration, and only with progressing conversion the effectiveness drops with decreasing substrate concentration and increasing product concentration. In many cases, such as penicillin hydrolysis, product(s) are inhibitors.

For a homogeneous system and Michaelis kinetics, the conversion $(S_0 - S_E)$ is obtained from the integrated Eq. (4) (it is easier to calculate the required residence time for a given conversion):

$$V_{max} \cdot \tau = S_0 - S_E + K_M \cdot \ln (S_0/S_E) \; . \qquad (4)$$

In general continuous processing is preferred in industrial application. However a continuous stirred tank reactor works at unfavourable substrate concentrations (exit concentration S_E):

$$(S_0 - S_E)/\tau = \frac{V_{max} \cdot S_E}{K_M + S_E} \qquad (5)$$

Thus, more enzyme or a longer residence time is required for a given conversion, higher by a factor of about 1,5 or 2,3 at 80% and 95% conversion, respectively (when $S_0 = 10K_M$) (see graph for different examples, Lilly, 1978). This effect becomes even more prominent when product is inhibiting and in immobilized systems when mass transfer limitation is significant (Figure 5 then applies to a batch or a continuous plug flow reactor).

Continuous plug flow systems offer the advantages of easy operation and high efficiency. However, problems of precipitate, pressure drop and clogging may prevent its application, and coenzyme regeneration may require approaches other than those based on porous carriers, e.g. membrane reactors (Wandrey et al., 1980). Figure 7 shows reactor principles which have been applied to bioconversion.

An example of optimization of a *membrane reactor* concerns lactose hydrolysis with soluble enzyme (Flaschel, this volume). Continuous reaction in a single stage

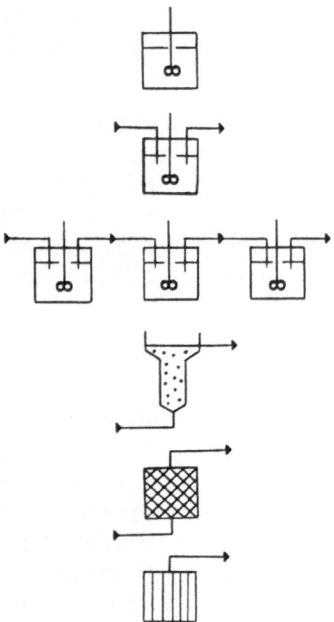

Figure 7. Reactor configurations: stirred tank, batch, and continuous, respectively cascade of stirred tanks, tubular fluidized bed and fixed bed, membrane reactor

(stirred vessel) reactor implies complete back mixing and rather low catalyst effectiveness. An elegant solution of this problem is to combine a tubular reactor with a membrane reactor where products leave the system, and recycling, at low rate, the enzyme of the tubular reactor. The catalyst efficiency thus can be higher by factors of 2 to 5 (or even more as compared to a CSTR), depending on the degree of conversion.

Tubular reactors with particulate (carrier bound) biocatalysts can also be operated as *fluidized bed* reactors in order to avoid problems of pressure drop or clogging. Backmixing, which in technical reactors would be very high, can be prevented by adequate installations. Flaschel et al. (1982) showed that available

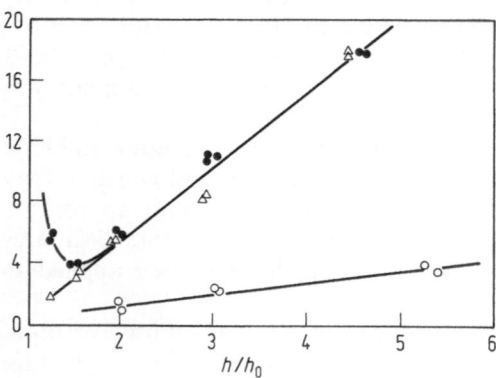

Figure 8. Equivalent stirred vessels (ordinate) in a fluidized bed reactor (○ empty tube), and with internals (● SMX, △ SMV, from Sulzer, CH). Intervals provide for much lower backmixing, notably with increasing bed expansion (h/h_0)

internals can provide much higher densities of biocatalysts (which may exhibit low settling rates) inside the reactor and reduce backmixing considerably (Fig. 8) which effect can be expressed by the rather high number of equivalent stirred vessels in a cascade. Fluidized bed reactors have been applied e.g. to glucose isomerization (Oestergaard et al., 1976).

It should furthermore be mentioned that the application of soluble enzymes which is inevitable with solid substrates such as cellulose can profitably make use of immobilization principles. Thus it is advantageous to have a longer residence time of the enzyme as compared to the fluid phase carrying the product in order to achieve an improved utilization. A simple principle works like a *reversible immobilization* when enzyme(s) is or are reversibly adsorbed on the solid substrate which is transported through the reactor counter current to the fluid. This results in a longer residence time of enzyme(s) in the reactor (Fig. 9) (Buchholz et al., 1982a).

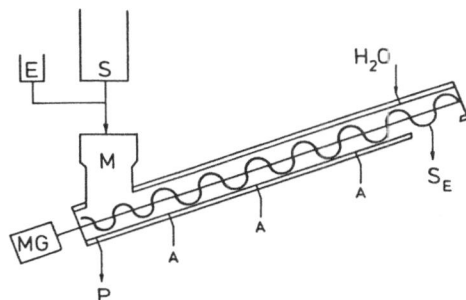

Figure 9. Continuous tubular reactor for cellulose hydrolysis, enzyme (E) and substrate (S) are mixed (M) and fed to the reactor, transport of the solid (S) is provided by a screw driven by a motor and gear (MG), probes for analysis can be taken at (A), product solution (P) and residual cellulose (S_E) are withdrawn

Conclusion

Immobilized biocatalysts are established in industrial processes, notably in food and pharmaceuticals production. A large range of research and development activities has widened the scope of potential application as well as contributed to improve on-stream processes. Nevertheless, many phenomena involved remain unknown and a deeper insight into these phenomena could lead to solution of existing problems (Buchholz et al., 1982b). Inactivation remains one of the bottle necks both for application of soluble and immobilized enzymes. Few is known on a quantitative basis concerning the stabilizing factors of native, thermophilic proteins. Obviously ionic interaction with metal ions and glycosidic residues play a major role, exhibiting tremendous stabilization effects. Stabilization of detoxification systems, cell organelles such as chloroplasts could possibly aid progress from a better understanding of inactivation and stabilization mechanisms. Application and stabilization of enzymes in non-aqueous systems is another topic of high interest. Genetic engineering offers a new pool of methods such as rapid analytical approaches for the investigation of structure — functional relationships.

With immobilized cells there are likewise several most important phenomena which are not understood. Obviously during immobilization, changes in cell walls may occur which affect the permeability. The generation time of some microorganisms can be reduced upon immobilization and even synchronous cell division has been observed. Basically, the effects of immobilization on cell physiology and regulation are not understood especially those phenomena concerned with secondary metabolites. The same is true with respect to stabilization where it remains unknown why several cell systems are more stable with respect to production in the immobilized state as compared to the free state. Progress in understanding of the inactivation mechanisms dependent on time and/or conversion would be highly important. It relates to productivity as one of the key factors in industrial application. These examples show that many fields are open for continuing research which could improve established and favour new applications.

Symbols

CSTR		continuous stirred tank reactor
D	$cm^2 s^{-1}$	diffusion coefficient
D_e	$cm^2 s^{-1}$	effective diffusion coefficient
d	cm, μm	diameter
E		enzyme
h	cm	height
I		ionic strength
K_M	$mol\, l^{-1}$	Michaelis constant
k_1	s^{-1}	kinetic rate constant (first order)
k_s	$cm\, s^{-1}$	mass transfer coefficient
L		length
n	s^{-1}	impeller rotation rate
Δp	Pa, bar	pressure drop
R	cm	radius
Re	$ud_p v^{-1}$	Reynolds number
Re_i	$nd_i^2 v^{-1}$	impeller Reynolds number
r		radial coordinate
S	$mol\, l^{-1}$	substrate concentration
Sc	vD^{-1}	Schmidt number
Sh	$k_s d_p D^{-1}$	Sherwood number
T	°C	temperature
t	s, min	time
u	$cm\, s^{-1}$	flow rate (superficial velocity)
V	cm^3	volume
V_{max}	$mol\, l^{-1} s^{-1}$	maximum reaction rate
x		local coordinate
Greek Symbols:		
δ		thickness of diffusion boundary layer
ε		voidage
η	$g\, cm^{-1} s^{-1}$	dynamic viscosity and effectiveness (dimensionless)
φ		Thiele modulus
ϱ	$g\, cm^{-3}$	density

References

Armstrong DE, Chesters G (1964) J. Soil Sci, vol 98:39
Borchert A, Buchholz K (1979) Biotechnol Lett, vol 1:15–20
Borchert A, Larsson PO, Mosbach K (1982) J Chromat, vol 244:49–56
Buchholz K (1979) Biotechnol Lett, vol 1:451–456
Buchholz K (1982a) Adv. Biochem Eng, vol 24:39–71
Buchholz K, Gödelmann B (1978) Enzyme Eng, vol 4:89–92
Buchholz K, Duggal SK, Borchert A (1979a) Enzyme Eng, vol 5:465–468
Buchholz K et al (1979b) Characterization of Immobilized Biocatalysts, Dechema Monograph, vol 84:1–43
Buchholz K, Borchert A (1982a) Biokonversion, Statusseminar, KFA, Jülich FRG, in press
Buchholz K, Klein J (1982b) Biotechnology in Europe, keynotes FAST-Biosociety workshop, Dechema, Frankfurt, 45–53
Carleysmith SW, Lilly MD (1979) Biotechnol Bioeng, vol 21:1057
Carleysmith SW, Dunnil P, Lilly MD (1980) Biotechnol Bioeng, vol 22:735
Fernandes PM et al (1979) Chem tech, July, 438
Flaschel E, Fauquex PF, Renken A (1982) Chem Ing Tech, vol 54:54–56
Goldstein L (1976) Methods in Enzymology, vol 44:397
Kasche V (1979) Characterization of Immobilized Biocatalysts, Dechema-Monograph, vol 84:224–241
Koch-Schmidt AC. Mosbach K (1977) Biochem, vol 16:2105–2109
Klein J, Eng H (1979b) Dechema-Monograph, vol 84:292–299
Klein J, Washausen P, Kluge M (1979a) loc cit, 277–291
Klein J, Vorlop KD, Eng H (1981) Poster 2nd Eur Congr Biotechnol, Eastbourne, GB
Krämer DM (1979) Dechema Monograph, Characterization of Immobilized Biocatalysts, vol 84:88–91
Krämer DM, Borchert A (1979) loc cit, 105–109
Lilly MD (1978), Biotechnology, Dechema-Monograph, vol 82:165–180
Manecke G, Ehrenthal E, Schlünsen J (1979) Dechema-Monograph, Characterization of Immobilized Biocatalysts, vol 84:73–87
Miller JR (1971) Progr Surface Membrane Sci, vol 4:299
Monsan P (1977/78) J Mol Catalysis, vol 3:371–384
Monsan P (1978) J Appl Microbiol Biotechnol, vol 5:1–11
Norsker O, Gibson K, Zittan L (1979) Stärke, vol 31:13–16
Oestergaard J, Knudsen S (1976) Stärke, vol 10:350
Plainer H, Sprössler GB (1981) Poster, 2nd Eur Congr Biotechnol, Eastbourne, GB
Poulson PB (1981a) Working Party on Immobilized Biocatalysts, European Federation of Biotechnology, Poster, 2nd Eur Congr Biotechnol, Eastbourne, GB
Poulson PB (1981b) Enz Microb Technol, vol 3:271
Reuss M, Buchholz K (1979) Biotechnol Bioeng, vol 21:2061
Rodriguez JLP, Weiss A, Lagaly G (1977) Clays and Clay Minerals, vol 25:243–251
Schlünsen J, Ehrenthal E, Manecke G, Dechema-Monograph, Characterization of Immobilized Biocatalysts, vol 84:145–168
Wandrey C, Wichmann R, Bueckmann AF, Kula MR (1980) Enzyme Engg, vol 5:453–465

Unusual Catalytic Properties of Usual Enzymes

Alexander M. Klibanov

Summary

The major objective of this paper is to stress that enzymes, in addition to their "normal" reactions (reflected in their names), often can catalyze other, sometimes quite different processes. The latter, although unimportant for the enzyme-producing organism, can be very valuable for biotechnological applications. The considered examples of "unnatural" catalytic activities of common enzymes, taken from the author's recent studies include: (I) reduction of aromatic compounds catalyzed by glucose oxidase; (II) prochiral and enantiomeric stereo-specificity of galactose oxidase in the oxidation of non-sugar, three carbon alcohols; (III) selective hydroxylation of aromatic compounds catalyzed by peroxidase; (IV) geometrically and positionally specific oxidation of aromatic aldehydes catalyzed by xanthine oxidase.

It is pointed out that along with conventional bacterial screening, a "chemical screening", i.e., the search for new, unconventional catalytic activities for known enzymes can be beneficial.

I. Introduction

More than 2,000 different enzymes have been described in the literature [1]. Out of these, about 200 are commercially available (most of them in miligram or gram quantities), and out of these two hundred only 16 are available in industrial amounts (kilograms) [2].

It is quite clear from the aforementioned data that if one is to employ an enzyme as a catalyst in a practical process, e.g. in preparative organic synthesis, it would be highly desirable if this enzyme is one of the sixteen or at least one of the two hundred. This is not only because in such a case the enzyme can simply be purchased from a commercial supplier as opposed to its laborious and time-consuming fermentative (or, even worse, animal or plant) production with no guarantee that the sought after enzyme will be found and will meet the requirements of the future process. The very fact that a given enzyme is commercially available usually implies that it is sufficiently active, stable and easy to handle.

The only objection this rather obvious reasoning is likely to raise is the following: What if none of the commercially available enzymes is capable of catalyzing the reaction one is interested in? It would seem then that a search for a new enzyme becomes inevitable.

At this point it appears necessary to make a general comment concerning the substrate specificity of enzymes. There is a vast misunderstanding among researchers, especially non-enzymologists with respect to what enzymes can and cannot do. This confusion is often caused by taking the names of enzymes literally and in the absolute sense. For example, if an enzyme is called ascorbate oxidase, this simply means that it can oxidize ascorbic acid with O_2 and that this reaction is probably a physiological function of this enzyme. The above name certainly *does not* mean that this enzyme cannot catalyze other reactions.

Despite the obviousness of this logic, it is usually not implemented in biotechnological strategy. Suppose one aims at finding an enzyme that can oxidize phenol(s). It is highly unlikely that one will test ascorbate oxidase to achieve this goal. The latter sounds almost ridiculous: to use ascorbate oxidase for oxidation of phenol(s). However, it turns out that actually this might not be such a bad idea, since, in fact, ascorbate oxidase does catalyze the oxidation of a number of different phenols [3].

The major trust of this paper is to promote biotechnological applications of such "unnatural", non-physiological, catalytic properties of enzymes. Using examples from our recent studies, it will be demonstrated that these properties, although perhaps unimportant for the organism that produced the corresponding enzyme, can be very valuable in biotechnology, in particular in preparative organic syntheses.

II. Glucose Oxidase for the Preparative Reduction of Aromatic Compounds

Glucose oxidase (EC 1.1.3.4) catalyzes the oxidation of β-D-glucose with molecular oxygen to produce D-gluconolactone (which is subsequently non-enzymatically hydrolyzed to D-gluconic acid) and H_2O_2 (Fig. 1) [4, 5]. This enzyme is currently manufactured in commercial quantities [2] because of its extensive use in the food industry [6].

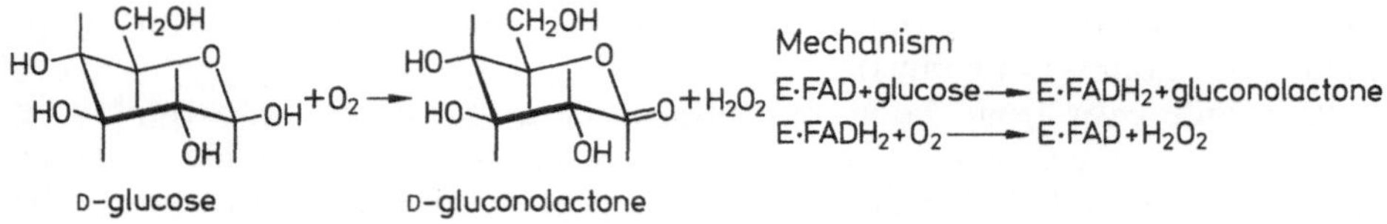

Figure 1. The reaction catalyzed and mechanism of glucose oxidase

From the viewpoint of a chemist, neither of the products of the glucose oxidase-catalyzed reaction (gluconic acid and hydrogen peroxide) is a coveted target and hence it would seem that glucose oxidase has a very limited, if any, potential for the chemical industry. This feeling persists if one reads text books on enzymology [5, 7], all of which assert that glucose oxidase is an extremely specific enzyme. This specificity is the reason for the wide use glucose oxidase has found in clinical and chemical analyses for selective determination of glucose in the presence of other compounds [8, 9].

These statements concerning the specificity, although correct, are somewhat misleading. Glucose oxidase, while indeed very specific with respect to the electron donor (D-glucose), is *not* specific at all with respect to the electron acceptor (O_2). For instance, molecular oxygen in this reaction can be replaced by some artificial dyes, in particular indophenols [10].

Glucose oxidase from *Aspergillus niger* has a high catalytic activity, is fairly stable (at least in the absence of H_2O_2) and inexpensive. The reducing substrate, D-glucose, is manufactured on a large scale from starch and is cheap. Therefore it occured to us that if O_2 can be replaced by another electron acceptor, the reduction of which will result in a commercially important product, such a process can be of practical interest.

We have illustrated this approach by the glucose oxidase-catalyzed production of hydroquinone (which is used in industry as a photographic developer and as an inhibitor of autoxidation and polymerization) from benzoquinone (Fig. 2) [11]. It has been found that (I.) the rate of the glucose oxidase-catalyzed oxidation of D-glucose under anaerobic conditions in the presence of 10 mM benzoquinone as an electron acceptor is 3 times as high as that under air in the absence of benzoquinone (the "normal" reaction, see Fig. 1). That is, benzoquinone is a very effective replacement for O_2 in the glucose oxidase catalysis; (II.) one molecule of hydroquinone is produced per one molecule of glucose consumed, i.e., the enzyme catalyzes a stoichiometric reduction of benzoquinone by glucose; (III.) the yield of hydroquinone in the enzymatic reaction (Fig. 2) is nearly 100%.

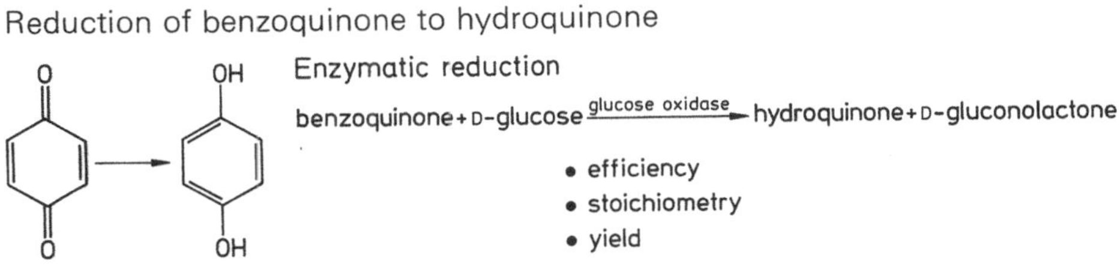

Reduction of benzoquinone to hydroquinone

Enzymatic reduction

benzoquinone + D-glucose $\xrightarrow{\text{glucose oxidase}}$ hydroquinone + D-gluconolactone

- efficiency
- stoichiometry
- yield

Figure 2. Glucose oxidase-catalyzed conversion of benzoquinone to hydroquinone

To make the production of hydroquinone preparative, we covalently attached glucose oxidase to alumina and packed a small column with the immobilized enzyme. A solution containing 0.1 M glucose and 0.1 M benzoquinone was passed through the column under anaerobic conditions. The immobilized glucose oxidase column was operated in this fashion for two weeks with no measurable decline in its catalytic efficiency and produced more than a hundred grams of hydroquinone [11].

It should be stressed that the enzymatic reduction of benzoquinone is just one example of the aforementioned approach. A number of substituted quinones (many of which, such as vitamin K derivatives, have biomedical applications) can replace O_2 in the glucose oxidase-catalyzed reaction, as well as various aromatic nitro compounds, e.g. nitrobenzene which is reduced to nitrosobenzene (H. H. Freedman, personal communication).

III. Galactose Oxidase for the Stereospecific Oxidation of Aliphatic Alcohols

Galactose oxidase (EC 1.1.3.9) is even more striking than glucose oxidase in terms of the diversity of the reactions catalyzed. This enzyme is not specific with respect to either electron acceptor (O_2) or electron donor (D-galactose).

In its "normal" reaction, *Dactilium dendroides* galactose oxidase oxidizes D-galactose with molecular oxygen to D-galacto-hexodialdose and H_2O_2 (Fig. 3) [12]. Again, as in the case of glucose oxidase, this normal reaction is of no interest from the standpoint of the chemical industry.

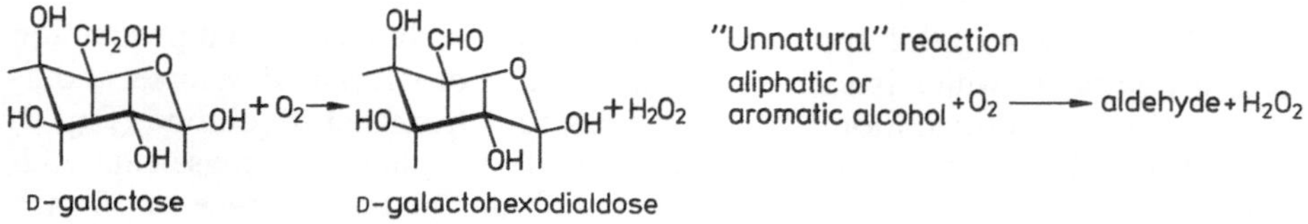

"Unnatural" reaction

aliphatic or aromatic alcohol $+ O_2 \longrightarrow$ aldehyde $+ H_2O_2$

D-galactose   D-galactohexodialdose

Examples of alcohols oxidized by galactose oxidase

1,3-propanediol	2-methylene-1,3-propanediol
glycerol	hydroxypyruvate
hydroxyacetone	salicyl alcohol
dihydroxyacetone	3,4-dimethoxybenzyl alcohol
glycolaldehyde	hydroxyacetophenone

Figure 3. "Normal" and "unnatural" reactions catalyzed by galactose oxidase

Fortunately, in addition to D-galactose, galactose oxidase can also oxidize a number of non-sugar, aliphatic and aromatic alcohols (some of which are listed in Fig. 3) to the corresponding aldehydes [12]. This reaction, although not breathtaking, might be useful because many alcohol-oxidizing chemical catalysts carry out the oxidation all the way to the acids instead of stopping at the aldehydes.

The "unnatural" reaction shown in Fig. 3 would really be attractive if galactose oxidase catalyzed the oxidation of alcohols *stereospecifically*, since nearly all conventional chemical catalysts fail to do that and there is a great demand for optically active compounds in the speciality chemicals area. Bearing this in mind, we have examined stereospecificity of the enzyme in the oxidation of simple three-carbon alcohols [13].

The simplest alcohol substrate of galactose oxidase possessing a prochiral carbon is glycerol. For this reason, glycerol was chosen to reveal whether the enzyme will oxidize it to glyceraldehyde asymmetrically. A mixture containing 2 M glycerol, galactose oxidase and catalase (to decompose H_2O_2 formed in the reaction) was incubated at 4 °C under air for 3 weeks after which time 105 mM glyceraldehyde was produced. Measurement of its optical rotation indicated that it was optically pure S(−)glyceraldehyde (Fig. 4). That is, galactose oxidase exhibits absolute *prochiral* specificity.

In order to elucidate the *enantiomeric* stereospecificity of the enzyme, the enzyme-catalyzed oxidation of 3-chloro-1,2-propanediol and 3-bromo-1,2-pro-

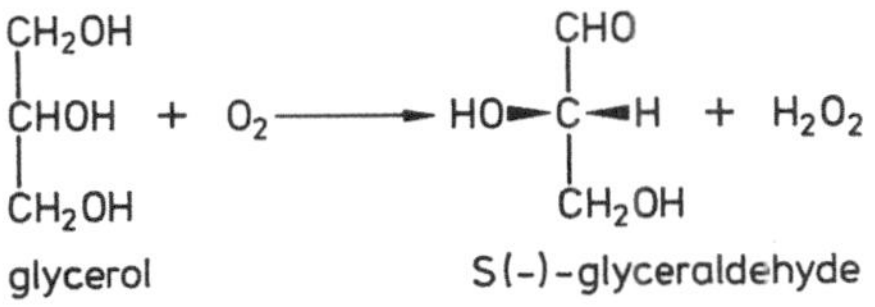

$$CH_2OH \quad\quad\quad CHO$$
$$|\quad\quad\quad\quad\quad |$$
$$CHOH + O_2 \longrightarrow HO\!\!-\!\!C\!\!-\!\!H + H_2O_2$$
$$|\quad\quad\quad\quad\quad |$$
$$CH_2OH \quad\quad\quad CH_2OH$$
glycerol   S(-)-glyceraldehyde

Figure 4. Galactose oxidase-catalyzed oxidation of glycerol to S(—) [L(—)] glyceraldehyde

panediol was studied. In contrast to glycerol, the C2 atom in these two alcohols is chiral. The enzymatic reaction conditions were the same as for glycerol. On the basis of both the experimental data obtained and conformational analysis, it was concluded that galactose oxidase reacts only with the R isomer of 3-halo-*1*,2-propanediol (Fig. 5).

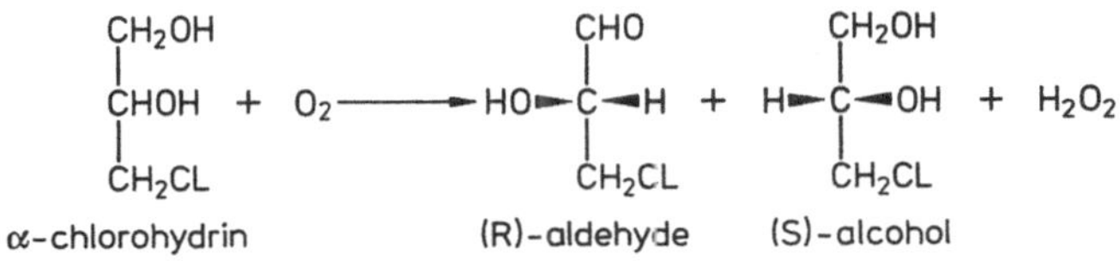

$$CH_2OH \quad\quad\quad CHO \quad\quad CH_2OH$$
$$|\quad\quad\quad\quad\quad |\quad\quad\quad |$$
$$CHOH + O_2 \longrightarrow HO\!\!-\!\!C\!\!-\!\!H + H\!\!-\!\!C\!\!-\!\!OH + H_2O_2$$
$$|\quad\quad\quad\quad\quad |\quad\quad\quad |$$
$$CH_2CL \quad\quad\quad CH_2CL \quad CH_2CL$$
α-chlorohydrin   (R)-aldehyde   (S)-alcohol

Figure 5. Galactose oxidase-catalyzed oxidation of 1-chloro-2,3-propanediol to the corresponding aldehyde. Only the R alcohol reacts with the enzyme

Hence, due to the dissymmetric environment of its active center, the enzyme stereoselectively oxidizes even its "unnatural" substrates.

IV. Peroxidase for the Selective Hydroxylation of Aromatic Compounds

Horseradish peroxidase (EC 1.11.1.7) catalyzes oxidation of various phenols and aromatic amines with H_2O_2 to form polyaromatic products and water [14]. This enzyme is widely used in biotechnology, in particular in enzyme immuno-assays and in clinical analyses (where it is coupled to a variety of other oxidases which form H_2O_2 as a product thereby affording easy colorimetric detection). Recently, we have developed a process using peroxidase for the removal of phenols, aromatic amines and other toxic organics from industrial aqueous effluents [15, 16].

In the late fifties, Mason and coworkers discovered that in addition to peroxidation of phenols and aromatic amines, horseradish peroxidase can also catalyze hydroxylation of aromatic compounds in the presence of dihydroxy-fumaric acid and molecular oxygen (e.g. see Fig. 6) [17]. Unfortunately, the yields of enzymatic hydroxylation were rather low and, even more importantly, several different products were formed in each reaction. This lack of specificity led Mason's group [17, 18] and others [19] to the conclusion that the peroxidase-catalyzed hydroxylation proceeds via a free radical mechanism which is responsible for multiple products.

We were not convinced by the above conclusions and re-examined the reaction in Fig. 6, initially using L-tyrosine as a model. In agreement with the

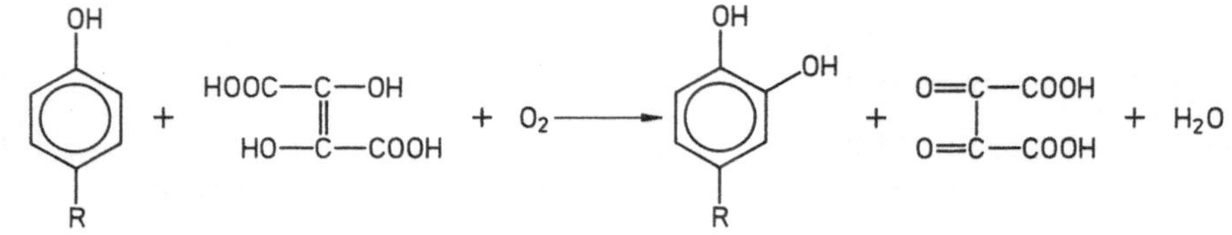

Figure 6. Peroxidase-catalyzed hydroxylation of *p*-substituted phenols in the presence of dihydroxy-fumaric acid and molecular oxygen

literature results, a yield not exceeding 50% was obtained and several different products were detected. However, we also found that even in the abscence of peroxidase, the hydroxylation takes place and its rate is comparable with that of the enzymatic reaction.

It occurred to us that maybe this non-enzymatic hydroxylation is responsible for the lack of selectivity of the overall process. Therefore, we undertook the search for conditions to eliminate the non-enzymatic hydroxylation by O_2 and dihydroxyfumaric acid.

We discovered [20] that a simple reduction in temperature from 25 °C to 0 °C nearly completely stops the non-enzymatic hydroxylation, while the rate of the enzymatic reaction is affected only slightly. Peroxidase-catalyzed hydroxylation of L-tyrosine at 0 °C (Fig. 7) was facile and specific: 70% degree of conversion was achieved after 3 hrs with L-3,4-dihydroxyphenylalanine (L-DOPA) being the *only* product.

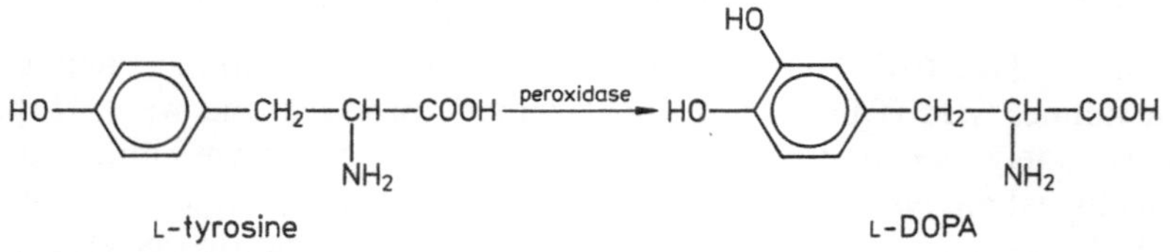

Figure 7. Peroxidase-catalyzed conversion of L-tyrosine to L-DOPA

To assess the generality of the peroxidase hydroxylation, we carried out the reaction in Fig. 6 with D(—)-4-hydroxyphenylglycine (an important intermediate in the synthesis of semisynthetic antibiotics). Again, after 3 hr reaction at 0 °C, up to a 70% degree of conversion was obtained with D(—)-3,4-dihydroxy-phenylglycine being the only product (Fig. 8).

In both of the above examples (Figs. 7 and 8), peroxidase catalyzed the hydroxylation of *para* substituted aromatic compounds at the *meta* position. It was of obvious interest to find out the direction of enzymatic hydroxylation in the case of *meta* substituted aromatic compounds. To address this problem, we studied peroxidase-catalyzed hydroxylation of L-phenylephrine (Fig. 9).

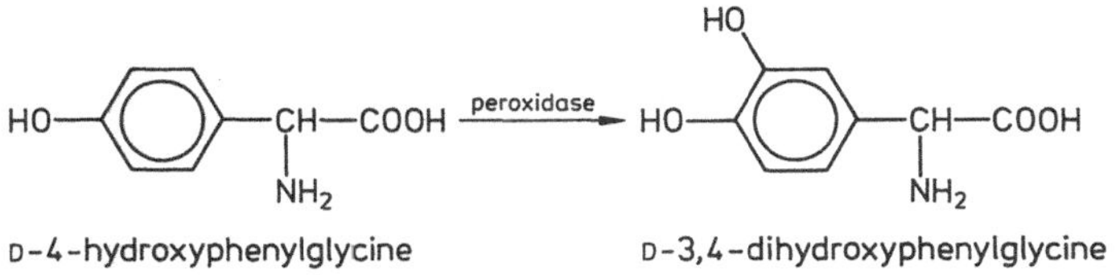

D-4-hydroxyphenylglycine   D-3,4-dihydroxyphenylglycine

Figure 8. Peroxidase-catalyzed conversion of D-4-hydroxyphenylglycine to D-3,4-dihydroxyphenyl-glycine

Ninety minutes after the beginning of the reaction, about 50% of the substrate was converted to a single product, L-epinephrine (Fig. 9). That is, the exclusive hydroxylation took place at the *para* position.

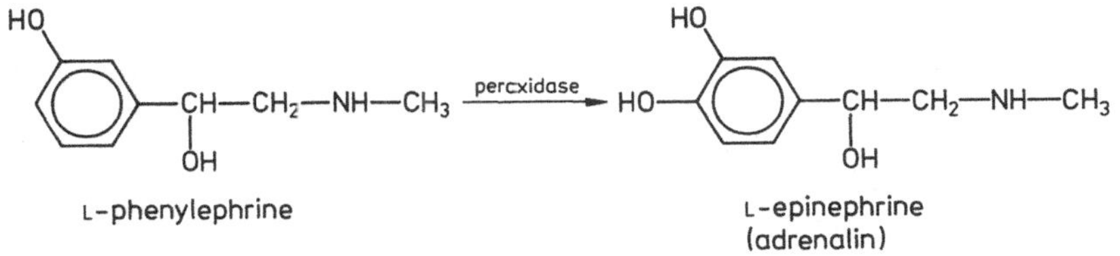

L-phenylephrine   L-epinephrine
                                         (adrenalin)

Figure 9. Peroxidase-catalyzed conversion of L-phenylphrine to L-epinephrine

The ability to selectively hydroxylate aromatic compounds seems to be a general feature of peroxidase: lactoperoxidase from milk and yeast cytochrome *c* peroxidase also convert L-tyrosine to L-DOPA in the reaction shown in Fig. 6 [20].

Selective hydroxylation of aromatic compounds is a difficult task in preparative organic chemistry. Therefore the simple and efficient hydroxylation catalyzed by peroxidase could be quite useful for preparative transformations of various pharmaceuticals and fine chemicals.

V. Xanthine Oxidase for the Geometrically and Regio-Specific Oxidation of Aldehydes

The natural function of cow's milk xanthine oxidase (EC 1.2.3.2) is to oxidize hypoxanthine and xanthine by O_2 to form uric acid and H_2O_2 (Fig. 10) [21]. In 1938, Booth discovered that the enzyme is also capable of doing something quite different — oxidizing a number of aliphatic and aromatic alcohols to the corresponding carboxylic acids [22].

hypoxanthine + O_2 ⟶ xanthine + H_2O_2 "Unnatural" reaction

xanthine + O_2 ⟶ uric acid + H_2O_2 aliphatic or aromatic aldehyde $+ O_2$ ⟶ carboxylic acid $+ H_2O_2$

Figure 10. "Normal" and "unnatural" reactions catalyzed by xanthine oxidase

Such oxidations, of course, can be carried out be chemical catalysts as well. Hence the enzymatic process can be of interest for preparative organic conversions only if it involves some unique features missing in chemical catalysts, e.g. selectivity. We have discovered that xanthine oxidase indeed exhibits both geometric (*cis/trans*) and regio (positional) specificity in the oxidation of aromatic aldehydes.

a) Geometric Specificity of Xanthine Oxidase in the Oxidation of β-arylacroleins [23]

Separation of *cis* and *trans* isomers of unsaturated compounds is a classical problem in organic chemistry. Functional properties (e.g. biological or pharmaceutical activities, odor, etc.) of geometric isomers are, generally speaking, different, since they are determined by the molecular geometry.

In order to take advantage of the properties of the geometric isomers of unsaturated, in particular β-arylacrylic compounds, one should be able to readily prepare both the *cis* and *trans* isomer separately. Unfortunately, nearly all conventional chemical syntheses of e.g. cinnamic compounds will yield almost exclusively *trans* isomers. A mixture of *trans* and *cis* isomers can be produced by ultraviolet irradiation of a solution of the *trans* isomer. However, separations of such mixtures are very laborious, time-consuming, inefficient and therefore not applicable on a large scale.

It occurred to us [23–25] that enzymes might be used for the preparative separation of *trans* and *cis* isomers of β-arylacrylic compounds. It is well established that active centers of most enzymes represent narrow clefts (openings). It is unlikely then that geometric isomers will fit into such an opening equally well and both in the correct way. Therefore, one can expect different reactivities of *cis* and *trans* isomers to a given enzyme. This, in turn, can be a basis for the enzymatic separation: since one of the isomers will be enzymatically converted to a new chemical functionality, it should be easy to separate it from the unreacted, second isomer.

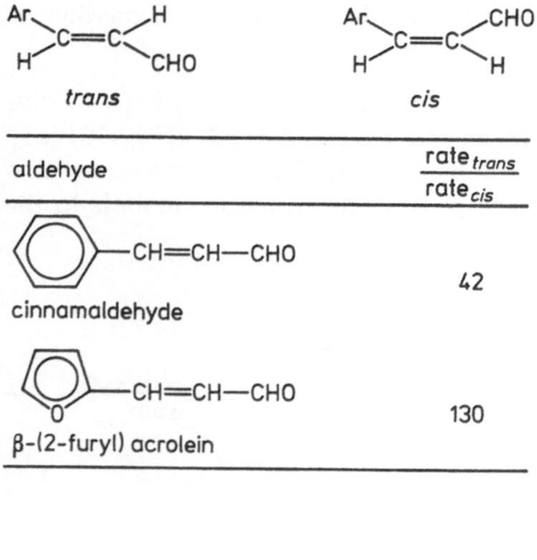

Figure 11. *Cis/trans* specificity of xanthine oxidase in the oxidation of β-arylacroleins

We verified the above approach in xanthine oxidase-catalyzed oxidation of *cis* and *trans* isomers of two β-arylacroleins, cinnamaldehyde and β-(2-furyl)-acrolein. In both cases, the *trans* isomer is oxidized by the enzyme about 2 orders of magnitude faster than its *cis* counterpart (Fig. 11).

This geometric specificity of xanthine oxidase can be used for preparation of the pure *cis* isomer from its commercially available *trans* counterpart, as exemplified by cinnamaldehyde: *trans*-cinnamaldehyde is irradiated with UV light, then the resultant mixture of geometric isomers is treated by the enzyme and the *trans*-cinnamic acid produced is absorbed on an anion-exchanger followed by extraction of the remaining *cis*-cinnamaldehyde with ether.

b) *Positional Specificity of Xanthine Oxidase in the Oxidation of Substituted Benzaldehydes [26]*

The introduction of a second substituent into monosubstituted benzene rings (e.g. in the reaction $ArXH + YZ \rightarrow ArXY + HZ$) usually results in the production of a mixture of positional isomers, *ortho*, *meta* and *para* (*I*, *II* and *III*, respectively) (Fig. 12). Preparative separation of such isomers, in particular *ortho* from *meta* and *ortho* from *para* (which is often of great practical significance) is in many cases rather difficult because most of their physico-chemical properties (e.g. boiling points, solubilities, etc.) are not sufficiently different.

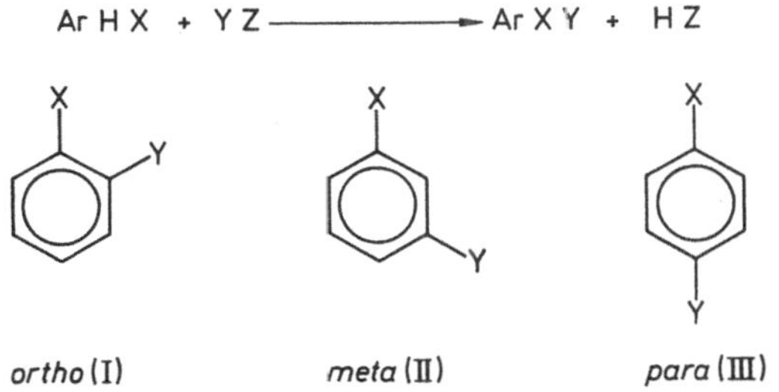

$$Ar\ H\ X\ +\ Y\ Z \longrightarrow Ar\ X\ Y\ +\ H\ Z$$

ortho (I)   *meta* (II)   *para* (III)

Figure 12. The introduction of a second substituent into a monosubstituted benzene ring resulting in the production of a mixture of *ortho*, *meta* and *para* isomers

It occurred to us that enzymes could be used for the separation of positional isomers. The rationale for this approach is as follows. Consider that an enzyme reacts with group X in *I*, *II* and *III*. Simple steric considerations suggest that, at least in the case of a bulky Y, such an enzymatic reaction will be hindered in *I*, but not in *II* or *III*, due to the proximity of the second substituent (Y). Leaving electronic effects aside, this should result in different reactivities to the enzyme of *I* and *II* or *III* which, in turn, can be the basis for their separation: since one of the isomers will be preferentially converted by the enzyme to a new chemical functionality, it should be easy to separate it from the other isomer. That is, the enzymatic treatment transforms in fact the problem of separation of positional isomers of a given chemical entity to the much easier problem of separation of two different chemical entities.

This rationale was confirmed in our studies on the reactions of xanthine oxidase with substituted benzaldehydes: we found that due to steric reasons the enzyme oxidizes *p*- (or *m*-) substituted benzaldehydes much faster than their *o*-counterparts in the case of bulky substituents.

Table I shows the initial rates of xanthine oxidase-catalyzed oxidation of *p*- and *o*-substituted halobenzaldehydes. One can see that while for fluorobenz-aldehyde the rates of the two isomers are comparable, upon further increase in the size of the halogen atom the *para/ortho* specificity factor greatly increases to exceed two orders of magnitude for iodobenzaldehyde. Comparison of the data for Cl-, Br- and I-benzaldehydes indicates that this regiospecificity of xanthine oxidase is due to diminishing reactivity of the *o*-isomer (as opposed to increasing reactivity of its *p*-counterpart). The observed positional specificity of the enzyme is not brought about by electronic effects since it takes place not only with electron-withdrawing halogen substituents but also with electron-donating alkoxy ones (Table 1).

Table 1: Xanthine Oxidase-Catalyzed Oxidation of Monosubstituted Benzaldehydes [26]

Benzaldehyde	$rate_{para}/rate_{ortho}$
fluoro-	0.7
chloro-	4
bromo-	11
iodo-	133
methoxy-	2
ethoxy-	14
propoxy-	34

Positional specificity of xanthine oxidase is truly remarkable in the oxidation of nitrobenzaldehydes: whereas *para* and *meta* nitrobenzaldehydes are enzymatically oxidized at the same rates (stressing once again a relative insignificance of electronic factors), the oxidation rate for *ortho* nitrobenzaldehyde is at least two and a half orders of magnitude lower. It is imperative for the observed regiospecificity that a bulky substituent be in close proximity to the carbonyl group. This conclusion follows from the fact that in contrast to nitrobenz-aldehyde, nitrocinnamaldehyde (which has an electronic structure similar to that of the former but in which the carbonyl moiety is separated from the benzene nucleus by a vinylene group) exhibits virtually no regiospecificity in the reaction with xanthine oxidase: *p*- and *o*-isomers have approximately the same reactivity.

Simple steric considerations suggest that a diminished reactivity of *ortho* (in comparison with *meta* or *para*) substituted benzaldehydes should be really profound only in interactions with a bulky reagent such as the active center of an enzyme but not with relatively small substances. We have tested this by studying the oxidation of *o*, *m*- and *p*-nitrobenzaldehydes by hydrogen peroxide (which at acidic pH is known to oxidize aromatic aldehydes to the corresponding carboxylic acids). In complete agreement with our model, *ortho* nitrobenzaldehyde reacts with H_2O_2 at the same rate as *meta* nitrobenzaldehyde, with the *para* isomer being only about twice as reactive as the other two.

We employed the regiospecificity of xanthine oxidase described above for the separation of a mixture of positional isomers. As an example, we have chosen a mixture of *m*-nitrobenzaldehyde (80%) and *o*-nitrobenzaldehyde (20%) which is routinely produced by the direct nitration of benzaldehyde, the procedure used industrially. A solution containing the above mixture and xanthine oxidase was stirred under air for 1 hr. Then the nitrobenzaldehydes remaining were extracted and analyzed gaschromatographically. This assay showed that more than 99% of *m*-nitrobenzaldehyde disappeared whereas the concentration of *o*-nitrobenzaldehyde did not change appreciably. That is, the treatment of a mixture of *o*- and *m*-NO_2-benzaldehydes with the enzyme converts it to a mixture of *o*-NO_2-benzaldehyde and *m*-NO_2-benzoic acid which can easily be separated by differential extraction.

VI. Concluding Remarks

In all of the examples discussed above, we took advantage of the ability of enzymes to catalyze "unnatural" reactions, often quite different from their "normal", physiological reactions reflected in their names. There is no doubt that this ability is not limited to the four enzymes considered herein but represents a rather general phenomenon. Recognition of this fact will greatly expand the arsenal of approaches at a biotechnologist's disposal. The existence of multiple activities of a given enzyme implies that in order to catalyze a new reaction one does not have to necessarily employ a new enzyme — sometimes a good look at one's "own backyard", at commercially available enzymes, can be extremely beneficial. Finally, the aforesaid indicates that along with conventional bacterial screening, it might be worthwhile to carry out a "chemical screening" — to systematically seek novel, unconventional activities for known enzymes.

Acknowledgement

The author is a recipient of the Henry L. Doherty Professorship.

References

1. Enzyme Nomenclature. 1979. Academic Press, New York.
2. Schmid RD 1979. Oxidoreductases — present and potential applications in technology. Process Biochem. 14(6):2–7
3. Malmström BG, Andréasson L-E, and Reinhammer B 1975. Copper-containing oxidases and superoxide dismutases, p. 574. *In* P. D. Boyer (ed.), The Enzymes, 3rd ed., vol. 12. Academic Press, New York.
4. Bright HJ and Porter DJT 1975. Flavoprotein oxidases, chapter 7. *In* P. D. Boyer (ed.), The Enzymes, 3rd ed., vol. 12. Academic Press, New York.
5. Whitaker JR 1972. Principles of enzymology for the food sciences, chapter 21. M. Dekker, New York.
6. Scott D 1975. Applications of glucose oxidase, chapter 19. *In* G. Reed (ed.), Enzymes in Food Processing, 2nd ed. Academic Press, New York.

7. Dixon M and Webb EC 1979. Enzymes, 3rd ed., pp. 243–244. Academic Press, New York.
8. Carr PW and Bowers LD 1980. Immobilized enzymes in analytical and clinical chemistry, chapter 5. Wiley, New York.
9. Guilbault GG 1982. Immobilized enzymes as analytical reagents. Appl. Biochem. Biotechnol. 7:85–98.
10. Keilin D and Hartree EF 1948. Properties of glucose oxidase. Biochem. J. 42:221–229.
11. Alberti BN and Klibanov AM 1982. Preparative production of hydroquinone from benzoquinone catalyzed by immobilized glucose oxidase. Enzyme Microb. Technol. 4:47–49.
12. Hamilton GA, DeJersey J and Adolf PK 1973. Galactose oxidase: the complexities of a simple enzyme, pp. 103–124. In T. E. King, H. S. Mason and M. Morrison (ed.), Oxidases and related redox systems, vol. 1. University Park Press, Baltimore.
13. Klibanov AM, Alberti BN and Marletta MA 1982. Stereospecific oxidation of aliphatic alcohols catalyzed by galactose oxidase. Biochim. Biophys. Acta, submitted for publication.
14. Saunders BC, Holmes-Siedle AG and Stark BP 1964. Peroxidase. Butterworth, London.
15. Klibanov AM, Alberti BN, Morris ED and Felshin LM 1980. Enzymatic removal of toxic phenols and anilines from waste waters J. Appl. Biochem. 2:414–421.
16. Alberti BN and Klibanov AM 1981. Enzymatic removal of dissolved aromatics from industrial aqueous effluents. Biotechnol. Bioeng. Symp. 11:373–379.
17. Mason HS, Onopryenko I and Buhler D 1957. Hydroxylation: the activation of oxygen by peroxidase. Biochim. Biophys. Acta 24:225–226.
18. Buhler D and Mason HS 1961. Hydroxylation catalyzed by peroxidase. Arch. Biochem. Biophys. 92:424–437.
19. Daly JW and Jerina DM 1970. Aerobic aromatic hydroxylation catalyzed by horseradish peroxidase: absence of NIH shift. Biochim. Biophys. Acta 208:340–342.
20. Klibanov AM, Berman Z and Alberti BN 1981. Preparative hydroxylation of aromatic compounds catalyzed by peroxidase. J. Amer. Chem. Soc. 103:6263–6264.
21. Massey V 1973. Iron-sulfur flavoprotein hydroxylases, pp. 301–360. In W. Lovenberg (ed.), Iron-sulfur proteins, vol. 1. Academic Press, New York.
22. Booth VH 1938. The specificity of xanthine oxidase. Biochem. J. 32:494–502.
23. Klibanov AM and Giannousis PP 1982. Geometrically specific oxidation of β-arylacroleins catalyzed by xanthine oxidase: the preparative potential. Biotechnol. Lett. 4:57–60.
24. Klibanov AM and Siegel EH 1982. Geometric specificity of porcine liver carboxyl esterase and its application for the production of cis-arylacrylic esters. Enzyme Microb. Technol. 4:No. 3.
25. Klibanov AM and Giannousis PP 1982. Geometric specificity of alcohol dehydrogenases and its potential for separation of trans and cis isomers of unsaturated aldehydes. Proc. Natl. Acad. Sci. USA 79:June.
26. Pelsy G and Klibanov AM 1982. Remarkable positional (regio) specificity of xanthine oxidase and dehydrogenases in the reactions with substituted benzaldehydes. Biochemistry 21:submitted for publication.

I. Enzyme Production

11. Enzyme Production

Purification and Characterization of Membrane-Bound Aldehyde Dehydrogenase from *Acinetobacter calcoaceticus* Grown on Long-Chain Alkanes*

H. Aurich, H. Sorger, R. Bergmann and J. Lasch

Summary

Acinetobacter calcoaceticus grown on alkanes synthesizes a membrane-bound aldehyde dehydrogenase which oxidizes aliphatic aldehydes with $NADP^+$ to the corresponding fatty acids. This enzyme is induced by exogenous alkanes, alcohols and aldehydes and repressed by intermediates of central metabolic pathways. It was shown ultracytochemically that it is located exclusively in the membranes which envelop intracellular hydrocarbon inclusions. The enzymic activity depends on cardiolipin.

Cytoplasma-free membranes were prepared by differential centrifugation after vibration with glass beads or lysozyme treatment of the microorganisms harvested in the *log*-phase. The enzyme was solubilized by detergents, purified in micellar form by chromatography on DEAE-cellulose and gel filtration on Sepharose CL-4B resulting in a 60fold enrichment. The enzyme is a tetrameric membrane protein with a molecular weight of 280000 daltons. Its monomeric form is enzymatically inactive. The aldehyde dehydrogenase oxidizes homologous aliphatic aldehydes with differing efficiency. The apparent K_M-values decrease with increasing chain length. At high substrate concentrations inhibition is observed. The aldehyde substrates are bound by hydrophobic interactions to the enzyme. Inhibition studies point to a functional important metal ion and an SH-group on the enzyme.

The enzyme can be incorporated into liposomal membranes prepared from bacterial lipids without loss of activity. Long-chain aldehyde substrates do not destroy the lipid vesicles. The reconstituted liposomal enzyme is very similar to its physiological form in the bacterial membrane as evidenced by an analogous behavior towards the action of proteases, phospholipases and detergents.

Introduction

The bacterial oxidation products of aliphatic hydrocarbons vary considerably. The occurence of certain oxidation products made it possible to postulate relatively early potential catabolic pathways of alkanes (Aurich, 1979). The most frequent encountered pathway leads from the alkane via primary alcohol and aldehyde to the corresponding fatty acid, i.e. alkane oxidation proceeds in 3 steps which are catalyzed by 3 enzymes or enzyme systems:

---

* Dedicated to Prof. Dr. med. Dr. phil. Dr. h. c. mult. Erich Strack, Leipzig, at his 85th birthday

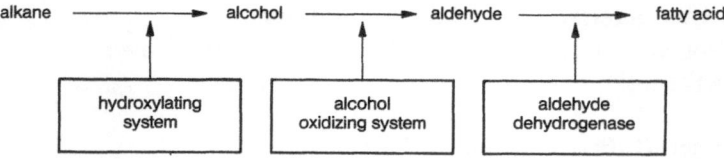

Pathway and enzymes of primary oxidation of n-alkanes in bacteria

In the extensively studied *Pseudomonas* species the *hydroxylation system* consists of 3 proteins (Peterson et al. 1966): the alkane-1-monooxygenase (a membrane-bound non-heme iron enzyme), the rubredoxin (a low molecular, soluble iron-sulfur protein) and the rubredoxin reductase (a cytoplasmic flavine enzyme).

The *alcohol dehydrogenase* catalyzing the second step of alkane oxidation falls into two types: a cytoplasmic, NAD(P)-dependent type (similar to the yeast alcohol dehydrogenase) and a membrane-bound type (seemingly akin to the bacterial methanol dehydrogenase). The membrane-bound alcohol dehydrogenase which is directly engaged in the alkane assimilation is coupled to an electron transport chain (the primary acceptor is a cytochrome) (Tauchert et al. 1978).

The *aldehyde dehydrogenase* is in all cases known as a NAD(P)-dependent enzyme. In *Acinetobacter* it is membrane-bound and $NADP^+$-specific (Aurich and Eitner 1977).

From the existence of these enzymes, their functional interplay, their tuned epigenetic regulation (Fennewald et al. 1979) and from their joint localization in membranes, it was inferred that a vectorial oxidation of the extracellular alkanes to intracellular fatty acids takes place during the permation through the membrane (Fig. 1).

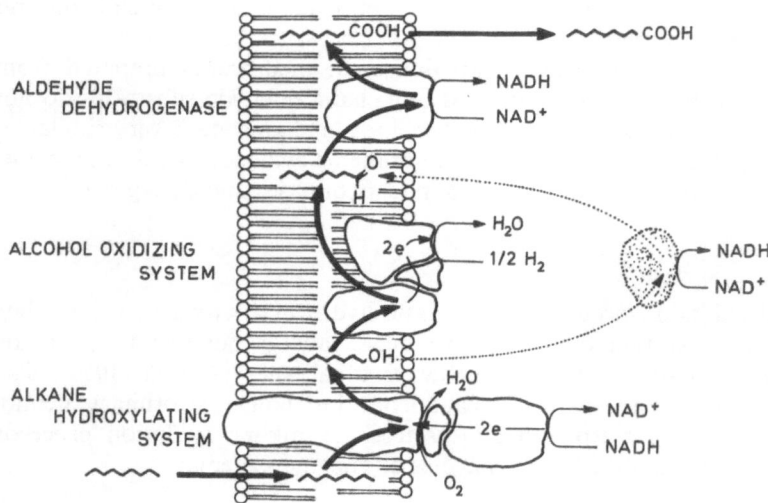

Figure 1. Model of vectorial alkane oxidation in bacterial membranes

It is assumed that the alkanes are oxidized in the lipid phase of the membrane (surface oxidation model) into which they will partition from the aqueous periplasmic space. We have shown, however, that enzymes can bind hydrophobic substrates directly from the aqueous phase by a hydrophobic region of the protein (Schöpp and Aurich 1973) as well as from interfaces of substrate particles (Rothe et al. 1976).

Up to now, little is known about the aldehyde dehydrogenase. This communication deals with the isolation and characterization of this inducible enzyme from *Acinetobacter*.

Results and Discussion

Intracellular Localization of Aldehyde Dehydrogenase

During alkane assimilation *Acinetobacter* species show distinct intracellular structures which are in contact with the cytoplasmic membrane (Fig. 2). These structures consist mainly of membrane-coated droplets (inclusions) containing preponderantly alcohols, wax esters and alkanes (Kennedy et al. 1975). The membraneous envelope of the inclusions differs markedly from the cytoplasmic membrane: 3fold lower protein contents and significantly lower phosphatidylethanolamine fraction. On the other hand, the cardiolipin content is twice as high as in the cytoplasmic membrane (Scott and Finnerty 1976).

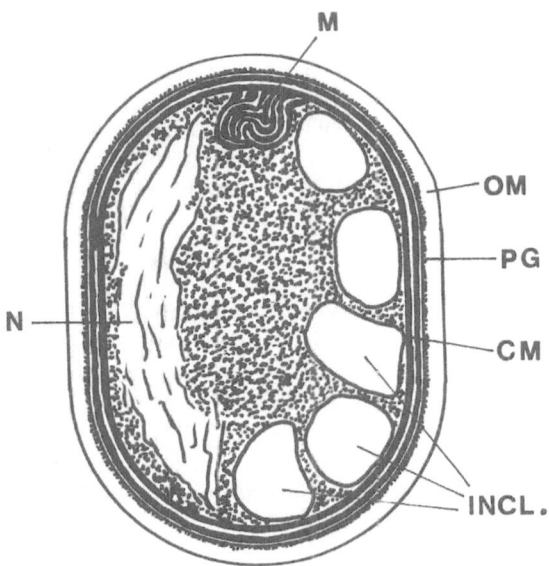

Figure 2. Model of an alkane-grown *Acinetobacter* cell

OM = outer membrane	PG = peptidoglycane
CM = cytoplasmic membrane	N = nuclear region
INCL. = alkane inclusions	M = intracellular membranes

The fraction of neutral lipids of the membranes (mainly alkanes and alcohols) increases during alkane assimilation. There is no lecithin in *Acinetobacter* membranes.

We were able to demonstrate ultracytochemically that the aldehyde dehydrogenase of *Acinetobacter* is located in the membraneous envelope of the inclusions (Fig. 3). The enzyme could not be detected in the cytoplasmic membrane.

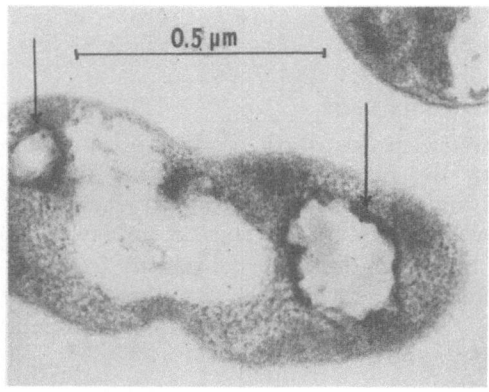

Figure 3. Electromicrograph of *Acinetobacter calcoaceticus* with cytochemical staining of aldehyde dehydrogenase in presence of butanol, NADP$^+$, phenacinmethosulphate, tetra nitro blue tetrazolium chloride, and sodium azide (Vorišek et al. 1982)

In order to isolate the enzyme the bacteria were disintegrated mechanically (glass bead vibrator) or enzymatically (lysozyme). This step was followed by differential centrifugation yielding cytoplasma-free particle fractions. The mechanical disintegration yields cell envelope fractions (Aurich et al. 1977), containing inclusions (partly free, partly associated with the cytoplasmic membrane), whereas by enzymic disintegration ghost-like cytoplasmic membranes are obtained (Sorger and Ylönen, unpublished results).

Purification of Aldehyde Dehydrogenase

Cell envelope preparations obtained by the mechanical disintegration were used as starting material for the purification of aldehyde dehydrogenase. The cell envelopes were prepared from bacteria cultivated in an optimal mineral salt medium containing 1 % alkanes (C_{13}–C_{18}) in a Biostat S fermentor (B. Braun/Melsungen, FRG) as published elsewhere (Fricke et al. 1982). The enzyme was solubilized from the cell envelopes by various detergents. The efficiency of the detergents to bring about the formation of enzymatically active detergent-enzyme mixed micelles differs considerably (Fig. 4).

In some cases an enzyme activation was observed at low detergent concentrations (e.g. with deoxycholate). At high detergent concentrations the enzyme is invariably inhibited (with concomitant increase of the apparent K_M-values of the aldehydes). The micellar form of the enzyme can be reverted to the membrane form by removal of the detergent by Bio-Beads SM-2 in the presence of enzyme-free bacterial cell envelopes.

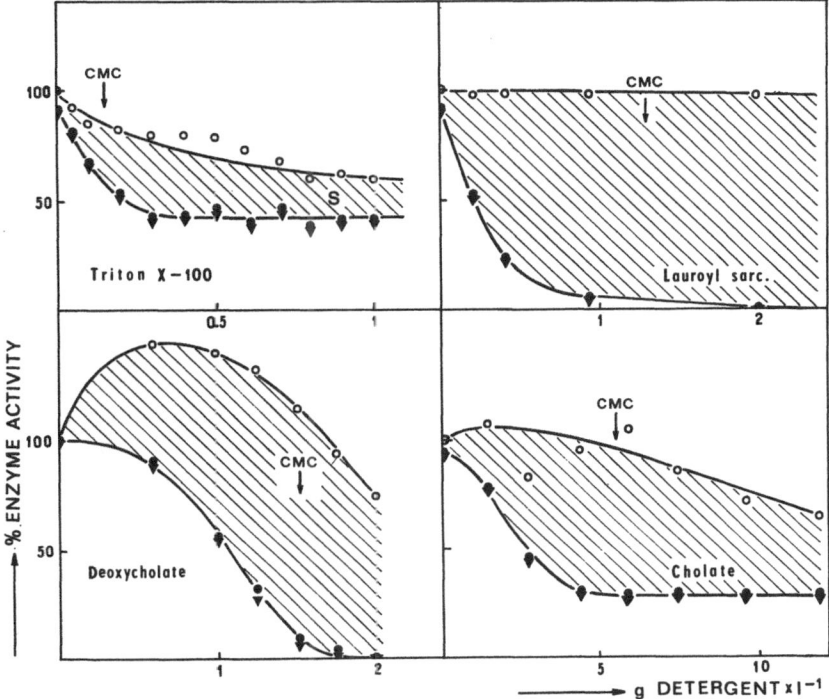

Figure 4. Effect of some detergents on the solubilization of aldehyde dehydrogenase from cell envelopes of *Acinetobacter calcoaceticus* CMC = critical micelle concentration, hatched areas = enzyme activity in the supernatant, white areas below them = enzyme activity in the sediment

The activity of the detergent form of the enzyme is enhanced by addition of bacterial lipid extracts. A detailed analysis revealed that of all studied lipids only cardiolipin had an activating effect. This led us to the assumption that the aldehyde dehydrogenase is physiologically located in a cardiolipin domain of the membraneous envelope of inclusions.

Table 1. Purification of aldehyde dehydrogenase from *Acinetobacter calcoaceticus*

	Protein tot. (mg)	Enzyme activity tot. (nkat)	Specific enzyme activity (nkat/mg)	Enrichment (x fold)
Crude extract	312	1203	4	1
Cell envelope preparation	149	1010	7	2
Solubilization by detergents	85	1039	12	3
DEAE-cellulose chromatography	5	720	133	33
Sepharose CL-4B gel filtration	0.7	161	229	57

The micellar enzyme form can be purified in the presence of detergents by conventional DEAE-cellulose chromatography and gel filtration on Sepharose CL-4B. A 60fold enrichment is achieved (Table 1).

Molecular and Kinetic Properties of Aldehyde Dehydrogenase

Aldehyde dehydrogenase in its detergent form has a molecular weight of 280000 daltons and an isoelectric point of 5.7. The enzyme is a tetramer. The dimer retains enzymic activity whereas the monomer is devoid of it. The pH optimum falls between pH 9 and 10. The highest stability, however, is found around pH 7.0.

The K_M-value with respect to NADP$^+$ is (non-dependent of the chain length of the aldehyde substrate) on the average $9 \cdot 10^{-5}$ (moles/l) (cf. Fig. 5). The reaction product NADPH is an inhibitor ($K_i = 2 \cdot 10^{-4}$ moles/l). Enzymic reduction of free fatty acids by NADPH could not be observed.

The following results were obtained, when homologous aldehydes with increasing chain length were used as substrates in molecular- and/or micellar-dispersed form:

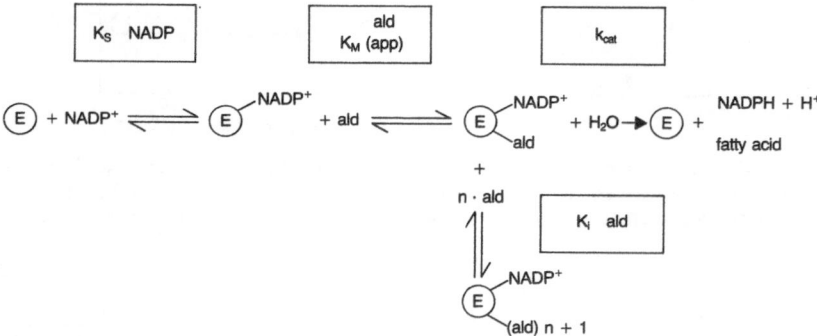

Kinetic reaction scheme of aldehyde dehydrogenase catalysis ⒺⒸ = aldehyde dehydrogenase, ald = aldehyde. Using graphic methods, n was determined to be 1

The enzyme shows substrate inhibition. The apparent V_{max}-values increase with increasing chain length and the optimal substrate concentration ([S] opt.) is shifted to lower values (Sorger and Aurich 1978). The substrate inhibition can be explained by the binding of a second aldehyde molecule per active site which acts as inhibitor. The reaction scheme can be formulated as follows:
The apparent K_M-values of the homologous aldehydes as well as their K_i-values decrease with increasing chain length (up to C_{10}–C_{12}) (Fig. 5).

The decrease of these parameters points to a better binding of the aldehydes to the enzyme as their chain length increases, i.e. the interaction is predominantly hydrophobic. The V_{max} effects indicate that this interaction interfers somehow with the catalytic step. The break in the linear relationships between the logarithms of the K_M- and K_i-values and the number of C-atoms at C_{10}–C_{12} is probably due to the limited dimension of the hydrophobic binding area on the enzyme

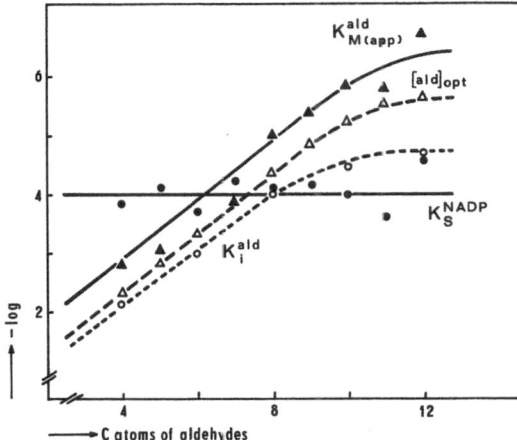

Figure 5. Dependence of K_s^{NADP}, $K_{M(app)}^{ald}$, K_i^{ald} and optimal aldehyde concentration ($[ald]_{opt}$) on the chain length of the aldehydes in aldehyde dehydrogenase reaction

surface and/or a restricted volume increment of the aldehyde molecules caused by curling back of the hydrocarbon chain.

The aldehyde dehydrogenase is inhibited by heavy metal ions (Fe^{3+}, Cu^{2+}, Sn^{2+}), molybdate and borate (inhibitor of the hydrid transfer of NAD(P)-enzymes), chloral hydrate (competitive with respect to the aldehydes; $K_i = 1 \cdot 10^{-2}$), EDTA and o-phenanthroline (non-competitive; $K_i = 3 \cdot 10^{-2}$ and $6 \cdot 10^{-3}$, respectively) and by 4-chloromercuribenzoate (non-competitive; $K_i = a \cdot 10^{-6}$). These results point to a catalytically essential SH-group and a metal ion in the enzyme. This is in keeping with properties of aldehyde dehydrogenases from other sources.

The reaction pathway postulated takes an intermediary thiohemiacetal and a thioester into account. These covalent intermediates should be bound additionally by hydrophobic contacts. For aldehyde dehydrogenases of other sources it is also discussed that SH-groups are necessary to bind the coenzyme:

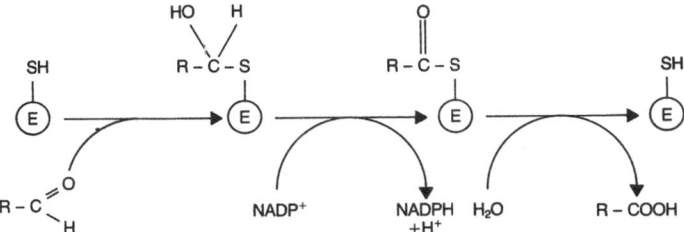

Assumed reaction pathway of aldehyde oxidation at the active center of aldehyde dehydrogenase

Reconstitution of Aldehyde Dehydrogenase in Liposomes

In order to study the interactions of the aldehyde dehydrogenase with membranes as well as the *in vitro* oxidation of aldehydes under *in situ* conditions, it was attempted to incorporate the detergent form of the enzyme into artificial membranes (liposomes). Liposomes were prepared from bacterial lipid extracts (total lipids) (Table 2). Egg lecithin proved to be unsuitable because cardiolipin is essential for the enzyme. Obviously, lecithin at high concentrations displaces cardiolipin from the enzyme.

Table 2. Preparation of aldehyde dehydrogenase liposomes

Preparation method	Phospholipid	Enzyme activity (nkat)			
		Initial activity	Enzyme lipid dispersion	Sephadex[a] G-75	Sepharose[b] CL-4B
Cholate dilution	egg lecithin	120	21	7	1
	bacterial lipids	18	19	4	2
Ultrasonication	egg lecithin	59	0.3	—	(+)
	bacterial lipids	13	4	—	2

[a] Gel filtration step to remove cholate
[b] Gel filtration to separate single-walled from multi-walled liposomes

   Both the cholate dilution method and the ultrasonication procedure were used to prepare liposomes (Koelsch et al. 1981). Homogeneous single-walled liposomes with a mean diameter of 25–30 nm into which more than 10 % of the added aldehyde dehydrogenase activity were incorporated could be obtained by both procedures. Visualization of the proteoliposomes by electron microscopy demonstrates that the protein is located in the liposomal membrane (Fig. 6).

   Turbidity measurements in dependence on the wavelength of light revealed no size changes of the enzymically active proteoliposomes when $NADP^+$ or long-chain aldehhydes are added.

   Obviously, the long-chain aldehydes partition rapidly takes place into the membrane in accordance with their high partition coefficient in the biphasic

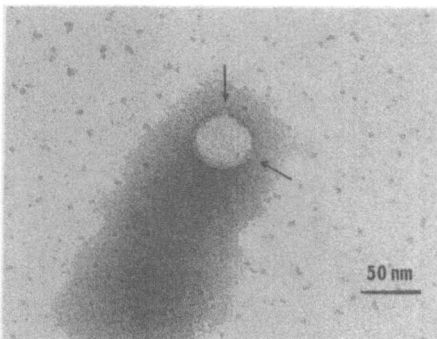

50 nm

Figure 6. Electromicrograph of aldehyde dehydrogenase containing liposomes produced from the purified enzyme and bacterial lipids by the ultrasonication method. Negative staining with 2 % ammonium molybdate. The micrograph was done by Dr. Ladhoff, Charité, Berlin

system membrane lipids/water. This partitioning proceeds faster than the turnover of the aldehyde substrates in the aqueous phase. On the other hand, $NADP^+$ could not be found in the liposomes.

We might thus conclude that the enzyme picks up the aldehyde substrates from the membrane pool and oxidizes them with $NADP^+$, taken from the aqueous phase.

The reconstituted liposomal aldehyde dehydrogenase has properties very similar to the enzyme in the physiological environment of the bacterial membrane. Proteinases inactivate it only slightly, but addition of detergents entails a total inactivation. The inactivation brought about by phospholipase A_2 can be abolished by addition of bacterial lipids.

The enzyme in its liposomal form as an *in situ* variant can oxidize even highly hydrophobic aldehydes. This provides a basis for the enzymatical transformation of hydrophobic substrates *in vitro* by bacterial membrane enzymes.

References

Aurich H (1979): Die Oxydation aliphatischer Kohlenwasserstoffe durch Bakterien. Sitzungsber. Akad. Wiss. DDR 16 N, Akad.-Verl. Berlin, pp. 1–24

Aurich H, Eitner G (1977): Induktion der NADP-abhängigen Aldehyddehydrogenase durch Kohlenwasserstoffe bei *Acinetobacter calcoaceticus*. Z. Allg. Mikrobiol. 17:263–266

Aurich H, Sorger H, Müller H (1977): Isolierung und Charakterisierung der Zellgrenzschichten von *Acinetobacter calcoaceticus*. Z. Allg. Mikrobiol. 17:333–338

Fennewald MS, Benson M, Oppici M, Shapiro J (1979): Insertion element analysis and mapping of the *Pseudomonas* plasmid *alk* regulon. J. Bact. 139:940–952

Fricke B, Bergmann R, Sorger H, Aurich H (1982): Optimierung von Kulturbedingungen für *Acinetobacter calcoaceticus* beim Wachstum auf n-Alkanen in einem Laborfermenter. Z. Allg. Mikrobiol. 22:365–372

Kennedy RS, Finnerty WR, Sudarsanan K, Young RA (1975): Microbial assimilation of hydrocarbons. I. The fine-structure of a hydrocarbon oxidizing *Acinetobacter sp*. Arch. Microbiol. 102:75–83

Koelsch R, Lasch J, Klibanov AL, Torchilin VP (1981): Incorporation of chemically modified proteins into liposomes. Acta biol. med. germ. 40:331–335

Peterson JA, Basu D, Coon MJ (1966): Enzymatic co-oxidation. I. Electron carriers in fatty acid and hydrocarbon hydroxylation. J. Biol. Chem. 241:5162–5164

Rothe U, Schöpp W, Aurich H (1976): Enzymatischer Umsatz von Tetradekanol in heterogener Phase durch Hefe-Alkoholdehydrogenase. Acta biol. med. germ. 35:7–14

Schöpp W, Aurich H (1973): Abhängigkeit der K_M- und V_{max}-Werte von der Kettenlänge des Substrates für die Reaktion der Alkoholdehydrogenase aus Hefe. Acta biol. med. germ. 31:19–28

Scott CCL, Finnerty RW (1976): Characterization of intracytoplasmic hydrocarbon inclusions from the hydrocarbon-oxidizing *Acinetobacter species H01-N*. J. Bact. 127:481–489

Sorger H, Aurich H (1978): Mikrobielle Aldehyddehydrogenasen und ihre Bedeutung für die Assimilation aliphatischer Kohlenwasserstoffe. Wiss. Z. KMU 27:35–46

Tauchert H, Schöpp W, Aurich H (1978): Pyridinnukleotid-unabhängige Alkoholdehydrogenase in alkanverwertenden Bakterien. Wiss. Z. KMU 27:25–34

Vořišek J, Sorger H, Aurich H, Lojda Z (1982): Ultracytochemical staining of aldehyde dehydrogenase activity in *Acinetobacter calcoaceticus* cultivated on n-alkanes. Symp. histochem. Karlovy Vary

Application of the Recombinant DNA Technique to Enzyme Technology

Jürgen Kreft and Werner Goebel

Summary

The recombinant DNA technique is at the threshhold of its practical application to enzyme technology. Recently a number of vector systems have been developed for this purpose, including plasmids and phage which allow molecular cloning of DNA in a variety of industrial microorganisms. In a few cases these systems have already been successfully used for enzyme production. Problems which arise during such experiments include non-expression of heterologous genes and instability of recombinant DNA. Efforts to overcome these problems are being made by construction and selection of specially designed host/vector systems, e.g. expression and integration vectors.

Introduction

Enzymes, in contrast to most other cell constituents, are primary gene products. This makes them particularly suitable for all kinds of genetic manipulations — only one or a few genes have to become altered to obtain strain improvement. This is in contrast to the situation with products synthezised by complex pathways, e.g. secondary metabolites such as antibiotics.

In general the improvement of organisms which are applied to enzymatic industrial processes can be aimed at different ends.

Desirable properties can be better yields and/or increased specific activities of an enzyme. On the other hand or in addition, facilitated enzyme purification or improved properties of the producing organisms during fermentation or bioreactor processes can be desired.

In the past this has been achieved by the classical methods of screening, mutation and selection. These methods are now sometimes complemented by the use of naturally occurring mechanisms of genetic exchange between microorganisms, e.g. conjugation.

The recent development of the recombinant DNA technique now offers an attractive alternative to the classical methods. In principle, all of the improvements mentioned above can be obtained by this methodology but recent results suggest that a variety of problems has to be overcome.

This article will only discuss work done with microorganisms including yeast. Developments in eukaryotic systems which are more or less in their infancy, despite the fact that some eukaryotic (fungal) enzymes are commercially important, will not be covered here. The *Bacillus* system will be given special emphasis since enzymes from this genus are of great interest (Priest, 1977) and the development of cloning systems is well advanced (Young, 1980; Kreft and Hughes, 1982).

Methods

See references given in the preceding chapter and Collins, 1977; Goebel, 1979, 1980; Kreft et al, 1978, 1982. Further references can be found in "Genetic engineering" (J. K. Setlow, A. Hollaender, eds.) 1979, 1980; Methods in Enzymology, Vol. 68, 1980; Current Topics Microbiol. Immunol., Vol. 96, 1982.

Results and Discussion

1. Cloning systems

As was already mentioned aboves vectors for molecular cloning of DNA in microorganisms comprise plasmids and phages. The most sophisticated systems have been developed for *E. coli*. A wide variety of plasmids with different genetic markers and recognition sites for restriction enzymes has been constructed, the plasmid pBR322 (Bolivar et al., 1977) and its derivatives being the most commonly used ones. Several vectors allow regulated expression of cloned genes from phage promoters or from promoters of inducible enzymes. Phage vectors for *E. coli* are based on the genome of labdoid phage; this type of vector allows very efficient introduction of recombinant DNA molecules into recipient cells (Hohn and Murray 1977).

Cosmids (Collins and Hohn 1978) combine the advantages of both systems by preferential cloning of very large fragments of DNA, highly efficient transformation and subsequent replication of the recombinant DNA in an autonomous state without cell lysis.

Plasmid vectors for the *Bacilli* comprise small antibiotic resistance plasmids originating from *S. aureus* (Ehrlich 1977; Gryczan and Dubnau 1978) and several *Bacillus* species (Bernhard et al. 1978; Bingham et al. 1979) as well as constructed vectors based on cryptic *Bacillus* plasmids (Tanaka and Kawano 1980). In addition bifunctional vectors capable of replication in both *E. coli* and *Bacillus* have been constructed (Ehrlich 1978; Kreft et al. 1978; Ehrlich et al. 1982; Kraft and Hughes 1982) (Fig. 1). Other bifunctional vectors allow cosmid cloning in *E. coli* and subsequent transformation of *Bacillus* with the recombinant DNA (Aubert et al. 1981). A cosmid-type cloning system for *B. subtilis* is under development (Marrero et al. 1981). Several bacteriophages from *Bacillus* have been used for cloning experiments (for a review see Kreft and Hughes 1982) but these systems are much less advanced than in the case of *E. coli*.

Recently a number of cloning vehicles has been constructed for *Streptomyces* (Chater et al. 1982).

Different types of vector plasmids are available for the yeast *Saccharomyces cerevisiae*. They are almost exclusively hybrids between *E. coli* plasmids and yeast DNA. Depending upon the ability of the yeast DNA sequence in these

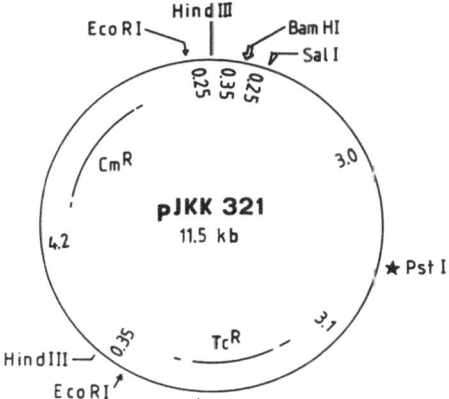

Figure 1. Restriction map of the bifunctional vector plasmid pJKK321. This plasmid can replicate in both *E. coli* and *B. subtilis*, thus allowing the easy transfer of cloned genes between these two species. It has been constructed by ligating *in vitro* the *Hind*III-cleaved vector pJKK3-1 (Kreft and Hughes 1982) and the *Staphylococcus* plasmid pC221. pJKK321 carries genes for resistance to tetracycline (Tc^R) and chloramphenicol (Cm^R) which are expressed in both host bacteria and single sites for the restriction enzymes *Bam*HI, *Sal*I and *Pst*I

hybrids to promote autonomous replication in yeast or not, integrating or extrachromosomally replicating vectors can be distinguished (for reviews see Hinnen and Meyhack 1982; Hollenberg 1982). Recently, plasmid vectors for yeast species other than *S. cerevisiae* are being developed (Heslot, 4th International Symposium on Genetics of Industrial Microorganisms — GIM 82 — Kyoto 1982).

Methods to introduce *in vitro* recombined DNA into cells of the desired host organism are, apart from the availability of suitable cloning vectors, a prerequisite for genetic manipulations by the recombinant DNA technique.

Such methods exist for all the microorganisms mentioned above and several others. Salt treatment in the case of *E. coli* (Mandel and Higa 1970) or enzymatic treatment resulting in removal of the cell wall combined with addition of polyethyleneglycol in the case of *Bacilli* and yeast (Chang and Cohen 1979; Hinnen et al. 1978) makes cells competent for the uptake of exogenous DNA. With *B. subtilis* the natural competence of some strains (Spizizen 1958) can also be used for this purpose.

Molecular cloning of chromosomal genes from one organism in the same or another species often involves a "shot gun"-type of experiment— fragment of genomic DNA are inserted at random into vector molecules, this mixture is then used for transformation. Direct selection for a desired clone is only possible in these few cases where successful cloning results in some resistance, complementation of host mutations or ability to grow on unusual substrates. This has been used for the cloning of the gene for an extremely heat-stable enzyme (3-Isopropylmalate dehydrogenase from *Thermus thermophilus* (Nagahari et al. 1980) and for sucrase (Fouet et al. 1982). Identification of clones is also relatively easy when a simple functional test is available, e.g. for α-amylase (Palva et al. 1982). In all cases a reduction in the number of clones to be analyzed is highly desirable. This can be achieved by techniques which allow enrichment of particular genes or by the cosmid technique mentioned above. Identification of clones can be done by using gene specific radioactive probes or by immunological methods.

2. *Expression of cloned genes*

The expression and the yield of an enzyme the gene for which has been cloned is influenced by several parameters such as copy number, appropriate signal sequences, stability of the recombinant DNA and the enzyme itself and eventually regulation mechanisms like repression. The situation is not always as favourable as in the case of α-amylase from *B. amyloliquefaciens* (Palva et al. 1981). In this case the cloned gene originated from a species closely related to the recipient strain, synthesis of the gene product is not under repression control, the enzyme is not harmful to the cell and is excreted into the medium. Therefore a dramatic increase in enzyme yield could rather easily be obtained. But already during the cloning of D-galactose dehydrogenase from *Pseudomonas* in *E. coli* (Buckel and Zehelein 1981) more sophisticated methods such as in vitro mutagenesis of the recombinant plasmid were necessary to obtain good expression of this enzyme. From our own results (Kreft et al. 1978; Kreft et al. 1982) and those of others (Ehrlich 1978) it is evident that expression of foreign genes in *B. subtilis* is limited. This is due to the fact that transcription and translation of genes in this species depends on the presence of strictly defined signal sequences (Moran et al. 1982; Kreft et al. in preparation) (Fig. 2). This problem can be circumvented by the use of vectors which carry the appropriate signals and allow insertion of foreign DNA in the right distance and orientation from these signals (Williams et al. 1981).

-35	-10	
T T G A C A	T A T A A T	σ^{55} recognition sequence
T T C A A A	C A C A A T	beta-lactamase
G G C A C G	C A T A A T	chloramphenicol acetyltransferase
T T G A C A	T T T A A T	tetracycline resistance

Figure 2. Comparison of the promoter sequence (—35 and —10 base pairs upstream to the start of transcription) recognized by the initiation factor σ^{55} of *Bacillus subtilis* (Losick and Pero 1981) with the promoters of three genes from *E. coli* (Russell and Bennett 1981; Le Grice and Matzura 1981). The sequence of the promoter for tetracycline resistance of the *E. coli* vector pBR322 has been communicated by J. Lowrie and J. Hedgpeth prior to publication. The tetracycline resistance promoter shows almost complete homology to the σ^{55} promoter, this gene from *E. coli* is readily expressed in *B. subtilis*, whereas the two other promoters with less homology do not function in *B. subtilis*. Sequences homologous to the ϱ^{55} promoter are underlined

Secretion of enzymes into the periplasmic space in the case of Gram-negative bacteria such as *E. coli* or directly into the medium as in the case of

Bacillus species greatly facilitates enzyme purification. In addition the secreted enzyme escapes from regulation mechanisms which otherwise may decrease enzyme yields. Attempts to use the cloned genes for the periplasmic enzyme alkaline phosphatase of *E. coli* and the exoenzyme α-amylase from *B. amyloliquefaciens* for the secretion of other enzymes the genes for which have been fused to these two genes were only partially successful (I. Palva, personal communication; Yoda et al. GIM '82).

Stability of recombinant DNA containing enzyme genes is an important prerequisite for the application of this technique to industrial production processes. Rapid loss of the entire recombinant DNA from a culture is caused by segregational instability. This can be prevented by the use of specially designed vectors (S. Chang, personal communication), multicopy integration vectors (Saito et al. GIM '82) being of particular interest for this purpose.

The problem is of course very important under conditions of continuous culture, e.g. in a chemostat where even vector plasmids with otherwise sufficient stability are lost (Noack et al. 1981). The loss of only parts of recombinant DNA molecules (structural instability) is a phenomenon often encountered, in particular with *B. subtilis* as a host (Kreft et al. 1982). Unfortunately no general method to solve this problem is available at present.

3. Conclusions

For several microorganisms relevant to enzyme production, model systems for the application of the recombinant DNA technique been exploited. In some cases these methods led to considerable improvements. The large scale application of these methods to enzyme technology will depend on further basic research to obtain a deeper insight into fundamental processes. The development of tailor-made cloning systems and a lot of project-specific research will be equally important. This is especially valid for enzymes involved in multistep biosynthetic pathways and for improvements in fermentation and bioreactor processes.

This work was supported by a grant from the Deutsche Forschungsgemeinschaft (SFB 105 — A-11).

References

Aubert E, Fargette F, Fouet A, Klier A, Rapoport G (1982) Use of a bifunctional cosmid for cloning large DNA fragments of *B. subtilis*. In: Ganesan AT, Chang S, Hoch JA (eds.) Molecular cloning and gene regulation in bacilli. Academic Press, New York, p. 11–24

Bernhard K, Schrempf H, Goebel W (1978) Bacteriocin and antibiotic resistance plasmids in *Bacillus cereus* and *Bacillus subtilis*. J. Bact 133:897–903

Bingham AHA, Bruton CJ, Atkinson T (1979) Isolation and partial characterization of four plasmids from antibiotic-resistant thermophilic bacilli. J Gen Microbiol 114:401–408

Bolivar F, Rodriguez RL, Greene PJ, Betlach MC, Heyneker HL, Boyer HW, Crosa JH, Falkow S (1977) Construction and characterization of new cloning vehicles. II. A multipurpose cloning system. Gene 2: 95–113

Buckel P, Zehelein E (1981) Expression of *Pseudomonas fluorescens* D-galactose dehydrogenase in *E. coli*. Gene 16: 149–159

Chang S, Cohen SN (1979) High frequency transformation of *Bacillus subtilis* protoplasts by plasmid DNA. MGG 168: 111–115

Chater KF, Hopwood DA, Kieser T, Thompson CJ (1982) Gene cloning in Streptomyces. Current Topics Microbiol Immunol 96: 69–95

Collins J (1977) Gene cloning with small plasmids. Current Topics Microbiol Immunol 68: 121–170

Collins J, Hohn B (1978) Cosmid: a type of plasmid genecloning vector that is packageable in vitro in bacteriophage λ heads. Proc Natl Acad Sci USA 75:4242–4246

Ehrlich SD (1977) Replication and expression of plasmids from *Staphylococcus aureus* in *Bacillus subtilis*. Proc Natl Acad Sci USA 74:1680–1682

Ehrlich SD (1978) DNA cloning in *Bacillus subtilis*. Proc Natl. Acad Sci USA 75:1433–1436

Ehrlich SD, Niaudet B, Michel B (1982) Use of plasmids from *Staphylococcus aureus* for cloning of DNA in *Bacillus subtilis*. Current Topics Microbiol Immunol 96:19–29

Fouet A, Klier A, Rapoport G (1982) Cloning and expression in *E. coli* of the sucrase gene from *Bacillus subtilis*. MGG 186:399–404

Goebel W (1979) Möglichkeiten und Gefahren bei der Anwendung moderner genetischer Techniken. Naturwiss Rdsch 32:265–273

Goebel W (1980) Plasmide als Kloniervehikel. Arzneim Forsch 30:533–540

Gryczan TJ, Dubnau D (1978) Construction and properties of chimeric plasmids in *Bacillus subtilis*. Proc, Natl, Acad Sci USA 75:1428–1432

Hinnen A, Hicks JB, Fink G (1978) Transformation of yeast. Proc Natl Acad Sci USA 75:1929–1933

Hinnen A, Meyhack B (1982) Vectors for cloning in yeast. Current Topics Microbiol Immunol 96:101–117

Hohn B, Murray K (1977) Packaging recombinant DNA molecules into bacteriophage particles in vitro. Proc Natl Acad Sci USA 74: 3259–3263

Hollenberg CP (1982) Cloning with 2-μm DNA vectors and the expression of foreign genes in *Saccharomyces cerevisiae*. Current Topics Microbiol Immunol 96: 119–144

Kreft J, Bernhard K, Goebel W (1978) Recombinant plasmids capable of replication in *B. subtilis* and *E. coli*. MGG 162: 59–67

Kreft J, Parrisius J, Burger KJ, Goebel W (1982) Expression and instability of heterologous genes in *B. subtilis*. In: Ganesan AT, Chang S, Hoch JA (eds.) Molecular cloning and gene regulation in bacilli. Academic Press, New York,, pp. 145–157

Kreft J, Hughes C (1982) Cloning vectors derived from plasmids and phage of Bacillus. Current Topics Microbiol Immunol 96: 1–17

Le Grice SFJ, Matzura H (1981) Binding of RNA polymerase and the catabolite gene activator protein within the *cat* promoter in *E. coli*. J Bact 150: 185–196

Losick R, Pero J (1981) Cascades of sigma factors. Cell 25: 582–584

Mandel M, Higa A (1970) Calcium-dependent bacteriophage DNA infection. J Mol Biol 53: 159–162

Marrero R, Chiafari FA, Lovett PS (1981) SPO2 particles mediating transduction of a plasmid containing SPO2 cohesive ends. J Bact 147: 1–8

Moran CP, Lang N, LeGrice SFJ, Lee G, Stephens M, Sonenshein AL Pero J, Losick R (1982) Nucleotide sequences that signal the initiation of transcription and translation in *Bacillus subtilis*. MGG 186: 339–346

Nagahari K, Koshikawa T, Sakaguchi K (1980) Cloning and expression of the leucine gene from *Thermus thermophilus* in *E. coli*. Gene 10: 137–145

Noack D, Roth M, Geuther R, Müller G, Undisz K, Hoffmeier C, Gaspar S (1981) Maintenance and genetic stability of vector plasmids pBR322 and pBR325 in *E. coli* K12 strains grown in a chemostat. MGG 184: 121–124

Palva I (1982) Molecular cloning of α-amylase gene from *Bacillus amyloliquefaciens* and its expression in *B. subtilis*. Gene 19: 81–87

Priest FG (1977) Extracellular enzyme synthesis in the genus *Bacillus*. Bacteriol Rev. 41: 711–753

Russell DR, Bennett GN (1981) Characterization of the β-lacatamase promoter of pBR322. Nucleic Acids Res 9: 2517–2533

Setlow JK, Hollaender A (eds) (1979 and 1980) Genetic engineering. Plenum Press, New York and London

Spizizen J (1958) Transformation of biochemically deficient strains of *Bacillus subtilis* by deoxyribonucleate. Proc Natl Acad Sci USA 44: 1072–1078

Tanaka T, Kawano N (1980) Cloning vehicles for the homologous *Bacillus subtilis* host vector system. Gene: 131–136

Williams DM, Schoner RF, Duvall EJ, Preis LH, Lovett PS (1981) Expression of *E. coli trp* genes and the mouse dihydrofolate reductase gene cloned in *B. subtilis*. Gene 16: 199–206

Wu R (ed) (1978) Methods in Enzymology, vol 68 Academic Press, New York

Young FE (1980) Impact of cloning in *Bacillus subtilis* on fundamental and industrial microbiology. J Gen Microbiol 119: 1–15

Induction of Cellulases in *Trichoderma reesei*

Hans-Peter Hohn and Hermann Sahm

Summary

Various synthetic derivatives of glucose and several different disaccharides were shaken with washed mycelia of *Trichoderma reesei* QM 9414 to test for CM-cellulase and for β-glucosidase activity. The extent of induction always increased with increases in the concentration of the inducers. Lactose and sophorose were good inducers, the latter being about as good as cellulose itself. Weak inducers were methyl-glucopyranoside, isopropyl-thioglucopyranoside, cellobiose and cellobionic acid, although cellobionic acid gave higher enzyme concentrations than cellobiose. One of the synthetic derivatives, tetramethyl-gluconolactone, induced only traces of cellulase and only when glycerol was also present but it did induce appreciable concentrations of β-glucosidase. The presence of glycerol affects the synthesis of cellulases and β-glucosidases in different ways. Cellulase formation is delayed except on sophorose, while β-glucosidase excretion is diminished. This different pattern indicates that there are different induction mechanisms for the two enzymes. Furthermore, sophorose seems to be a direct inducer for cellulases.

Introduction

During the last few years world-wide the interest in renewable resources of chemicals and liquid fuels has grown and is now very strong because everyone is aware of the progressive exhaustion of conventional sources of energy. Cellulose is such a renewable resource, one which is always produced widely in nature and by only the transformation of solar energy, water and carbon dioxide in photosynthesis. That is why there is such growing interest in the enzymatic hydrolysis of cellulose.

Cellulases are inducible enzymes synthesized and mostly excreted into the environment by a number of fungi and bacteria during their growth on cellulosic materials. They also can be induced by many oligomeric and dimeric sugars including some of their derivatives, all of which are metabolized by the organisms. Up to now, however, very little is known about the induction mechanisms for the enzymes. It would be a big step forward, especially for the commercial utilization of cellulases, if a strong inducer could be found which is not degraded by the organisms.

The best known example for so-called "gratuitous induction" is the induction of β-galactosidase by isopropyl-thiogalactoside in *E. coli*. The synthetic analogue of lactose — the real substrate of the enzyme — is an even better inducer than lactose itself. Furthermore, it could be shown that inside the cell small amounts of lactose must first be converted to another galactoside, called allolactose, which is the real inducing compound. Therefore — in contrast to the more active synthetic analogues — lactose is not the direct inducer of β-galactosidase synthesis.

From this work of J. Monod (1956) just cited, one can surmise that cellobiose and other oligomers of cellulose are not direct inducers since they have been found to be very weak inducers compared to cellulose itself. Starting from the thesis that these oligomers have to be converted to a directly inducing compound by the organism, we have tried to find a synthetic analogue of cellobiose which is able to induce large amounts of cellulases, but which cannot be metabolized itself.

The experiments were carried out with the imperfect fungus *Trichoderma reesei* strain QM 9414 which is known to be one of the most effective producers of cellulases.

Materials and Methods

Chemicals — All substances tested for induction, except cellobionic acid and tetramethylgluconolactone were commercial products of analytical grade from E. Merck AG and Serva Fine Chemicals GmbH. Cellobionic acid was synthesized in our Institute by Dr. K. Hamacher, as described by Lindgren and Nilson (1973). Also tetramethyl-gluconolactone was synthesized as described by P. Moeller (1972).

Preparation of Washed Mycelia — *Trichoderma* was grown on a mineral salts medium at pH 5.0 according to Mandels et al. (1962) containing 0.075% protease peptone (Difco) and 0.2% Tween 80. The carbon source was 1% glycerol since it is reported to induce no cellulase (Vaheri et al. 1979). Precultures of 50 ml were inoculated from agar slants and grown at 27 °C with rotary shaking at 110 rpm for 48 hours. They were then added to 500 ml of culture medium in 2 l baffled Erlenmeyer flasks. After incubation for 24 hours under the same conditions, the mycelium was harvested by sterile centrifugation at 15,000 g. It was then washed several times with sterile mineral salts medium at pH 3.0 containing only Tween 80. The pellets were finally dispersed in this solution at concentrations of about 2 mg dry weight per ml.

Conditions for Inductive Enzyme Formation — 50 ml portions of washed mycelia were dispersed in sterilized 500 ml Erlenmeyer shake flasks. The inductions were started by adding the substances to be tested, concentrated in 1 ml solution and sterilized by ultrafiltration. The flasks were then incubated at 27 °C on a rotary shaker at 110 rpm. All test substances were employed at concentrations from 10^{-6} M to 10^{-2} M. One series contained only the test substance, another series contained in addition 0.2% glycerol which would enable the cells to produce enzyme protein in case that a putative inducer could not be metabolized by the organism. There was always one flask in each series which contained either no test substance or contained only glycerol to serve as controls. Samples were taken at different times and assayed for cellulase, β-glucosidase and concentration of added test substance.

Enzyme Assays — Cellulase concentrations were assayed by the determination of reducing sugars liberated from carboxymethylcellulose (CM-cellulase) during incubation with culture filtrate (Mandels and Weber, 1969). Enzyme activities are expressed as international units per ml. One unit is that amount of enzyme that liberates 1 μmole of glucose equivalents per minute.

β-glucosidase was determined according to Wood (1971) as aryl-β-glucosidase by the hydrolysis of p-nitrophenyl-β-glucopyranoside. The concentration of p-nitrophenol was estimated from the extinction at 405 nm under alkaline conditions. One enzyme unit is defined as that amount of enzyme that catalyzes the cleavage of 1.0 μmole per min.

Concentrations of additives — Reducing compounds were measured by the method of Nelson (1944). Cellobionic acid, isopropyl-thioglucoside and -galactoside were determined with the anthrone-method according to Herbert et al. (1971). Tetramethyl-gluconolactone was assayed as described by Moeller (1972). The concentration of glycerol was estimated from the intensity of the spots on thin layer chromatographs in accordance with Gübitz et al. (1976).

Results and Discussion

One of the principal observations was that in every case of induction the extent was closely correlated to the concentration of inducer; i.e., the higher the concentration of inducer, the higher was the level of induction. An optimal concentration such as described by Nisizawa et al. (1971) and by Sternberg and Mandels (1979) for 10^{-3} M sophorose was never found. Therefore, we report here only on results at the highest concentrations tested. Furthermore, we never detected cellulase activities in the controls, neither with nor without glycerol. In both types of controls β-glucosidase appeared after a certain time, but this must be ascribed to the release of constitutive cell-bound enzymes by cell lysis (Fig. 14/15). This must also be the reason for the second rise in the concentration of this enzyme after a first peak — a phenomenon that was observed in most of the experiments.

Synthetic Derivatives of Glucose

Three synthetic additives were tested: Isopropyl-thioglucopyranoside, methyl-β-D-glucopyranoside and tetramethyl-gluconolactone. None was very effective as an inducer of enzymes, but none was appreciably metabolized either.

Isopropyl-thioglucopyranoside is the derivative of glucose corresponding to isopropyl-thiogalactoside. Therefore this compound was tested first as inducer for cellulases presuming similar mechanisms of induction for those for β-galactosidase. Fig. 1 shows some induction of enzyme activity, but the excretion occurred only very slowly and in relatively small amounts measured in milli-units. None of the added thioglucoside was metabolized; the apparent increase in its concentration was probably due to liberation of other sugars by cell lysis. Failure to induce could not be attributed to lack of a metabolizable carbon source, however, since Fig. 2 shows that enzyme levels were essentially unchanged when glycerol was added at the start of the experiment. Slight differences in the rate of enzyme production between Fig. 1 and Fig. 2 may or may not be significant. In any case, this derivative should be regarded as unsuitable for gratuitous induction. Formation of enzymes was so slow that in both series of experiments we observed a progressing cell lysis beginning at about the same time as enzyme production.

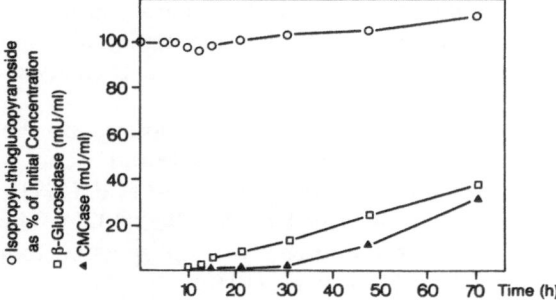

Figure 1. Induction of cellulases by 10^{-2} M isopropylthioglucopyranoside of *Trichoderma reesei*. Washed mycelia (about 2.0 mg dry weight per ml) were incubated with the inducer at 27 °C on a rotary shaker at 110 rpm. Samples were assayed for CM-cellulase, β-glucosidase and the concentration of the compound to be tested for induction

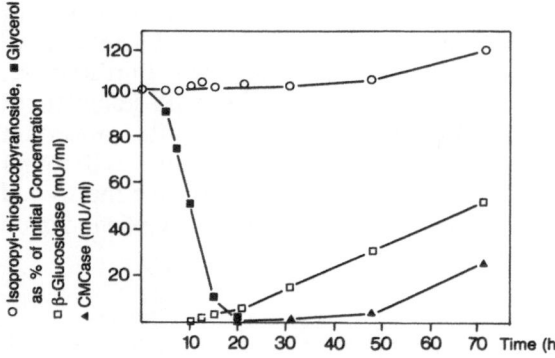

Figure 2. Induction of cellulases by 10^{-2} M isopropyl-thioglucopyra-noside $+0.2\%$ glycerol. Experimental conditions as described in Fig. 1

For a second experiment we chose methyl-β-D-glucopyranoside, also a derivative of glucose but with another substituent in the C_1-position. Actually it underwent a slow but steady degradation by the mold both in the presence and in the absence of glycerol; after 96 hours only about 10% was still left in the culture fluid. Both of the tested enzymes were induced, but only to a level of 70 mU/ml after 96 hours (Fig. 3). An initial dose of 0.2% glycerol stimulated the synthesis of the two enzymes in different ways (Fig. 4). The level of CM-cellulase was raised to almost double its value, 140 mU/ml in 96 hours, after a retardation of induction for 5 hours. The β-glucosidase, however, was produced

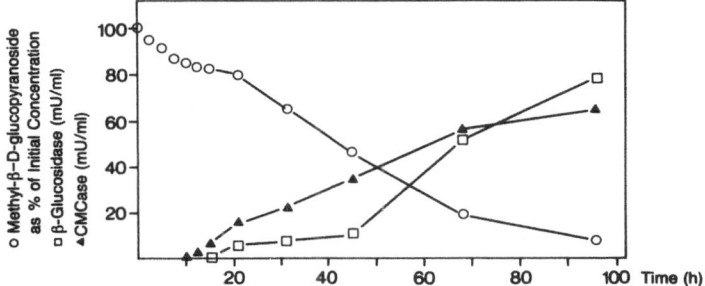

Figure 3. Induction of cellulases by 10^{-2} M methyl-β-D-glucopyranoside. Experimental conditions as described in Fig. 1

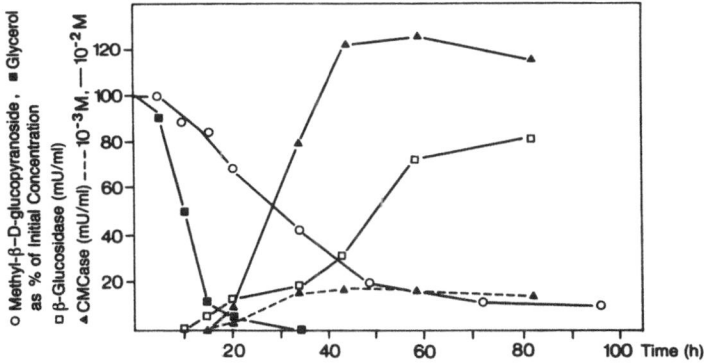

Figure 4. Induction of cellulases by 10^{-2} M methyl-β-D-glucopyranoside +0.2% glycerol. Experimental conditions as described in Fig. 1

earlier but with the same concentrations as without glycerol. From these results methyl-β-D-glucopyranoside also seems to be unsuitable as a gratuitous inducer.

Another possibility for a synthetic inducer is suggested in the work of Bruchmann (1978) who investigated the induction of cellulases by lactones. In our experiments tetramethyl-gluconolactone induced no cellulases at all when given as the sole substrate. When glycerol was employed as a supporting carbon source, however, the concentration of β-glucosidase rowe to a peak of 100 mU/ml after the rapid consumption of the glycerol (Fig. 5). CM-cellulase was only detectable in trace amounts of about 7 mU/ml. This latter result substantiates the observation of Bruchmann that only an exo-glucanase is induced by tetramethyl-gluconolactone using another strain of *Trichoderma reesei*. Only about 10% of the added lactone were metabolized in our experiments.

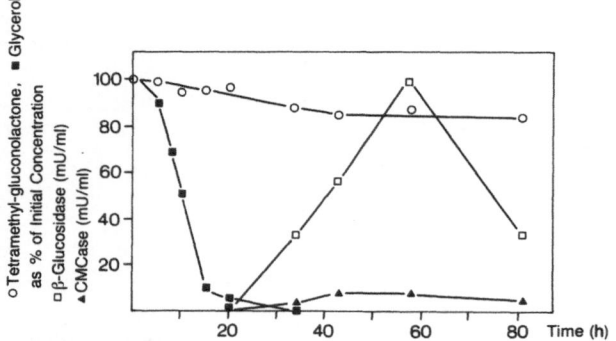

Figure 5. Induction of cellulases by 10^{-2} M tetramethyl-gluconolactone $+ 0.2\%$ glycerol. Experimental conditions as described in Fig. 1. No induction was observed without glycerol

Disaccharides and their derivatives

Since all the synthetic substrates failed to induce high levels of cellulolytic enzymes, we next cønsidered the early opinion of Mandels and Reese (1960) that cellobiose can be the natural inducer of cellulases since it is a natural degradation product of cellulose. On cellobiose the fungus synthesized cellulase much more rapidly than on any synthetic material tested so far (Fig. 6). There was a primary peak of 90 mU/ml after an incubation period of 10 hours. The β-glucosidase shows a peak already at 7 hours, but the activity was not as high as that on various synthetic substrates. Cellobiose was degraded very quickly; after 15 hours about 90% had been metabolized. This rapid disappearance of the inducer may be the reason why the enzyme concentrations decline. Their subsequent rise could be ascribed to cell lysis and growth on leakage products from dead cells.

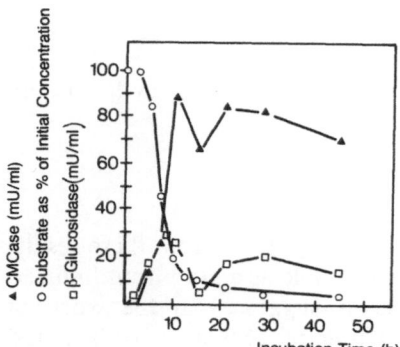

Figure 6. Induction of cellulases by 10^{-2} M cellobiose. Experimental conditions as described in Fig. 1

From Fig. 7 it appears that cellobiose is preferred to glycerol as a substrate for assimilation. Consequently, we find a similar course of enzyme activities as before. There are some differences, however. The cellulase excretion was delayed, but the maximum was appreciably higher probably because on the basis of glycerol the organism grew to a greater extent. The β-glucosidase was affected in another way. The peak of activity occurred at the same time but was essentially lower. Furthermore, the enzyme disappeared for about

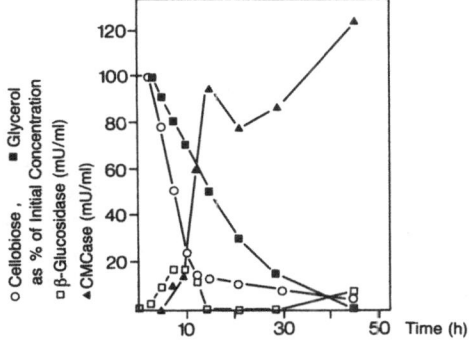

Figure 7. Induction of cellulases by 10^{-2} M cellobiose + 0.2% glycerol. Experimental conditions as described in Fig. 1

15 hours. These facts may be interpreted as a repression by glycerol, but there seem to exist different induction mechanisms for every kind of enzyme.

If we consider the enzyme levels induced with cellulose itself — about 10 U/ml of cellulase and about 0.5 U/ml of β-glucosidase — all these above mentioned levels are disappointingly low. The reason must be the rapid degradation of cello-

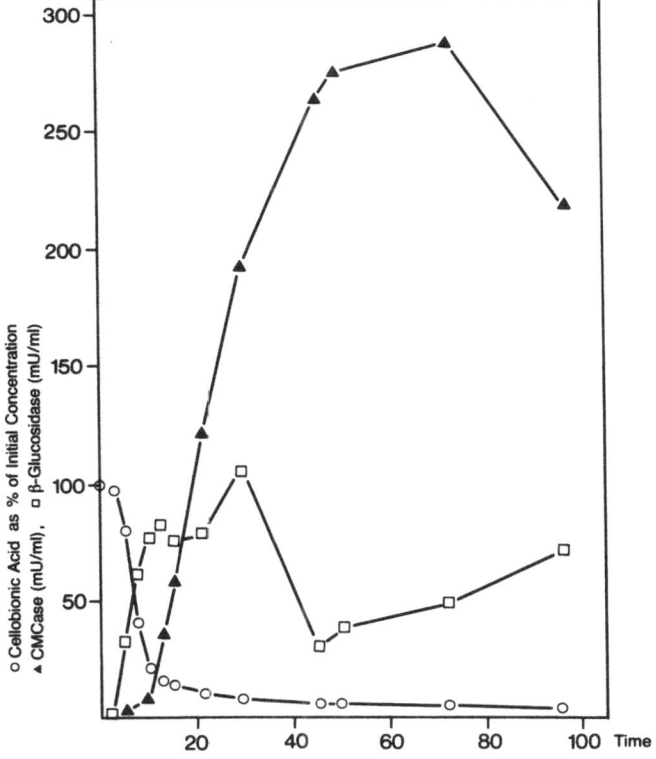

Figure 8. Induction of cellulases by 10^{-2} M cellobionic acid. Experimental conditions as described in Fig. 1

biose and its hydrolysis by β-glucosidases to glucose which is a potent repressor of cellulase synthesis (Nisizawa et al. 1972). During the degradation of cellulose, cellobiose is liberated only at low levels so that its cleavage does not provide enough glucose for repression.

To alleviate possible glucose repression, it was decided to replace the cellobiose with cellobionic acid which would give only one glucose instead of two per molecule, if it could be hydrolyzed at all. Furthermore, the acid was reported to be formed in some cellulolytic fungi from cellobiose by oxidases or dehydrogenases (Ayers et al. 1978; Westermark and Eriksson 1975; Dekker 1980; Coudray et al. 1982). In addition, first results of Eriksson and Westermark (1974) indicate the existence of such an enzyme in *Trichoderma reesei*. Also Canevascini et al. (1979) observed induction of cellulases by cellobionic acid in *Sporotrichum thermophile*.

When the washed mycelia of *Trichoderma reesei* were incubated with cellobionic acid, induction relationships were similar to those for growth on cellobiose (Fig. 8), but with much higher enzyme activities. Glycerol, when added

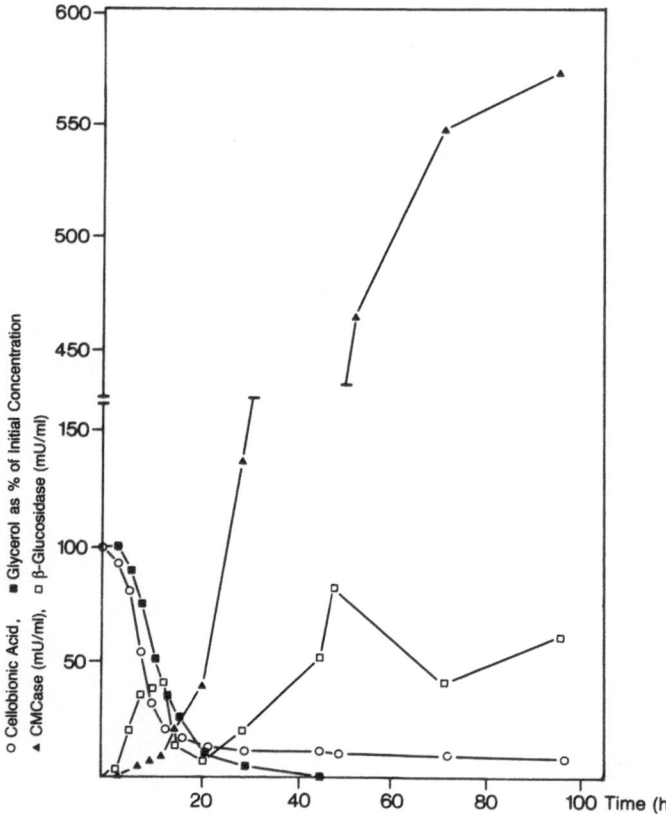

Figure 9. Induction of cellulases by 10^{-2} M cellobionic acid + 0.2 % glycerol. Experimental conditions as described in Fig. 1

as an additional carbon source (Fig. 9), caused a further rise in the CM-cellulase concentration from 290 mU/ml to 570 mU/ml. Enzyme synthesis was initially delayed until glycerol and cellobionic acid had been metabolized. Also the formation of β-glucosidase was repressed by glycerol as had been also observed with cellobiose as an inducer. Both metabolites seem to be consumed at almost the same time with a slight preference for the sugar acid.

Since the rapid utilization of cellobionic acid was a surprise, we tried to elucidate the fate of this compound in the metabolism of *Trichoderma*. It appears to be hydrolyzed by several β-glucosidases having different affinities for cellobiose, cellobionic acid and carboxymethylcellulose. The enzymes were found in crude extracts of mycelia as well as in the culture fluid. Lactonase, as reported for *Aspergillus niger* (Bruchmann 1978), appears to be unnecessary for their hydrolytic action on cellobionic acid.

Even the amounts of enzymes induced by cellobionic acid are not sufficient to be of practical interest, and it cannot be regarded as a gratuitous inducer because of its rapid metabolism. An important aspect is that the critical concentration for induction is not maintained for a sufficient period if the additive is metabolized as easily as cellobiose or its acid. Lactose, however, was reported by Mandels and Reese (1960) to be a rather poor growth substrate for *Trichoderma*, but a better inducer than cellobiose. Therefore, lactose was also tested using the present experimental conditions. Addition of lactose without glycerol (Fig. 10) yielded 2 U of CM-cellulase per ml, a value higher than on any

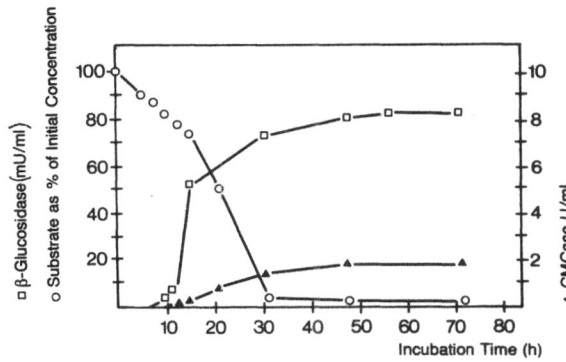

Figure 10. Induction of cellulases by 10^{-2} M lactose. Experimental conditions as described in Fig. 1

other compound tested so far. The β-glucosidase reached levels similar to induction by cellobionic acid, in the range of 80 mU/ml. If glycerol is employed together with lactose (Fig. 11) it is metabolized before lactose. There is only a slight disappearance of lactose during the rapid utilization of most of the glycerol. We assume that this delay in uptake of lactose is the reason why in this culture the synthesis of both enzymes is delayed. But the formation of the cellulase is retarded much longer than that of β-glucosidase. Growth on glycerol and lactose combined results in cellulase activity raised to 4 U/ml.

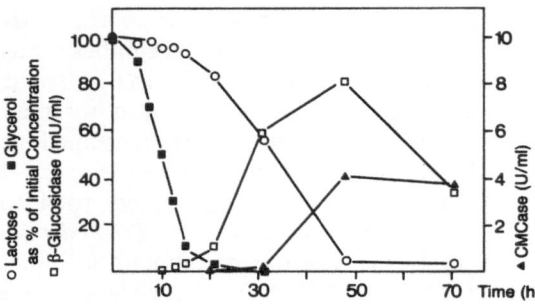

Figure 11. Induction of cellulases by 10^{-2} M lactose + 0.2% glycerol. Experimental conditions as described in Fig. 1

The concentration of β-glucosidase is not affected, perhaps because of a repression by remaining low levels of glycerol.

The best inducer of CM-cellulase was found to be sophorose, whereas the β-glucosidase showed congruent results with the other experiments (Figs. 12, 13, 14). Cellulase activities of about 6 U/ml were achieved after the rapid degradation of the sophorose. With the combination of sophorose and glycerol levels of even 11 U of CM-cellulase per ml were reached (Fig. 13). These enzyme concentrations are as high as those induced by cellulose itself. Glycerol is attacked only after the consumption of sophorose and presence of glycerol does not delay induction of the CM-cellulase. The synthesis of β-glucosidase, however, is repressed in the same manner as in other experiments where glycerol was used (Fig. 15). Perhaps the greatest difference between sophorose and other inducers is the fact that the induction at a concentration of 10^{-2} M occurs later than at a concentration of 10^{-3} M (Fig. 12, 13). The reason for the delay in the induction of the cellulase at higher concentrations may lie in the fact that sophorose itself is cleaved to glucose by β-glucosidases (Sternberg and Mandels, 1980). Higher concentrations of glucose will in turn repress the enzyme formation. This delay as a function of sophorose concentration may also explain why others have reported 10^{-3} M to be the most effective concentration for induction

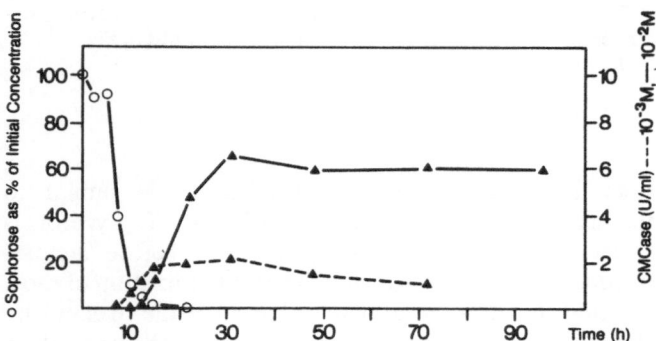

Figure 12. Induction of CM-cellulase by 10^{-3} M and 10^{-2} M sophorose. Experimental conditions as described in Fig. 1

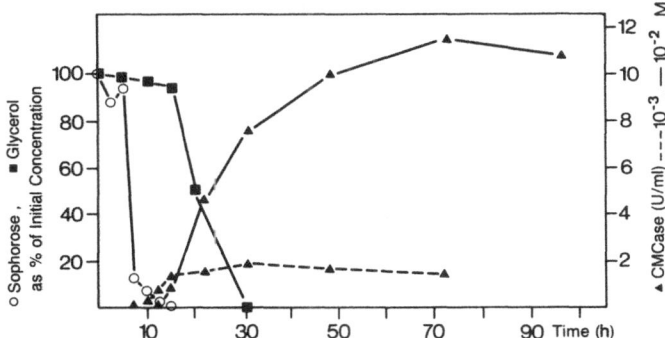

Figure 13. Induction of CM-cellulase by 10^{-3} M and 10^{-2} M sophorose $+2.8\%$ glycerol. Experimental conditions as described in Fig. 1

(Nisizawa et al. 1974a, b; Sternberg and Mandels, 1979). For example, observations made at 10–12 hours (see Figs. 12, 13) might lead to the conclusion that 10^{-2} M is inferior to 10^{-3} M sophorose. It is necessary to observe the course of enzyme formation over a sufficient time for repressive effects to disappear.

In most of the experiments with inducers, the formation of CM-cellulase and the formation of β-glucosidase are affected in different ways by addition of glycerol. The production of cellulase is delayed for a few hours by glycerol whereas the β-glucosidase is excreted in the same time pattern with or without glycerol; only its levels are lower in the latter case. This indication of different regulation patterns for the two enzymes is expressed for sophorose in another way. For sophorose, the induction of cellulase is not delayed after addition of glycerol while induction of β-glucosidase parallels its behaviour with other additives (Fig. 14, 15).

The conclusion from these results might be that sophorose is the natural or at least a direct inducer for cellulases, since if it is incubated along with glycerol it induces as much cellulase formation as cellulose itself. Furthermore, in this case the cellulase appears without any time lag as compared to the culture without glycerol. However, the question arises how sophorose can induce cellulolytic

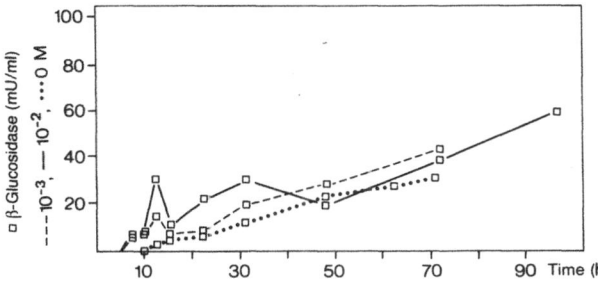

Figure 14. Induction of β-glucosidase by 10^{-3} M and 10^{-2} M sophorose and formation of the enzyme without substrate. Experimental conditions as described in Fig. 1

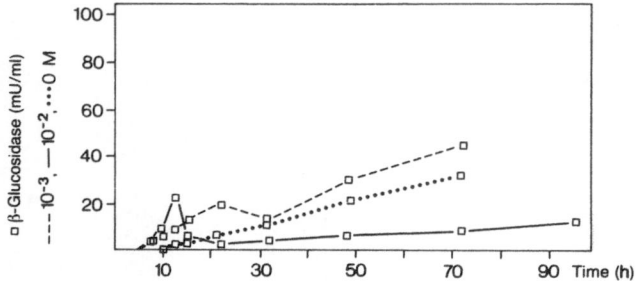

Figure 15. Induction of β-glucosidase by 10^{-3} M and 10^{-2} M sophorose + 0.2% glycerol and formation of the enzyme with glycerol as the only substrate. Experimental conditions as described in Fig. 1

Table 1. Maximum Activities, for all substrates at 10^{-2} M Concentrations

Substrate	Without Glycerol		+0.2% Glycerol	
	CMCase	β-Glucosidase	CMCase	β-Glucosidase
Sophorose	6.5	0.032	11.6	0.022
Lactose	1.8	0.082	4.1	0.102
Cellobionic Acid	0.290	0.107	0.570[a]	0.080
Cellobiose	0.084	0.028	0.124	0.017
Isopropyl-thiogalactoside	0	0.026[a, b]	0	0.060[a]
Isopropyl-thioglucoside	0.031[a]	0.024[a, b]	0.025[a]	0.031[a, b]
Methyl-β, D-glucopyranoside	0.061[a]	0.051	0.127	0.080[a]
Tetramethylgluconolactone	0	0.023[a, b]	0.008	0.099
Control without Substrate	0	0.022[a]	0	0.033[a]

[a] increasing activities during the cultivation without maximum
[b] no deviation from the control all activities expressed as U/ml

enzymes if it is not a natural degradation product of cellulose. An explanation may come from the finding that *Trichoderma* cellulase possesses transglycosylation activity enabling it to form sophorose from cellulose and cellobiose (Vaheri et al. 1979). This reaction could take place at the beginning of cellulose attack from constitutive levels of cellulolytic enzymes in *Trichoderma* (Nisizawa et al. 1971 a).

If sophorose is the sole substrate, it must serve as an enzyme inducer and as a carbon source at the same time. By this means the effective concentration of inducer is diminished. The combination of sophorose with glycerol as an additional carbon source may simulate the substrate relationships of cellulose degradation according to the following model: While sophorose is consumed it is adsorbed to the effector that mediates the start of DNA-transcription. After this substrate is exhausted, the adsorbed inducer is utilized as a carbon source. In the presence of glycerol, however, the inducer remains adsorbed and the organism maintains its anabolism at the expense of glycerol. For induction with cellulose the role of glycerol is replaced by the continuously released degradation products of cellulose itself. The proof for this model and for sophorose being the real inducer of cellulases in *Trichoderma* is missing at the moment, but it might be provided if sophorose can be shown to arise from transglycosylation in all mycelia grown on any inducing compound. If sophorose is the natural inducer for cellulases, then the induction of β-glucosidases must be ascribed to another degradation product of cellulose since sophorose yields only a weak induction of this enzyme as compared to cellulose. Sternberg and Mandels (1980) even report repression of β-glucosidase formation by sophorose for the *Trichoderma reesei* — wild type QM 6a strain.

In further experiments we intend to investigate the inducing capacity of thiocellobiose on *Trichoderma*-cellulases. Thiocellobiose was recently reported to be a good inducer of cellulases in *Schizophyllum commune* without being metabolized (RHO et al. 1982).

Acknowledgements

We would like to thank Dr. K. Hamacher for the synthesis of 2.3.4.6-tetra-methyl-gluconolactone and cellobionic acid, and Prof. Dr. R. K. Finn for the correction of the manuscript.

References

Ayers AR, Ayers SB, Eriksson K-E (1978) Cellobiose Oxidase, Purification and Partial Characterization of a Hemoprotein from *Sporotrichum pulverulentum*, Eur J Biochem 90:171–181

Bruchmann EE (1978) Lactone, Reduktone und enzymatische Celluloseverzuckerung, Chemiker-zeitung 102/11:387–389

Canevascini G, Coudray M-R, Rey J-P, Southgate RJG, Meier H (1979) Induction and Catabolite Repression of Cellulase Synthesis in the Thermophilic Fungus *Sporotrichum thermophile*, J Gen Microbiol 110:291–303

Coudray M-R, Canevascini G, Meier H. (1982) Characterization of a Cellobiose Dehydrogenase in the Cellulolytic Fungus *Sporotrichum (Chrysosporium) thermophile*, Biochem. J. 203:277–284.

Dekker RF (1980) Induction and Characterization of a Cellobiose Dehydrogenase Produced by a Species of *Monilia*, J Gen Microbiol 120:309–316

Eriksson K-E, Pettersson B, Westermark U (1974) Oxidation: An Important Enzyme Reaction in Fungal Degradation of Cellulose, FEBS Letters 49/2:282–208

Gübitz G, Frei R.W., Bethke H (1976): Fluorescence Densitometric Method for the Determination of Gluconic and Lactobionic Acids ("Sugar Acids") in Pharmaceutical Preparations, J Chromatogr 117:337–343

Herbert D, Phipps PJ, Strange RE (1971) Chemical Analysis of Cells. In "Methods in Microbiology", 5b, Norris and Ribbons (ed), Academic Press, Inc Ltd London

Lindgren BO, Nilsson T (1973) Preparation of Carboxylic Acids from Aldehydes (Including Hydroxylated Benzaldehydes) by Oxidation with Chlorite, Acta Chem Scand 27:888–890

Mandels M, Reese ET (1960) Induction of Cellulase in Fungi by Cellobiose, J Bacteriol 79:816–826

Mandels M, Parrish FW, Reese ET (1962) Sophorose as an Inducer of Cellulase in *Trichoderma viride*, J Bacteriol 83:400–408

Mandels M, Weber J (1969) The production of Cellulases. In "Cellulases and their Applications". Adv Chem Ser 95:391–414

Moeller P (1972) Darstellung von 2.3.4.6-Tetramethyl-D-glucono-1,5-lacton, Liebigs Ann Chem 755:191–193

Monod J (1956) Remarks on the Mechanism of Enzyme Induction. In: A Woff and A Ullmann (ed), Selected Papers in Molecular Biology by Jaques Monod, Academic Press, Inc New York, 313–334

Nelson N (1944) A Photometric Adaption of the Somogyj Method for the Determination of Glucose, J Biol Chem 53:375–380

Nisizawa T, Suzuki H, Nakayama M, Nisizawa K (1971) Inductive Formation of Cellulases by Sophorose in *Trichoderma viride*. J Biochem 70:375–385

Nisizawa T, Suzuki H, Nisizawa K (1971) "De novo" Synthesis of Cellulase Induced by Sophorose in *Trichoderma viride*

Nisizawa T, Suzuki H, Nisizawa K (1972) Catabolite Repression of Cellulase Formation in *Trichoderma viride*. J Biochem 71:999–1007

Rho D, Desrochers M, Jurasek L, Driguez J, Defaye J (1982) Induction of Cellulase in *Schizophillum commune*: Thiocellobiose as a New Inducer. J Bacteriol 149:47–53

Sternberg D, Mandels GR (1979) Induction of Cellulolytic Enzymes in *Trichoderma reesei* by Sophorose. J Bacteriol 139:761–769

Sternberg D, Mandels GR (1980) Regulation of the Cellulolytic System in *Trichoderma reesei* by Sophorose: Induction of Cellulase and Repression of β-glucosidase. J Bacteriol 144:1197–1199

Vaheri MP, Vaheri MEO, Kauppinen VS (1979) Formation and Release of Cellulolytic Enzymes During Growth of *Trichoderma reesei* on Cellobiose and Glycerol. European J Appl Microbiol Biotechnol 8:73–80

Vaheri MP, Leisola M, Kaupinen VS (1979) Transglycosylation Products of Cellulase System of *Trichoderma reesei*. Biotechnol Letters 1:41–46

Westermark U, Eriksson K-E (1975) Purification and Properties of Cellobiose: Quinone Oxidoreductase from *Sporotrichum pulverulentum*. Acta Chem Scand B 29:419–424

Wood TM (1971) The Cellulase of *Fusarium solani*, Purification and specifity of the β-(1,4)-Glucanase and β-D-Glucosidase Components. Biochem J 121:353–362

Extracellular Acid Protease of *Rhizopus rhizopodiformis*

J. Schindler, R. Lehmann, H. Pfeiffer and R. Schmid

Summary

A fungal strain, identified as *Rhizopus rhizopodiformis*, capable of producing extracellular acid protease has been isolated in the course of a special screening program. The strain reaches its maximum of protease synthesis after 48 hrs of submerged fermentation. The process of fermentation and enzyme recovery is described. The protease shows a pH-optimum of about 3, is stable between pH 3 and pH 6 at temperatures of 40 °C and has a temperature optimum of 60 °C.

Introduction

Microbial enzymes find increasing use on a large technical scale. The alkaline *Bacillus* protease for detergents and cleansing agents constitute an illustrative example for this development (Aunstrup 1979, 1980; Fogarty and Kelly 1979; Ruttloff 1981).

The requirements of technologically usable enzymes with respect to substrate specificity, pH and temperature optimum or stability vary widely. The search for new enzymes for specific applications therefore implies frequently the search for completely new microorganisms. The goal of the described effort was a screening for acid proteases suitable as digestive enzymes for animal nutrition which might, e.g., facilitate a better nutritional yield. The proteases should be characterized by a low pH optimum and a high temperature stability. The enzyme source for the process to be developed should be microbial, the enzyme to be isolated should be extracellular, and the production strain should be suitable for submerged fermentations without losing its enzymatic properties.

Classification of Microorganisms and their Proteases

The ability to produce extracellular proteases is wide-spread among microorganisms. The enzymatic properties of these proteases of widely different taxonomic origin also vary widely. Thus one distinguishes between alkaline serine proteases (EC. 3.4.21), SH proteases (EC. 3.4.22), acid proteases (EC. 3.4.23), and neutral metalloproteases (EC. 3.4.24). Often certain relationships may be deduced from the taxonomy of a strain and the properties of the enzymes produced by it. These may serve as a first orientation aid when screening for new protease producers. Thus, alkaline proteases are preferably formed by bacilli and acid proteases by fungi. However, one species produces frequently more than one enzyme type (Table 1).

Among the fungi one distinguishes again between strains — such as those of the genus *Aspergillus, Penicillium, Rhizopus, Paecilomyces, Alternaria,* and *Trametes* — secreting pepsin-like proteases and others — such as *Aspergillus, Mucor, Endothia,* and *Byssochlamys* — which form rennin-like proteases. Thermostable enzymes are to be expected from microorganisms growing preferably at higher temperatures, i.e., representatives of the genus *Humicola, Mucor, Sporotrichum, Talaromyces, Thermoascus* etc. (Adams 1978; Cooney and Emerson 1967; Ong and Gaucher 1972; Ong and Gaucher 1976).

Table 1. Microbial sources of different protease types

Organisms	Proteases
Alternaria, Aspergillus, Byssochlamys Endothia, Mucor, Paecilomyces, Penicillium, Rhizopus, Trametes	Acid Proteases
Aeromonas, Bacillus, Pseudomonas Streptomyces Aspergillus	Neutral Proteases
Arthrobacter, Bacillus Streptomyces Alternaria, Aspergillus, Gliocladium, Penicillium, Sorangium	Alkaline Proteases

Screening and Strain Properties

The proper choice of selective enrichment and isolation techniques is of decisive importance for a successful screening program for microbial specialists. The above mentioned systematic relationships may serve as a first starting point in such an effort.

Acidic enrichment media containing fungus-specific nutrients and proteins as a nitrogen source were inoculated with soil samples of widely varying origin. The cultures were incubated at 45 °C and the mixed populations subsequently on casein, azocasein or fibrin blue agar plates. Pure cultures were obtained from colonies exhibiting strong caseolytic haloes. The demonstration of extracellular proteases by the observation of caseolysis on casein agar plates is a well known technique for the isolation of proteolytic microorganisms.

A few hundred fungal strains forming extracellular proteases were isolated in this manner. All strains were grown in shake cultures for the selection of the most active protease producers. One of the most interesting strains was identified as *Rhizopus rhizopodiformis.*

Literature studies on *R. rhizopodiformis* are rendered difficult by the fact that this species has a number of synonyms of which *R. chinensis* or *R. cohnii* are probably the best known (Inui et al. 1965).

The majority of lower fungi usually grow optimally at temperatures about 30 °C. Especially a few representatives of the genus *Rhizopus* positively prefer higher temperatures (Wolf and Wolf 1969). *R. rhizopodiformis* thus tolerates temperatures of up to 50 °C for its growth.

Thus, this species does not only fulfill the temperature stability requirements but also seems interesting as a protease producer as was already demonstrated by a broad screening program by Ellis et al. (1974) (Table 2).

Table 2. Proteolytic and amylotytic activities in strain of *Rhizopus Species*

Species	Number of strains demonstrating activity				
	Milk clotting		Amylolytic		Strains
	Rice*	Wheat*	Rice*	Wheat*	tested
R. rhizopodiformis (Cohn & Lichth.) Zopf	7	62	4	62	65
R. microsporus van Tiegh.	29	79	12	79	82
R. chinensis Saito	1	32	0	33	35
R. oligosporus Saito	0	22	0	28	31
R. hangchao Yamazaki	1	5	0	6	6
R. niveus Yamazaki	3	7	3	7	7
R. arrhizus Fischer	9	81	8	79	82
R. oryzae Went & P. Geerligs	3	13	2	14	14
R. chungkuoensis Yamazaki	0	2	0	2	2
R. stolonifer (Ehrenb. ex. Fr.) Vuill.	0	0	0	0	23
C. oryzae Went & P. Geerligs	7	6	8	9	9
Total strains	60	309	37	319	356

*with 2 % rice flour or 2 % wheat flour as growth substrate

Fermentation

The isolated strain *R. rhizopodiformis* has been deposited at the Centraalbureau voor Schimmelcultures in Baarn (CBS 227.75). The culture conditions for this strain were first optimized in shake culture experiments and subsequently scaled up to a semitechnical scale in 1,000 l fermenters.

Since the strain synthesizes amylases as well, it grows on corn starch as a carbon source. However, a prehydrolysis of the starch results in an accelerated fermentation. Increased concentrations of the hydrolized starch obviously lead to a catabolite repression of protease formation. The strain requires proteins such as casein or gelatine as nitrogen source and protease inductor. The strain only synthesizes insignificant amounts of extracellular proteases with inorganic nitrogen compounds alone.

The main fermenter is inoculated with 2% by volume of a 24 hr old seed culture. Although *R. rhizopodiformis* is able to grow at temperatures of up to about 50 °C, incubation temperatures of about 33 °C have proved more favourable for optimal protease yields. At an aeration rate of 1 vvm and a stirrer speed of 300 rpm the production of the extracellular protease starts in the late growth phase and reaches its maximum during the stationary phase. *Rhizopus* strains grow relatively fast and, with 48 hr fermentation time, exhibit clear advantages with respect to volume and time yields in comparison with most strains of *Aspergillus* and *Mucor* (Fig. 1).

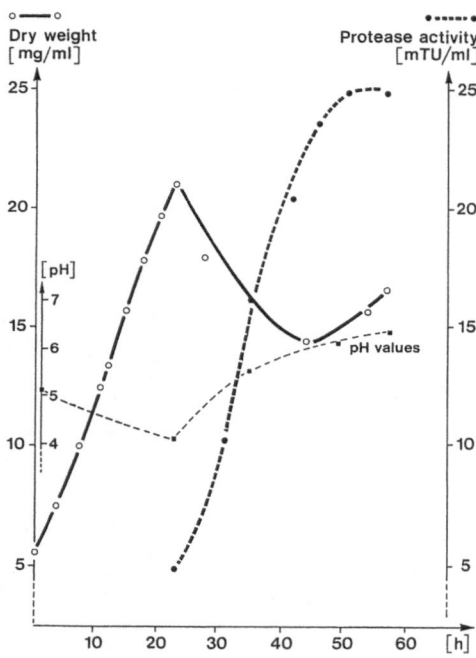

Figure 1. Protease fermentation — 1 m³ scale

Isolation and Properties of the Protease

After the fermentation is finished, filter aids are added to the culture and the mycelium separated by means of a vacuum drum filter. The clear filtrate subsequently concentrated in a thin layer evaporator or by ultrafiltration. The enzyme may be precipitated from the concentrate by the addition of salts or organic solvents. The precipitated enzyme is separated by means of a filter press and the wet enzyme cake dried in a vacuum pedal dryer. The dry raw protease may already be used in this form as a feed additive or it may be converted into pelleted and coated forms.

The *Rhizopus* protease was further concentrated and purified for its enzymatic characterization. Depending on the test substrate used, pH optima of 3 and 6 were determined (Fig. 2). The temperature optimum of this protease is at 60 °C (Fig. 3). The enzyme is very stable in the pH range from 3 to 5.5 even at a temperature of 50 °C (Fig. 4).

Although the genus *Rhizopus* is a member of the large family of the *Mucoraceae*, it has so far become comparatively less known as an industrially important protease producer than have other representatives of this family or the *Aspergillaceae*. Acid proteases have hitherto been mainly obtained from strains of the genus *Aspergillus* and *Mucor*. Especially strains of the groups "*Aspergillus niger*" and "*Aspergillus flavus-oryzae*" have gained industrial importance (Cohen 1977; Ichishima 1972). Mucor strains have long been used industrially, mainly for the production of rennin-like proteases. According to Aunstrup the rennin-like proteases industrially produced in 1980 by *Mucor* strains represented a value of more than five million US dollars (Aunstrup 1980; Federici 1981; Sternberg 1976).

The preparation of typical oriental foods — e.g., Koji, Miso, Tempeh etc. — employing fungal metabolic processes has a long tradition in Japan.

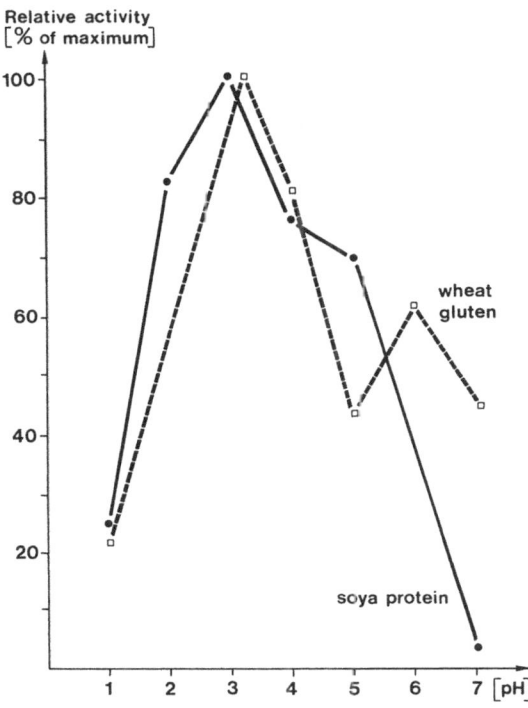

Figure 2. Protease activity at different pH-values

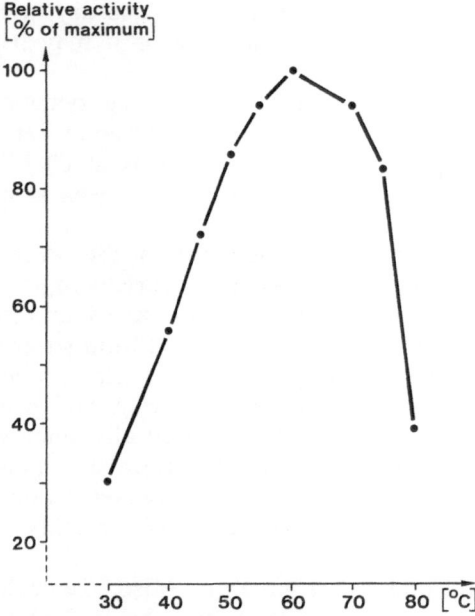

Figure 3. Protease activity at different temperatures

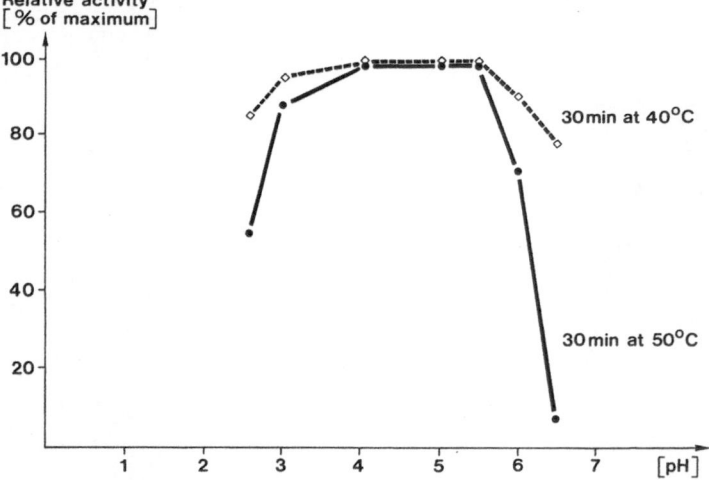

Figure 4. Protease stability in relation to pH-values and temperatures

The methods employed for this purpose were early translated into technical processes for enzyme production in this country. This might be the reason why even today numerous industrial processes for the production of fungal proteases are based on solid state or surface cultures. Quite a number of industrially employed *Aspergillus* strains give higher protease yields in surface than in submerged cultures (Aunstrup 1979; Yoshida and Ichishima 1964). The technical production of acid proteases from *A. saitoi* employed either surface or submerged cultures (Ichishima 1972). *A. oryzae* and *A. sojae* strains were grown on wheat bran (Tsujita and Endo 1977).

For the technical protease production, *Mucor pusillus* is cultured on 60% wheat bran (Arima 1964). Somkuti and Babel (1967, 1968) investigated the protease production by *M. pusillus* under submerged culture conditions. *M. miehei* and *M. renninus* grow in solid state cultures or fermenters, however, the protease production often requires long fermentation times (Aunstrup 1980; Belyauskaite et al. 1980). The same species required four days in shake cultures before maximum protease synthesis was established (Wang and Hesseltine 1965; Wang et al. 1974). *R. chinensis* was cultured for 70 hrs at 25 °C on wheat bran for protease production (Fukumoto et al. 1967).

The protease from *R. chinensis* — probably comparable with the present enzyme — has been extensively characterized. It belongs to the pepsin type, has a pH optimum between 2.9 and 3.2 and its temperature optimum at 60 °C. The molecular weight was determined to be 35,000. The amino acid sequence, active centers and tertiary structure of the enzyme molecule have been largely elucidated. It is similar to proteases such as pepsin, rennin, penicillopepsin etc. (Graham et al. 1973; Mizobe et al. 1973; Nakamura 1977, 1978; Sepulveda et al. 1975; Subramanian 1978; Subramanian et al. 1977; Takahashi et al. 1972; Tsuru et al. 1969).

Technical Application

The majority of fungal acid proteases exhibits extensive similarities with pepsin, rennin or papain. Therefore they have been mainly used in the nutritional realm.

Aspergillus proteases are mainly employed as digestive enzymes or in the production of protein hydrolysates. In Japan the main part of the *Aspergillus* protease is used for soy sauce production (10^9 l/year). The commercial importance of the *Mucor* and *Endothia* proteases sold as rennin substitutes has already been mentioned. Further possibilities for application are cheese production, pastry and meat preparation ("tenderizer"), beer brewing ("chillproofing") etc. (Aunstrup 1977, 1979, 1980; Blain 1975).

The protease from *R. rhizopodiformis* has initially been developed as a digestive enzyme for animal nutrition. The goal is an improved efficiency with respect to the protein fraction of the feed and thus protein and cost savings in cattle fattening. Beyond that a growth stimulation and an accelerated weight gain of the treated animals is hoped for (Hiller 1980, 1980b).

The *R. rhizopodiformis* protease should be applicable wheresoever its enzymatic properties correlate with those of other, already known and utilized fungal proteases. For instance, our *Rhizopus* protease may — similar to already known *Aspergillus* enzymes — also be employed in the enzymatic softening of hides (Asbeck et al. 1980).

References

Adams PR, Deploey JJ (1978) Enzymes produced by thermophilic fungi. Mycologia 70:906–910

Arima K (1964) US Patent 3, 151, 039

Asbeck A, Pfeiffer HF, Plapper J, Schmid R (1978) DE-OS 2, 836, 824

Aunstrup K (1977) Production of industrial enzymes. In: Meyrath J, Bu'lock JD (eds.) Biotechnology and fungal differentiation. Academic Press, London. p 157

Aunstrup K (1979) Production, isolation, and economics of extracellular enzymes. Appl Biochem Bioeng 2:27–69

Aunstrup K (1980) Proteinases. In: Rose AH (ed.) Economic microbiology. vol 5. Academic Press, London. p 50

Belyauskaite JP, Palubinskas VJ, Anchenko OE, Vesa VS, Glemzha AA (1980) Purification and some properties of the extracellular acid proteases from Mucor renninus. Enzyme Microb Technol 2:37–44

Blain JA (1975) Industrial enzyme production. In: Smith JE, Berry DR (eds.) The filamentous fungi. Vol I. Industrial mycology. Edward Arnold, London. p 193

Cohen BL (1977) The proteases of Aspergilli. In: Smith JE, Patemann JA (eds.) Genetics and physiology of Aspergillus. Academic Press, London. p 281

Cooney DG, Emerson R (1967) Thermophilic fungi. Freeman and Company. San Francisco London

Ellis JJ, Wang HL, Hesseltine CW (1974) Rhizopus and Chlamydomucor strains surveyed for milk-clotting, amyloytic and antibiotic activities. Mycologia 66:593–599

Federici F (1981) Proteasi fungine per l'industria alimentare. Ind Aliment 20:679–694

Fogarty WM, Kelly CT (1979) Developments in microbial extracellular enzymes. In: Wiseman A (ed.) Topics in enzyme and fermentation biotechnology. Vol. 3. John Wiley & Sons, London, p 45

Fukumoto J, Tsuru D, Yamamoto T (1967) Studies on mold protease. Part I. Purification, crystallization and some enzymatic properties of acid protease of Rhizopus chinensis. Agric Biol Chem 31:710–717

Graham JES, Sodek J, Hofmann T (1973) Rhizopus acid proteinases (Rhizopus-pepsins): properties and homology with other acid proteinases. Can J Biochem 51: 789–896

Hiller G (1976) DE-OS 2, 633, 105

Hiller G (1977) DE-OS 2, 728, 850

Ichishima E (1972) Acid proteinase and acid carboxypeptidase of molds of the genus Aspergillus. Proc. IV IFS: Ferment Technol Today: 259–270

Inui T, Takeda Y, Iizuka H (1965) Taxonomical studies on genus Rhizopus. J Gen Appl Microbiol 11 (Suppl.): 1–119

Mizobe F, Takahashi K, Ando T (1973) The structure and function of acid proteases. I Specific inactivation of an acid protease from Rhizopus chinensis by diazoacetyl-DL-norleucine methyl ester. J Biochem 73: 61–68

Nakamura S, Takahashi K (1977) Amino acid sequences around 1,2-epoxy-3-(p-nitrophenoxy)-propane-reactive residues in Rhizopus chinensis acid protease: homology with pepsin and rennin. J Biochem 81: 805–807

Ong PS, Gaucher GM (1972) Production, purification, and partial characterization of an extracellular protease from a thermophilic fungus. Proc. IV IFS: Ferment Technol Today: 271–278

Ong PS, Gaucher GM (1976) Production, purification and characterization of thermomycolase, the extracellular serine protease of the thermophilic fungus Malbranchea pulchella var. sulfurea. Can J Microbiol 22: 165–176

Ruttloff H (1981) Stand der Entwicklungstendenzen der mikrobiellen Enzymproduktion. Abh Akad Wiss DDR, Abt Math, Naturwiss, Techn (3 N, Mikrob Enzymprod): 11–21

Sepulveda P, Jackson KW, Tang J (1975) The amino terminal sequences of acid proteases — human pepsin and gastriscin and the protease of Rhizopus chinensis. Biochem Biophys Res Com 63: 1106–1111

Somkuti GA, Babel FJ (1967) Conditions influencing the synthesis of acid protease by Mucor pusillus Lindt. Appl Microbiol 15: 1309–1312

Somkuti GA, Babel FJ (1968) Acid protease synthesis by Mucor pusillus in chemically defined media. J Bacteriol 95: 1415–1418

Sternberg M (1976) Microbial rennets. Adv Appl Microbiol 20: 135–157

Subramanian E (1978) Molecular structure of acid proteases. Trends Biochem Sci 3: 1–3

Subramanian E, Swan IDA, Liu M, Davies DR, Jenkins J, Tickle IJ, Blundell TL (1977) Homology among acid proteases: comparison of crystal structures at 3 A resolution of acid proteases from Rhizopus chinensis and Endothia parasitica. Proc Natl Acad Sci USA 74: 556–559

Takahashi K, Mizobe F, Chang WJ (1972) Inactivation of acid proteases from Rhizopus chinensis, Aspergillus saitoi and Mucor pusillus, and calf rennin by diazoacetylnorleucine methylester. J Biochem 71: 161–164

Tsujita Y, Endo A (1977) Extracellular acid protease of Aspergillus oryzae grown on liquid media: Multiple forms due to association with heterogeneous polysaccharides. J Bacteriol 130:48–56

Tsuru D, Hattori A, Tsuji H, Yamamoto T, Fukumoto J (1969) Studies on mold proteases. Part II. Substrate specificity of acid protease of Rhizopus chinensis. Agric Biol Chem 33: 1419–1426

Wang HL, Hesseltine CW (1965) Studies on the extracellular proteolytic enzymes of Rhizopus oligosporus. Can J Microbiol 11: 727–732

Wang HL, Vespa JB, Hesseltine CW (1974) Acid protease production by fungi used in soybean food fermentation. Appl Microbiol 27: 906–911

Wolf FA, Wolf FT (1969) The fungi. Vol II. Hafner Publishing Comp. New York London

Yoshida F, Ichishima E (1964) US Patent 3, 149, 051

Enzymatic Halogenation of Allyl Alcohol to 2,3-Bromochloro-, Bromoiodo-, and Fluoroiodo-1-propanols: A Study at the Interface of Chemical and Enzymatic Catalyses

Saul L. Neidleman, John Geigert, Demetrios J. Dalietos, and Susanne K. DeWitt

Summary

The enzymatic conversion of allyl alcohol to 2,3-bromochloro-, bromoiodo-, and fluoroiodo-1-propanols by various haloperoxidases is reported. This represents the first enzymatic transformation of an unhalogenated organic substrate to heterogeneous dihalide derivatives. Synthesis of the fluoroiodo-compound involves the first reported enzyme-associated fluorination. The mechanism appears to involve the oxidation and incorporation of a halide into allyl alcohol, followed by a non-enzymatic cascade reaction to insert the second and less reactive halide.

Introduction

Many future advances in enzyme technology will occur at the interface of chemical and enzymatic catalyses where principles of chemistry and enzymology coalesce. Improvements in enzyme performance by carrying out reactions in organic solvents (Klibanov et al. 1977, Patterson et al. 1979, Antonini et al. 1981, Oyama et al. 1981, Tanaka et al. 1981, Lilly 1982) as well as increases in enzyme activity and changes in substrate specificity after metal exchange at the active site (Vallee 1980, Gilles et al. 1981) are examples of such interface studies. We wish to add our studies on enzymatic heterogeneous dihalogenation as an additional example.

We and others (Neidleman 1975, Morrison and Schonbaum 1976, Hager 1982) have been investigating enzymatic halogenation of a variety of substrates. Amoung our latest studies have been those concerned with the conversion of alkenes to halohydrins, such as propylene to propylene bromohydrin (Neidleman 1980, Neidleman et al. 1981a and b):

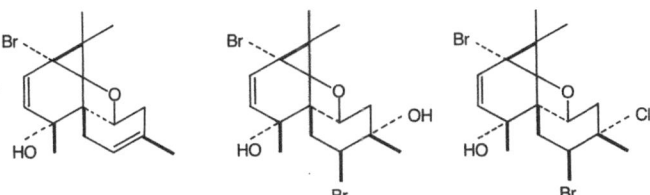

Figure 1. Sesquiterpene alcohols found in the marine alga *Laurencia nipponica* (Suzuki 1980)

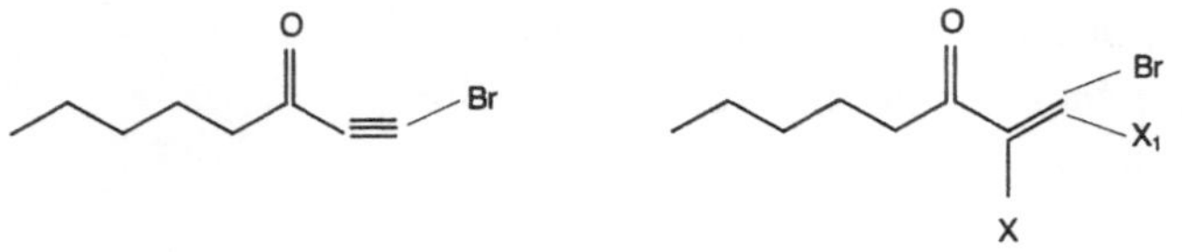

S. L. Neidleman et al.

Figure 2. Halogenated ketones found in marine algae (Rose et al. 1977; McConnell and Fenical 1977)

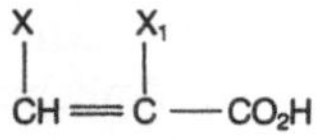

Figure 3. Halogenated acrylic acids found in the marine alga *Asparagopsis taxiformis* (Woolard et al. 1979)

For some time, we have been intrigued by a diverse family of metabolites, produced by marine organisms that contain mixed or heterogeneous halides such as those in Fig. 1–3 (Rose et al. 1977, McConnell and Fenical 1977, Woolard et al. 1979, Suzuki, 1980). We have wondered how mixed halides were incorporated into these compounds. Biomimetic studies involving chemically generated bromonium ions have been presented as models to explain heterogeneous dihalogenation in marine systems (Faulkner 1976, Wolinsky and Faulkner 1976, Fenical 1979), but no enzymatic studies have supported the model in Fig. 4.

Recently, we discovered that with allyl alcohol (or allyl chloride) the ratio of the 2,3-dibromo/2,3-bromohydrin-derivatives produced in enzymatic reactions could be controlled by varying Br^- concentration (details below). The ease of chemical and enzymatic halide ion oxidation increases in the order Cl^-, Br^-, I^- (Hager 1982). Enzymatic activation of F^- has never been demonstrated although fluorinated metabolites occur in plants (Marais 1944, Hall 1977). Therefore, we wondered whether, with halogenating enzymes, high concentrations of a less reactive halide ion might favor its insertion into allyl alcohol by a non-enzymatic cascade reaction associated with the enzyme-catalyzed incorporation of a more reactive halide ion. For example, could Cl^- insertion accompany Br^- incorporation? Could Br^- insertion accompany I^- incorporation? Most excitingly, could F^- insertion accompany I^- incorporation? Thus, we might form heterogeneous dihalide derivatives of allyl alcohol and support the model of Fig. 4.

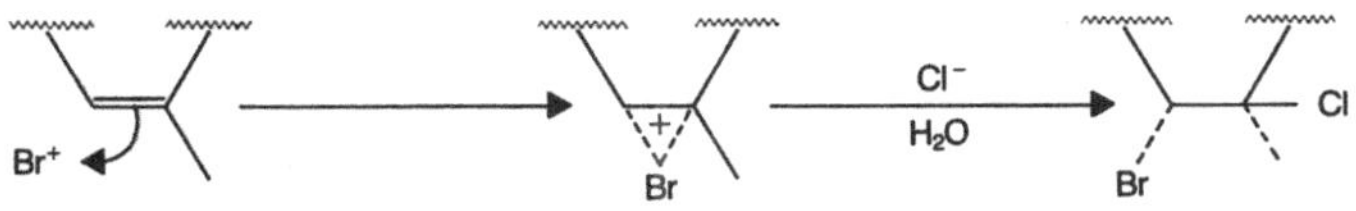

Figure 4. Biomimetic studies on bromonium ion-induced biosynthesis of marine-derived terpenoids (Fenical 1979)

The ability of various haloperoxidases to oxidize halide ions in the presence of hydrogen peroxide is shown in Table 1. The synthesis of heterogeneous dihalides by any of these enzymes was presumed possible, but the most convincing support for a non-enzymatic cascade reaction dependent upon an enzyme-mediated event would be in cases where a not normally incorporated halide ion was inserted in the presence of an appropriate mixed halide reaction. Examples would be Cl^- insertion by lacto- or bromoperoxidase, Br^- or Cl^- insertion by horseradish or thyroid peroxidase, and F^- insertion by any of these enzymes.

Table 1. Known Haloperoxidases

Halide Oxidation	Enzyme Source	Common Enzyme Name
Cl^-, Br^-, I^-	human/animal	
	leukocytes	myeloperoxidase (MPO)
	fungal	
	Caldariomyces fumago	chloroperoxidase (CPO)
Br^-, I^-	human/animal	
	milk, saliva, tears	lactoperoxidase (LPO)
	algal	
	Bonnemaisonica hamifera	bromoperoxidase (BPO)
	Rhipocephalus phoenix	
	Penicillus capitatus	
	Rhodomela larix	
	> 50 others	
I^-	human/animal	
	thyroid	thyroid peroxidase (TPO)
	plant	
	horseradish	horseradish peroxidase (HRPO)

This paper describes the first successful enzymatic syntheses of mixed dihalides of an unhalogenated organic substrate, specifically allyl alcohol. Included are examples of enzymatically synthesized new compositions of matter, among these the first enzyme-associated formation of a carbon-fluorine bond.

Materials and Methods

Enzymes

Chloroperoxidase (EC 1.11.1.10, CPO from *Caldariomyces funago*, 4–8×10^6 units/mg protein), lactoperoxidase (EC 1.11.1.7, LPO, from bovine milk,

70–100 units/mg protein) and horseradish peroxidase (EC 1.11.1.7, HRPO, from horseradish, 150–200 units/mg solid) were obtained from Sigma Chemical Co. The haloperoxidase of the seaweed *Coralina* sp. was prepared according to U.S. Patent No. 4,247,641 (2 units/ml).

Chemicals

Allyl alcohol; allyl chloride; 2,3-dibromo-1-propanol; 2,3-dichloro-1-propanol were obtained from Aldrich Chemical Company. Hydrogen peroxide (30%), sodium chloride, potassium bromide, potassium iodide and sodium fluoride were obtained from Merck, Darmstadt, FRG. Seawater was obtained off the Monterey, CA, coast.

Enzymatic Reactions

The reaction mixtures were incubated at room temperature in 100 ml Pyrex flasks equipped with a magnetic stir bar and stirrer. Each mixture contained 2 mg haloperoxidase (or in the case of *Coralina* haloperoxidase, 2 units), varying levels of two halide salts, and 10 ml of 300 mM potassium phosphate buffer at either pH 3.5 (for CPO) or 7.0 (for LPO, HRPO, and *Coralina* enzyme). Alkene substrate was set at a concentration of 20 mM at the start of the reaction. Hydrogen peroxide was the last reagent added, to give a final concentration of 30 mM.

After initiation, the reaction was allowed to proceed for 15 minutes. All reactions were run at room temperature and atmospheric pressure. Appropriate controls in the absence of enzyme, H_2O_2, halide ion or substrate showed no halogenated products or non-enzymatic halide exchange. Reactions were not necessarily run under optimized conditions nor were reactions run to complete conversion of the substrate.

Reaction Mixture Analysis

Aliquots of reaction mixtures (10 μl) were injected into a Finnigan 4021 gas chromatograph-mass spectrometer (GCMS) equipped with a 1.8×4 mm coiled, glass column packed with Tenax-GC (80/100 mesh). The carrier gas was helium, set at 25 ml/minute. For rapid analysis of the different reaction mixtures, the following temperatures were employed: the column temperature was programmed from 100 °C to 250 °C at a rate of 10 °C/minute, and then held at 250 °C for 10 minutes; the injection and jet separator temperatures were set at 260 °C. The mass spectrometer was operated in electron impact (EI) mode at 70 eV. The mass range from m/z 40 to 400 was scanned every 2 seconds.

Confirmation of the identity of the reaction products were made by GC retention time and mass spectral comparison with authentic standards, whenever available. Product positional isomers were not separately quantitated for purposes of this study.

Results

Synthesis of 2,3-Dibromo-Derivative and Bromohydrins of Allyl Chloride with Chloroperoxidase

The effects of Br^- concentration on the ratio of 2,3-dibromo/2,3-bromohydrin derivatives of allyl chloride and the total yield of products are shown in Table 2. With increasing Br^- concentration the ratio increased, while the total yield of product decreased. In forming the 2,3-dibromoderivative, the incorporation of the first bromine atom is presumed to occur via an enzyme-mediated oxidation of Br^- to Br^+, while insertion of the second bromine atom occurs in a non-enzymatic cascade reaction. In analogous fashion the 2,3-bromohydrins are formed by non-enzymatic insertion of OH^-.

Table 2. Homogeneous dihalogenation with haloperoxidase: Product control as a function of halide ion content

$$CICH_2CH{=}CH_2 \xrightarrow[Br^-, H_2O_2]{chloroperoxidase} \underset{\text{bromohydrin}}{CICH_2\overset{OH}{C}H{-}\overset{Br}{C}H_2} + \underset{\text{dibromo}}{CICH_2\overset{Br}{C}H{-}\overset{Br}{C}H_2}$$

allyl chloride

[Br⁻]	Ratio dibromo: bromohydrin	Relative Yield[a]
20 mM	<0.01	0.9
85	0.03	1.0
170	<0.09	1.0
250	0.13	0.5
1,050	0.65	0.3
3,045	2.14	0.3

[a] A relative yield of 1.0 equals $\sim$40 μmoles of product formed, $\sim$20% conversion of substrate.

Synthesis of 2,3-Dichloro-, Dibromo-, and Bromochloro-Derivatives, Chlorohydrins, Bromohydrins of Allyl Alcohol with Chloroperoxidase

The effects of altering Br^- and Cl^- concentrations on product distribution are shown in Table 3. At 200 mM halide ion, the ratio 2,3-dihalo-1-propanols/2,3-halohydrins was $\sim$1; at $\geqq$2,000 mM, the ratio was between 7 and 13. Allyl alcohol was more readily converted to dihalo-derivatives than allyl chloride. In both cases, increasing halide ion concentration, increased the ratio. In this study, maximum formation of the 2,3-bromochloro-1-propanols was at a Cl^-/Br^- ratio of $\sim$100, although substantial formation occurred at ratios between 1 and 333.

Synthesis of 2,3-Dibromo-, Bromochloro-Derivatives, and Bromohydrins of Allyl Alcohol with Lactoperoxidase

The effects of altering Br^- and Cl^- concentrations on product distribution are shown in Table 4. At 200 mM Br^-, the ratio 2,3-dibromo-1-propanol/2,3-bromohydrins was $\sim$1; at 2,400 mM, the ratio was $\sim$16. These results were

Table 3. Heterogeneous dihalogenation with chloroperoxidase: Product control as a function of mixed halide ion content

$$\underset{\text{allyl alcohol}}{CH_2-CH=CH_2}\ \overset{\text{OH}}{|} \quad \xrightarrow[\text{Br}^-,\ \text{Cl}^-,\ \text{H}_2\text{O}_2]{\text{chloroperoxidase}} \quad \text{Halogenated Products}$$

Product	Product Composition, %						
	mM KBr 200	2,400	1,200	20	6	0	0
	mM NaCl 0	0	1,200	2,000	2,000	2,000	200
OH OH Br $\ \ \mid$ $CH_2-CH-CH_2$	52	7	4	4	1	0	0
OH Br Br $\ \ \mid\quad\mid\quad\mid$ $CH_2-CH-CH_2$	48	93	28	7	2	0	0
OH Cl Br $\ \ \mid$ $CH_2-CH-CH_2$	0	0	59	82	73	0	0
OH Cl Cl $\ \ \mid\quad\mid\quad\mid$ $CH_2-CH-CH_2$	0	0	6	5	20	86	41
OH OH Cl $\ \ \mid$ $CH_2-CH-CH_2$	0	0	3	2	4	14	59
Relative Yield[a]	0.7	1.0	0.8	0.8	0.6	0.6	0.3

[a] A relative yield of 1.0 equals $\sim$100 µmoles of product formed, $\sim$50% conversion of substrate

similar to those with chloroperoxidase and Br^-. As expected there was no Cl^- insertion in the absence of Br^-, in contrast to chloroperoxidase. In the presence of Cl^- and Br^-, the major products were the 2,3-bromochloro-1-propanols. As the ratio of Cl^-/Br^- increased from 1 to 333, the dominance of the 2,3-bromochloro-compounds increased as well.

Synthesis of 2,3-Dibromo, Bromochloro-Derivatives, and Bromohydrins of Allyl Alcohol in Seawater with a Crude Algal Bromoperoxidase from Coralina sp.

Several natural niches occur where the ratio of Cl^-/Br^- is in the range we have found to favor heterogeneous dihalogenation (Table 5). We carried out a study in seawater with a crude, algal bromoperoxidase, adding increasing levels of Br^-. The results are shown in Fig. 5. In seawater traces of the 2,3-bromohydrins and bromochloro-1-propanols were detected. With increasing concentrations of Br^- more total product was formed with relatively more 2,3-bromohydrins and 2,3-dibromo-1-propanol as compared to 2,3-bromochloro-

1-propanols. Dibrominated and bromohydrin metabolites are synthesized by marine organisms (for example, Figures 1–2). Dichlorinated and chlorohydrin metabolites are to be expected if marine chloroperoxidases exist or if the bromine in marine products is displaced enzymatically or non-enzymatically by Cl^- or OH^-.

Table 4. Heterogeneous dihalogenation with bromoperoxidase: Product control as a function of mixed halide ion content

OH
|
$CH_2-CH=CH_2$ $\xrightarrow[Br^-,\ Cl^-,\ H_2O_2]{lactoperoxidase}$ Halogenated Products
allyl alcohol

Product	Product Composition, %						
	mM KBr 200	2,400	1,200	20	6	0	0
	mM NaCl 0	0	1,200	2,000	2,000	2,000	200
OH OH Br | $CH_2-CH-CH_2$	54	6	5	6	3	0	0
OH Br Br | | | $CH_2-CH-CH$	46	94	34	8	5	0	0
OH Cl Br | $CH_2-CH-CH_2$	0	0	61	86	92	0	0
OH Cl Cl | | | $CH_2-CH-CH_2$	0	0	0	0	0	0	0
OH OH Cl | $CH_2-CH-CH_2$	0	0	0	0	0	0	0
Relative Yield[a]	0.6	1.0	1.0	0.8	0.7	0	0

[a] A relative yield of 1.0 equals ~80 µmoles of product formed, ~40% conversion of substrate

Table 5. Natural niches of high halide ion concentrations

	Approximate Concentration, mM		Approximate Ratio
	Cl^-	Br^-	$Cl^-:Br^-$
Seawater	500	1	500
Bitterns (Leslie Salt)	5,400	30	180
Dead Sea	5,700	60	95

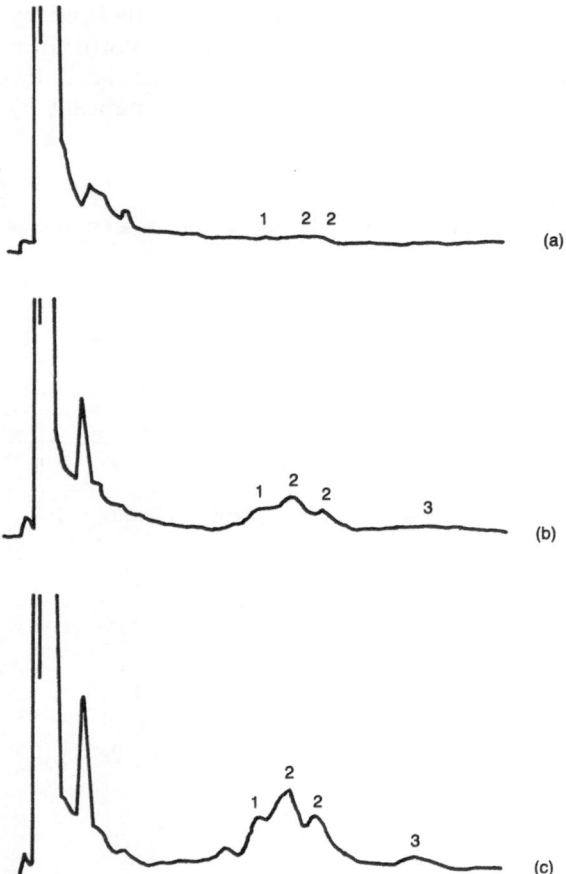

Figure 5. Production of halogenated derivatives of allyl alcohol with a marine bromoperoxidase from *Coralina* sp. and seawater (pH 7.2): **a** without added bromine ion; **b** with 42 mM final concentration of added bromide ion; **c** with 105 mM final concentration of added bromide ion. Products: (1) 2,3-bromochloro-1-propanols; (2) 3-bromo-1,2-propanediol and 2-bromo-1,3-propanediol; and (3) 2,3-dibromo-1-propanol

Synthesis of 2,3-Diiodo-, Bromoiodo-Derivatives, Iodohydrins of Allyl Alcohol with Horseradish Peroxidase

The effects of altering I^- and Br^- concentrations on product distribution are shown in Table 6. At 200 mM I^-, the ratio 2,3-diiodo-1-propanol/2,3-iodohydrins was ~2; while, as expected, there was no Br^- insertion in the absence of I^-. In the presence of Br^-/I^- at a ratio of 100, the 2,3-bromoiodo-1-propanols were the major products. These compounds are new compositions of matter.

Table 6. Heterogeneous dihalogenation with horseradish peroxidase: Bromoiodination

$$
\underset{\text{I}^-,\text{Br}^-,\text{H}_2\text{O}_2}{\overset{\text{horseradish peroxidase}}{\longrightarrow}}
$$

OH
|
CH$_2$–CH=CH$_2$ $\xrightarrow[\text{I}^-,\,\text{Br}^-,\,\text{H}_2\text{O}_2]{\text{horseradish peroxidase}}$ Halogenated Products

Product	Product Compositions, %		
	mM KI 200	20	0
	mM KBr 0	2,000	200
OH OH I | CH$_2$–CH–CH$_2$	67	30	0
OH I I | | | CH$_2$–CH–CH$_2$	33	28	0
OH Br I | CH$_2$–CH–CH$_2$	0	42	0
OH Br Br | | | CH$_2$–CH–CH$_2$	0	0	0
OH OH Br | CH$_2$–CH–CH$_2$	0	0	0
Relative Yield[a]	1.0	1.2	0

[a] A relative yield of 1.0 equals $\sim$20 µmoles of product formed, $\sim$10% conversion of substrate.

Synthesis of 2,3-Diiodo-, Fluoroiodo-Derivatives, and Iodohydrins of Allyl Alcohol with Horseradish Peroxidase

The effects of altering I$^-$ and F$^-$ concentrations on product distribution are shown in Table 7. At 200 mM I$^-$, results were as in the preceeding experiment. As expected, there was no F$^-$ insertion in the absence of I$^-$. In the presence of F$^-$/I$^-$ at a ratio of 100,9% of the total products were 2,3-fluoroiodo-1-propanols. This is the first enzyme-associated fluorination of an organic molecule of which we are aware and the compounds are new compositions of matter.

Discussion

The catalytic versatility of marine and terrestrial organisms in producing halogenated metabolites has been amply demonstrated and reviewed (Siuda and DeBernardis 1973, Fenical 1975, Burreson et al. 1976, Faulkner 1977, Thomson 1978, Moore 1979, Fenical 1982, Hager 1982). The enzymologist, biochemist, and chemist are only beginning to understand and reproduce some of these

Table 7. Heterogeneous dihalogenation with horseradish peroxidase: Fluoroiodination

OH
|
$CH_2-CH=CH_2$ $\xrightarrow[\text{I}^-, \text{F}^-, \text{H}_2\text{O}_2]{\text{horseradish peroxidase}}$ Halogenated Products

Product	Product Composition, %		
	mM KI	200	20	0
	mM NaF	0	2,000	200
OH OH I | |__| $CH_2-CH-CH_2$		67	52	0
OH I I | | | $CH_2-CH-CH_2$		33	39	0
OH F I | |__| $CH_2-CH-CH_2$		0	9	0
OH F F | | | $CH_2-CH-CH_2$		0	0	0
OH OH F | |__| $CH_2-CH-CH_2$		0	0	0
	Relative Yield[a]	1.0	0.2	0

[a] A relative yield of 1.0 equals ~20 μmoles or product formed, ~10% conversion of substrate.

biosynthetic achievements in the laboratory. In addition, there are biological questions that relate to the significance in nature of the wide spectrum of halogenated compounds, especially among marine forms. Do the heterogeneous dihalides have special roles that differ from monohalogenated derivatives or homogeneous dihalides? Our results in seawater and synthetic halide mixtures have shown how ratios of Cl^-/Br^- and other halide ions influence product distribution. Do marine organisms, by selective halide concentration and within specialized compartments, control with delicate balance this complex chemistry as seasoned chemists?

We have also realized with greater clarity the possible relationship between the catalytic properties of various haloperoxidases and the biosynthetic pressures that face marine organisms (Table 8). If it were important for marine organisms to produce brominated and bromochlorinated metabolites rather than chlorinated compounds in a mixture of Cl^- and Br^-, then the bromoperoxidase was a brilliant invention based on chemical principles. If a marine organism needs to synthesize iodinated, bromoidoinated, or chloroiodoinated metabolites and exclude brominated, chlorinated, and bromochlorinated derivatives, an iodoperoxi-

Table 8. Synthetic expectations for haloperoxidases

Enzyme Class	Monohalogenated	Homogeneous Dihalogenation	Heterogeneous Dihalogenation	
Chloroperoxidase	Cl	Cl—Cl	Br—Cl	Cl—F
	Br	Br—Br	I—Cl	Br—F
	I	I—I	I—Br	I—F
Bromoperoxidase	Br	Br—Br	Br—Cl	Br—F
	I	I—I	I—Cl	I—F
			I—Br	
Iodoperoxidase	I	I—I	I—Cl	I—F
			I—Br	

dase such as horseradish or thyroid peroxidase would be required. On the other hand, to take full advantage of the bountiful Cl⁻ supply of the marine and terrestrial spheres, a chloroperoxidase would be the choice. Nature was not casual in applying chemical principles to biological problems and creating appropriate enzymes.

Furthermore, it is exciting for the scientist to take these biocatalysts, synthesized to solve particular biological problems and study what the effect of artificially imposed chemical pressures on them might be: reactions in organic solvents, metal exchange at active sites, varying halide ion ratios. How will the enzyme respond?

In this paper we have begun to examine the effects of mixed halide ions on reactions catalyzed by various haloperoxidases. The first heterogeneous dihalide derivatives of an unhalogenated organic substrate have been synthesized, including the first example of enzyme-associated fluorination. We have shown that the biomimetic approach to bioheterogeneous dihalogenation (wherein chemistry models for enzymology, Fig. 4) has a counterpart that is chemomimetic (wherein enzymology models for chemistry, Fig. 6).

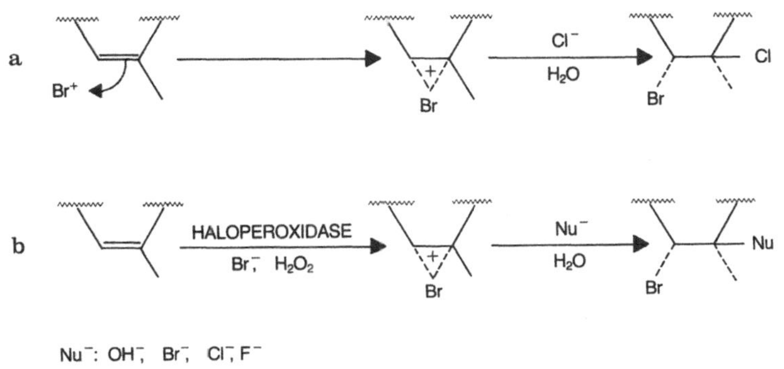

Nu⁻: OH⁻, Br⁻, Cl⁻, F⁻

Figure 6. Mimetic studies of bromonium-induced synthesis: (a) biomimetic and (b) chemomimetic

References

Antonini E, Carrea G, Cremonesi P (1981) Enzyme catalysed reactions in water-organic solvent two-phase systems. Enzyme Microb. Technol. 3:291–296

Burreson BJ, Moore RE, Roller PP (1976) Volatile halogen compounds in the alga *Asparagopsis taxiformis* (Rhodophyta). J. Agric. Food Chem. 24:856–861

Butler LG (1979) Enzymes in non-aqueous solvents. Enzyme Microb. Technol. 1:253–259

Faulkner DJ (1976) Biomimetic synthesis of marine natural products. Pure Appl. Chem. 48:25–28

Faulkner DJ (1977) Interesting aspects of marine natural products chemistry. Tetrahedr. 33:1421–1443

Fenical W (1975) Halogenation in the *Rhodophyta*; a review. J. Phycol. 11:245–259

Fenical W (1979) Molecular aspects of halogen-based biosynthesis of marine natural products. Recent Adv. Phytochem. 13:219–239

Fenical W (1982) Natural products chemistry in the marine environment. Science 215:923–928

Gilles I, Loffler HG, Schneider F (1981) Co^{+2}-substituted acylamino acid amido hydrolase from *Aspergillus oryzae*. Z. Naturforsch. 36c: 751–754

Hager LP (1982) Mother nature likes some halogenated compounds. Basic Life Sci 19:415–429

Hall RJ (1977) in Trace Substances in Environmental Health (DD Hemphill, ed) The presence and biosynthesis of carbon-fluorine compounds in tropical plants and soils. pp. 156–163

Klibanov AM, Samokhin GP, Martinek K, Berezin IV (1977) A new approach to preparative enzymatic synthesis. Biotechnol Bioeng 19: 1351–1361

Lilly MD (1982) Two-liquid-phase biocatalytic reactions. J Chem Tech Biotechnol 32: 162–169

Marais JSC (1944) Monofluoroacetic acid, the toxic principle of "gifblaar in *Dichapetalum cymosum*" (Hook) Engl. Onderstepoort J Vet Sci Anim Ind 20: 67–73

McConnell OJ, Fenical W (1977) Polyhalogenated 1-octene-3-ones, antibacterial metabolites from the red seaweed *Bonnemaisonia asparagoides*. Tetrahed. Lett., pp. 1851–1854

Moore RE (1979) Marine aliphatic natural products. Aliphatic Related Nat Prod Chem 1:20–67

Morrison M, Schonbaum GR (1976) Peroxidase-catalyzed halogenation. Ann. Rev. Biochem. 45: 861–888

Neidleman SL (1975) Microbial halogenation. CRC Crit. Rev. Microb., pp. 333–358.

Neidleman SL (Nov., 1980) Use of enzymes as catalysts for alkene oxide production. Hydrocarbon Proc., pp. 135–138

Neidleman SL, Amon WF Jr., Geigert J (1981) Method for producing epoxides and glycols from alkenes. U.S. Pat. 4, 247, 641

Neidleman SL, Amon WF Jr, Geigert J (1981) Method for producing epoxides and glycols from gaseous alkenes. U.S. Pat 4, 284, 723

Oyama K, Nishimura S, Nonaka Y, Kihara K, Hashimoto T (1981) Synthesis of an aspartase precursor by immobilized thermolysin in an organic solvent. J. Org. Chem. 46: 5241–5242

Patterson JDE, Blain JA, Shaw CEL, Todd R, Bell G (1979) Synthesis of glycerides and esters by fungal cell-bound enzymes in continuous reactor systems. Biotechnol. Lett. 1: 211–216

Rose AF, Pettus JA, Sims JJ (1977) Marine Natural Products XIII. Isolation and synthesis of some halogenated ketones from the red seaweed *Delisea fimbriata*. Tetrahed. Lett. 22: 1847–1850

Siuda JF, DeBernardis JF (1973) Naturally occurring halogenated organic compounds. Llyodia 36: 107–143

Suzuki T (1980) Two new sesquiterpene alcohols containing bromine from the marine alga *Laurencia nipponica* Yameda. Chem Lett 541–542

Tanaka T, Ono E, Ishihara M, Yamanaka S, Takinami K (1981) Enzymatic acyl exchange of triglyceride in n-hexane. Agric Biol Chem 45:2387–2389

Thomson RH (1978) Halogenated metabolites from marine animals and plants. J Indian Chem Soc 55:1209–1215

Vallee BL (1980) Zinc and other active site metals as probes of local confirmation and function of enzymes. Carlsberg Res Commun 45: 423–441

Wolinsky LE, Faulkner DJ (1976) A biomimetic approach to the synthesis of *Laurencia* metabolites. Synthesis of 10-bromo-α-chamigrene. J Org Chem 41:597–600

Woolard FX, Moore RE, Roller PP (1979) Halogenated acetic and acrylic acids from the red alga *Asparagopsis taxiformis*. Phytochem 18:617–620

II. Cell Disintegration

Disintegration of Cells by Extrusion Under Pressure

L. Edebo

In order to extract intracellular constituents from microorganisms the barrier of the cell envelope has to be disrupted. Since this barrier is usually resistant and the intracellular components with enzymatic or similar activities are often unstable, methods for disintegration are sought that destroy the barrier and preserve the desired constituents. Several reviews exist that deal with different chemical, physical and mechanical methods [26, 34, 36, 60, 122, 129]. As suggested in an earlier review [34] certain methods are usually preferred for particular purposes. For example agitation with abrasive particles has often been employed in the preparation of cell walls. Other methods are avoided when certain materials are treated. Sonic disintegration was not so efficient in the disintegration of mold mycelium, presumably due to impaired cavitation (Table 1). Certain enzymes are inactivated by sonic treatment, also when the cooling is efficient [23, 44, 106] such that sonic treatment was not considered a good method for primary cell breakage [122]. In order to prepare microbial extracts with enzymatic activity pressure extrusion methods are often employed since a minimal amount of enzyme inactivation occurs during this disruption relative to other procedures [26]. Also the disintegrative efficiency of such methods may be great. In a recent review Dobrogosz [26] ranked the commonly used physical methods freeze-pressing > vibration mills > French press (extrusion of non-frozen material) > ultrasonication > alumina grinding. This presentation aims to review pressure extrusion methods, their different modifications, the treatment of different microbes and other biological materials, and release of different

Table 1. Susceptibility of various kinds of cells in suspension to disintegration by various methods. The numbers show the ranking within the group. A high number indicates that the cells are sensitive. Parenthesis indicates that the number is very uncertain

	Liquid pressing	Freeze-pressing	Agitation	Sonic
Animal cells	7	7	7	7
Gram-negative bacilli and cocci	6	6	5	6
Gram-positive bacilli	5	4	(4)	5
Yeast	4	2.5	3	3.5
Gram-positive cocci	3	2.5	(2)	3.5
Spores	2	1	(1)	2
Mycelium	(1)[a]	5	6	1

[a] clogging of orifice may occur

constituents. Some aspects will be given on the treatment of larger quantities of microbial material.

Since microbial cell walls are giant bag-shaped macromolecules, the disruption of microbes inolves the breaking of covalent bonds. Extrusion of microorganisms through an orifice by pressure initially involves a stress on the microbial cell envelope from the outside. Only when the cell has become disrupted and the intracellular contents released will the stress affect the protoplasmic constituents. Since in extrusion under pressure the disintegrating activity is focused at the orifice and probably instantaneous for the individual cell, disruption of the cell wall should be greatly favoured over disintegration of protoplasmic constituents. In contrast, methods involving continuous exposure of suspensions of microbes for a length of time will gradually subject increasing quantities of intracellular constituents to the disintegrating stress.

Pressure Extrusion of Non-frozen Material

Buchner already used a press to prepare extracts from yeast to show fermentative activity in the absence of cells. In later designs (42, 51, 62, 86, 87, 108) different valve designs have been employed to press microbes though narrow orifices at pressures 40–350 MPa (400–3500 kp cm^{-2}). Many different microbes have been disrupted in suspension by these methods. The pressure required for disintegration was different for different kinds of cells.

A recent investigation [62] used an apparatus with an extrusion orifice formed from a conical seating with a hardened ball loaded into it. Flow of fluid occurred only after a given pressure was exceeded. Different pressures up to 270 MPa were tested for a number of different microorganisms and the cell disruption estimated by release of UV-absorbing material and reduction of viable counts. The pattern found for all the microbes was that little or no disruption occurred until a certain level of applied pressure was reached, and that subsequently the proportion of disrupted cells increased rapidly with increasing pressure until almost all the cells were disrupted. In each case disruption followed a sigmoid curve. Different pressures were required to disrupt 50% of the cells in each strain (Table 2). When plotted against the Log of the applied pressure the cell viability followed a Log Normal disruption. This rela-

Table 2. Pressures required for disintegration of different microbes (ref. 62)

Strain	Gram stain	Dimension (μm)	Pressure for 50% disruption (Pa)
Escherichia coli	negative	2 –4 × 0.5	1.5×10^7
Bacillus subtilis	positive	1.5–3 × 0.5–0.8	2.4×10^7
Lactobacillus casei	positive	up to 4 × 0.4–0.7	3.1×10^7
Streptococcus faecalis	positive	∅1.02	1.5×10^8
Staphylococcus aureus	positive	∅1	1.9×10^8
Saccharomyces cerevisiae	positive	7 –12 × 5–8	1.5×10^8

tionship was suggested to be due to different resistance against disruption among the population of cells, possibly related to the statistical size distribution.

The ranking of the different kinds of cells within the different disintegration methods is very similar (Tables 1 and 2). The one exception is mycelium which is easily disintegrated by agitation with beads and by freeze-pressing but comparatively resistant to sonic vibration. At least part of this effect is due to impaired cavitation in suspensions of mycelia, such that actually mycelium may not be more resistant to sonic cavitation. These results as well as studies on the molecular structure of the microbial cell wall [113] indicate that small size, spherical rather than rod shape, and the toughness of the cell envelope, e.g. as caused by several peptidoglycan layers in the murein sacculus, contribute to resistance towards mechanical disintegration. It follows then that the choice of method will be greatly influenced by the mechanical efficiency and dependability of the equipment provided that inactivation of the constituents can be avoided. Heat generated by pressing may inactive the enzyme.

When an industrial homogenizer, the Manton-Gaulin homogenizer, was adopted for disintegration of microorganisms [42, 51] it was observed that a knife-edge valve-seat improved the disintegration over a flat valve-seat. This might be due to a steeper pressure-gradient with greater acceleration and stress on the microbes. The disruption efficiency was independent of the suspension concentration in the range 300–600 g packed yeast cells/liter. The rate of release of protein could be described by the relationship

$$\log \frac{R_m - R}{R_m} = - KNP^{2.9} \qquad \text{(eq. 1)}$$

where R = amount of soluble protein released, R_m = maximum amount of soluble protein that can be released, K = dimensional temperature-dependent rate constant, N = number of passages through valve, and P = operating pressure. At a temperature of 30 °C and a pressure of 55 MPa, 62% of the water-soluble protein was released in a single pass through the homogenizer.

The rate of release of seven different enzymes was studied in relation to the release of protein. The rate was dependent on the enzyme location in the cell: the release of acid phosphatase and invertase which are supposed to be located primarily outside the cell membrane was faster than the overall protein release. The dehydrogenases which belong to the cell sap were released slightly faster or at the same rate as the overall protein. Alkaline phosphatase and fumarase which in the intact cell probably are mainly located in plasma membranes and mitochondria, respectively, were released more slowly.

At operating pressures above 80 MPa and high cell concentrations (750 g packed yeast/liter) the disruption rate was reduced and no longer first-order. Increase of temperature of the yeast suspension from 5 °C to 30 °C increased the disruption rate 1.5 times. Some of these effects might be due to reduced flow of material and disintegration at higher viscosity.

Considerable investigations have been directed towards the elucidation of the hydrodynamic processes during pressure extrusion that are important for disintegration [20, 21, 28, 37, 42, 51, 62]. All results emphasize the importance of the shape of the orifice for efficient disruption at a given pressure. The role of turbulence as a means of disintegration in liquid extrusion homogenizers has often been pointed out [20, 21, 28]. The turbulent motion of the liquid passing through the orifice contains great velocity fluctuations due to the random motion of numerous eddies of different scales. These eddies were considered important in yeast disruption [28].

The mechanism of disintegration by liquid extrusion was recently reinvestigated by Engler and Robinson [37] using suspensions of *Candida utilis* cells. They used a newly constructed disruption device capable of producing pressures of up to 300 MPa and compared the disruption achieved when the suspension could flow freely after passing through an outlet nozzle with that observed, when the outlet jet struck an impact plate situated near the orifice discharge. The disruption seen after the free pass was considered to be mainly caused by shear stresses, whereas impingement stresses were considerably when the jet struck the impact plate. The impingement disruption was 70–90 % for different batches of *C. utilis* at an operating pressure of 100 MPa. Their data indicated that shear stress was only about 20 % as effective as impingement for disruption. However, the most effective impingement disruption of *C. utilis* in their device was less efficient than that calculated for *S. cerevisiae* in the Manton-Gaulin homogenizer published by Hetherington et al. [51]. This effect was still greater at higher pressures since the pressure exponent with the Manton-Gaulin homogenizer [2.9 see Eq. 1] was considerably greater than that in the impingement device (1.17–1.77). Because of the hydrodynamic similarities between the impingement device and the Manton-Gaulin homogenizer valves. Engler and Robinson suggested that the disruption mechanisms are similar.

At temperatures above 35 °C, significant losses of enzyme activity were observed [42, 51]. At lower temperatures, no loss in activity on prolonged treatment was observed except in the cases of fumarase and invertase. Optimization in the recovery of a labile intracellular enzyme was achieved at 60–80 % disintegration [3].

Since the energy consumed in a pressing operation (P × V) is largely converted into heat, higher pressures involve risks of heat denaturation. Pressing one ml (cm^3) at 200 MPa (= 2000 atm. = 2000 kp cm^{-2}) requires 20 kpm which is equivalent to 47 cal and a temperature rise of 47 °C. The Ribi press [108] was provided with a cooling device employing liquid nitrogen to balance the heat generated.

It is a general experience that during cell disruption there is an increase in the viscosity as the protoplasmic material is released [30, 31]. Liquid extrusion disruption of baker's yeast increased the suspension viscosity and its non-Newtonian behaviour [89]. The physical properties of the homogenate were such that solid-liquid separation by centrifugation or filtration presented problems. The advantage of cell disruption methods that minimize the degradation of cell walls was emphasized both with respect to the reduction of cell fragments and to the restriction of the increase in viscosity due to colloidal cell-wall glycan [89].

The denaturation of proteins by the kind of shear encountered in liquid extrusion disruption were considered less severe than supposed earlier [138].

Only a partially active preparation of the enzyme yeast alcohol dehydrogenase was precipitated by the shear, whereas the fully active molecule was unaffected. The interfacial denaturation that was held responsible for the precipitation [138] is probably of greater magnitude in sonic and agitation devices.

Freeze-pressing (X-press, 29)

Extrusion of material that has been frozen does not require a valve to adjust the pressure [59] but utilizes the phase changes in water under pressure (Fig. 1; [17, 18]) to regulate the pressure at which the material starts to flow. In principle, a pressure chamber with an outlet ca 1 mm wide suffices. The standard X-press consists of two identical cylindrical axial chambers separated by a disk with a central round hole about one mm wide (Fig. 2). As long as the material is frozen (ice I) pressure may be applied without flow. Under pressure transformation of the crystal structure of water occurs which reduces the cohesiveness of the material such that it will flow through the orifice. This process is related to the phase transitions in water described long ago as the phase diagram of water (Fig. 1).

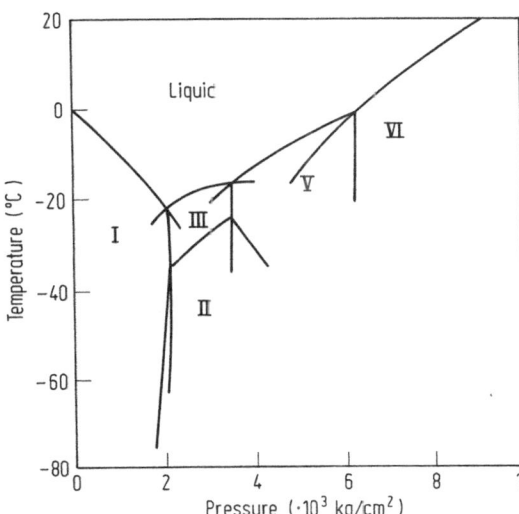

Figure 1. A part of the phase diagram of water in the liquid-solid region. (From P. W. Bridgman, 17)

The X-press was originally designed for laboratory disintegration of microorganisms and other tough biological materials with the aid of a manual hydraulic press. With the equipment, several different kinds of cells, tissues and other materials have been disintegrated and different constituents extracted. A great number of different enzymes involved in nucleic acid and protein synthesis and in substrate hydrolysis have been extracted from *E. coli* and other microbes and purified after freeze-pressing (Table 3). Some of the references selected illustrate lines of successful biochemical investigation [106] starting

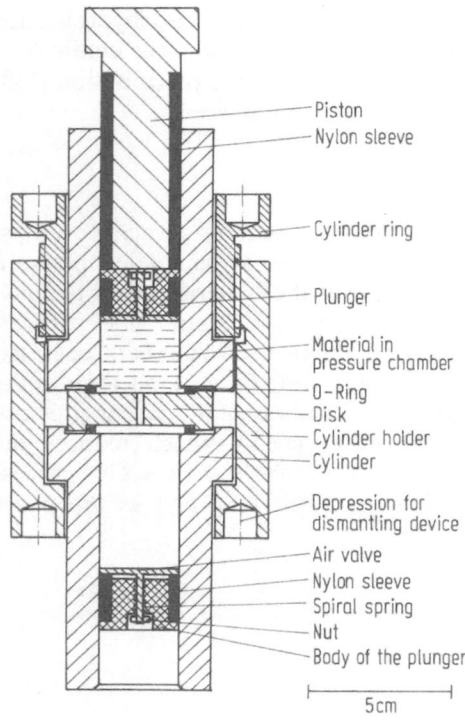

Figure 2. Section through the X-press

Table 3. Preparation of cellular constituents from gram-negative bacteria by freeze-pressing (X-press)

Cell	Constituent	Reference
E. coli	amino acid: tRNA ligases [7]	1
	deoxycytidine phosphate forming enz.	106
	DNA N-glycosidases	71, 107
	nucleoside diphosphate kinase	135
	ribonucleoside diphosphate reductase	22, 40
	ribonucleotide reductase	116
	thioredoxin	55, 56
	thioredoxin reductase	132
	soluble enzyme preparations for synthesis of nucleotide substrate	R. Plapp[a]
	L-asparaginase	124
	protein synthesis in vitro	123
	β-galactosidase	P. M. Bhargava[a]
	tRNA	88
	ribosomes	70
Enterobacteria (5 genera)	proteinase inhibitors	43
Salmonella (5 species)	precipitinogens	54
Haemophilus influenzae	precipitinogens	16
Haemophilus parainfluenzae	'endotoxin'	94
Salmonella cholerae-suis	'endotoxin'	95
Pseudomonas aeruginosa	'endotoxin'	90, 91
Pseudomonas fluorescens	vanillate-demethylase	24

Table 3. (Forts.)

Cell	Constituent	Reference
Bordetella pertussis	cell wall immunogen, cytoplasmic neurotoxin	12
Leptotrichia buccalis	cell walls	52
Fusobacterium sp	precipitinogen	64
Bacteroides (3 species)	'endotoxin'	53
Bacteroides fragilis	hydrolases [6]	10
Neisseria gonorrhoeae	antigen	137

[a] personal communication

Table 4. Preparation of cellular constituents from grampositive cocci and bacilli by freeze-pressing (X-press).

Cell	Constituent	Reference
Staphylococcus aureus	antigen, cell wall	46, 73, 100, 139
Staphylococcus epidermidis	cell wall	45
Micrococcus sp.	cell wall mucopeptide	45, 48
Diplococcus pneumoniae	antigen	11
Streptococcus pyogenes	antigen	98, 99
Streptococcus faecalis	tyrosine decarboxylase	2
Streptococcus sanguis	α-glucosidase	93
Streptococcus mutants	α-galactosidase	93
Streptococcus salivarius	β-galactosidase	93
Streptococcus mitis	β-N-acetylglucosaminidase	93
Streptococcus mitis	β-fructofuranosidase	127, 128
Streptococcus mitis	α-glucosidase	128
Ruminococcus flavefaciens	succinic acid production	57
Lactobacillus fermenti	cell wall peptidoglycan	136
Corynebacterium (10 species)	precipitins	109
Bacillus stearothermophilus	asparagine: tRNA ligase	117
Bacillus stearothermophilus	lysine tRNA ligase	117
Bacillus stearothermophilus	serine tRNA ligase	117
Bacillus stearothermophilus	valine tRNA ligase	117

Table 5. Preparation of cellular constituents from mycobacteria, actinomycetes and mycoplasma by freeze-pressing (X-press)

Cell	Constituent	Reference
Streptomyces aureofaciens	citrate synthase	58
	aconitate hydratase	58
	isocitrate dehydrogenase	58
	fumarate hydratase	58
	malate dehydrogenase	58
Nocardia sp.	O-demethylases	19, 23
Nocardia 9 species	precipitinogens	110, 112
Mycobacterium 11 species	precipitinogens	5, 96, 111
Nycobacterium friedmannii	precipitinogens	125
Mycobacterium ranae	precipitinogens	126
Micropolyspora faeni	precipitinogens	8
Thermoactinomyces vulgaris	precipitinogens	8
Mycoplasma agalactiae	precipitinogens	4

Table 6. Preparation of cellular constituents from yeasts by freeze-pressing (X-press).

Cell	Constituent	Reference
Saccharomyces cerevisiae	valyl RNA synthetase	67, 68, 69
	phenylalanyl RNA synthetase	67
	leucyl RNA synthetase	67
	lysyl transfer RNA synthetase	114
	valine: tRNA ligase	115
	lysine: tRNA ligase	115
	nucleoside diphosphate kinase	135
	superoxide dismutase	L. Ringstedt[a]
	RNA-fractions	63
	dolicholmonophosphates: mannosyl acceptors	131
Saccharomyces carlsbergensis	α-acyl-α-hydroxy acid synthetase	47
Candida utilis	glucose-6-phosphate dehydrogenase	27
Candida albicans	antigens	Å. Frisk[a]
Cryptococcus neoformans	antigens	Å. Frisk[a]

[a] personal communication

Table 7. Preparation of cellular constituents from fungi by freeze-pressing (X-press) (Yeasts see Table 6)

Cell	Constituent	Reference
Aspergillus niger	glucose oxidase	140
Blakeslea trispora	β-carotene	25
Cephalosporium acremonium	D-amino acid oxidase	9
Neurospora crassa	aryl-aldehyde: NADP oxido-reductase	G. G. Gross[a]
Trichosporon cutaneum	cathechol 1:2-oxygenase	134
Trichosporon cutaneum	phenol hydroxylating enzymes	92
Alternaria alternata	precipitinogens	7
Aspergillus fumigatus	precipitinogens	7
Botrytis cinerea	precipitinogens	7
Cladosporium herbarum	precipitinogens	7
Paecilomyces variotii	precipitinogens	7
Penicillium frequentans	precipitinogens	7
Pullularia pullulans	precipitinogens	7
Mucor hiemalis	precipitinogens	7
Rhizopus rhizopodiformis	precipitinogens	7
Claviceps purpurea ⎫ Claviceps fusiformis ⎬ Claviceps paspali ⎭	DAHP synthetase	81, 118, 121
	chorismate mutase	81
	prephenate dehydrogenase	81
	prephenate dehydratase	81
	anthranilate synthetase	81, 120
	tryptophan synthetase	81
	chanoclavin-I-cyclase	38
Claviceps purpurea	alkaloids	102
Coprinus lagopus	anthranilate synthetase	50
	indol-glycerolphosphate synthetase	50

[a] personal communication

Table 8. Extraction of precipitinogens from pollen by freeze-pressing (X-press)

Latin name	English name	Kind of plant	Reference
Alnus glutinosa	alder	deciduous tree	6, 39
Betula verrucosa	birch	deciduous tree	6, 39
Corylus avellana	hazel	deciduous tree	6, 39
Fagus grandifolia	beech	deciduous tree	39
Populus tremula	aspen	deciduous tree	6, 39
Quercus alba	oak	deciduous tree	39
Ulmus americana	elm	deciduous tree	39
Phleum pratense	timothy	grass	6, 39
Phragmites communis	common reed	grass	39
Ambrosia artemisilfolia	ragweed	weed	29
Chrysanthemum leucanthemum	marguerite	weed	29
Taraxacum vulgare	dandelion	weed	29

Table 9. Disintegration of virus by freeze-pressing (X-press)

Coliphage	Inactivation rate (K)[a]	Reference
$\emptyset$X-174, T1, T3	0.6–1.8	35
T5, P2, λ vir	4.0–5.2	35
T2L, T2Hr13h, T4D, T6	6.6–7.6	31, 35
Animal virus	Constituent	Reference
Pox virus	precipitinogens	82[b]
Vaccinia virus	precipitinogens	82

[a] The inactivation rate was described by $S = e^{-NK}$, where S is the fraction of surviving plaque-forming units, N is the number of pressings, and K describes the sensitivity of the phage.
[b] personal communication

Table 10. Preparation of cellular constituents from animal tissue and cells by freeze-pressing (X-press).

Tissue	Species	Constituent	Reference
Artery	human, rabbit	lipids	14, 15
Brain	human	lactate dehydrogenase	105, 106
	human	hydroxybutyrate dehydrogenase	105, 106
Hair	human	antigen	65
Ileum	rat	phospholipidases	13
Liver capsule	bovine	heparin proteoglycan	72
		heparan sulfate proteoglycan	61
Mastocytoma	mouse	heparin, glycosaminoglycan	49, 97
Skin	human	antigen	65, 66, 101
	rat	hyaluronate (100%)	83
Synovial tissue	human	antigen	85
Spermatozoa	human	antigen	41

before the first description of the X-press was published [29] and still going on at Karolinska Institutet [116]. It was originally found that extracts prepared after freeze-pressing were superior to those obtained after sonic treatment by the formation of deoxycytidine phosphate in extracts from *Escherichia coli*, since sonic treatment inactivated the activity [29]. Also recently alternative disintegration methods have been inferior with respect to yield and enzymatic activity as determined by antigenicity and enzymatic activity of the enzyme protein [56]. Furthermore, enzymes and antigens have been extracted from grampositive cocci and bacilli (Table 4), from mycobacteria and actinomycetes (Table 5), from yeast cells (Table 6), and from molds (Table 7). Antigens (allergens) have been extracted from pollens (Table 8) and from the protozoa *Plasmodium falciparum* (malaria, 133), *Trypanosoma gambiense* (African sleeping sickness, 84) and *Trypanosoma musculi* [103]. Soluble precipitinogens were prepared from pox virus (Table 9). The sensitivity of coliphages to freeze-pressing was closely related to the size of the virus particle (Table 9). Enzymes and antigens have been extracted from animal tissues (Table 10). Minimally degraded hyaluronic acid was recently extracted quantitatively from rat skin by freeze-pressing [83]. Provided the cells and tissues are surrounded by enough water, such that large pieces with tough structure are avoided, most biological materials are fragmented by freeze-pressing. Tough tissues like skin and tendon may be cut into slices, a few mm wide, mixed with an equal volume of water before freezing and then frozen and pressed. Soft tissues like liver and kidney may be just frozen and pressed. The applications chosen above have been achieved with the original X-press [29] or slight modifications of it [33].

Few systematic comparisons of different disintegration methods are available. Some involving freeze-pressing with the X-press will be referred to. The yield of glucose oxidase from *Aspergillus niger*, a heat sensitive enzyme, was greater after freeze-pressing than after sonic treatment, hard-grinding with quartz sand, treatment with a high-speed vibratory tissue disintegrator or with a laboratory homogenizer. After shaking with glass beads (7.5 × wet mycelia) for 40 min the same enzyme yield was achieved as after freeze-pressing in the X-press [140]. Enzyme extracts from *Saccharomyces carlsbergensis* showed higher activity of the enzyme α-acyl-α-hydroxy acid synthetase after freeze-pressing than after sonic treatment or grinding with quartz sand in a mortar [47]. The specific activity of the particulate fraction of *Saccharomyces cerevisiae* as dolichol-monophosphates: mannosyl acceptors was higher after freeze-pressing than after disruption in the French press [131]. Bacterial mixed-function oxidases prepared by freeze-pressing seemed to consist of only two components. Claims for the existence of separate reductases and redoxins might be based on artifacts from the use of ultrasonic methods for cell disruption [19]. Soluble enzymes for the synthesis of nucleotide substrate in the synthesis of murein showed the highest activity in preparations made by freeze-pressing as compared to grinding with alumina, shaking with glass beads and sonic treatment (R. Plapp. pers. commun.). *Ruminococcus flavefaciens* was resistant to all methods tried except freeze-pressing [57].

Development of Freeze-pressing to Optimize Disintegration

In order to improve the conditions for disintegration, the influence of different factors such as pressure, the suspension medium and concentration of cells, the temperature, the width and length of the extrusion orifice, and the speed of flow through the orifice have been investigated. In general, baker's yeast has been used as model cells. The extent of disruption was estimated with an electronic particle counter, the Coulter Counter, on the basis of reduced electrical resistance of individual yeast cells on disruption [77].

Freeze-pressing with a Mechanical Valve

The original X-press with an open orifice in the disk may be considered as a pressure extrusion apparatus with a physical rather than a mechanical pressure regulating device. In this way pressures up to 200 MPa (2000 kp cm^{-2}) may be selected by pre-cooling at temperatures down to -20 °C [29]. Pressures above 200 MPa were achieved by closing the orifice with a spring-loaded piston. In this way 90 % disintegration was achieved by pressing *Saccharomyces cerevisiae* at -22 °C (13.5 % dry weight, 400–480 MPa), with yeast paste (27 % dry weight) 50 % disintegration was achieved at 400–520 MPa [36]. The impaired disintegration of yeast paste may be due to its higher viscosity.

Influence of Suspending Medium on Flow of Material and Disintegration

It has often been observed that when dilute biological material is pressed through an open orifice in the X-press disk, the material flows through the orifice discontinuously in short pulses accompanied by sharp bangs, whereas concentrated microbial paste at the same temperature passes the orifice more smoothly [29, 34]. Similar flow patterns were achieved by pressing water with different concentrations of gelatin. They were described by recording the pressure in the pressing chamber during the process [74, 78]. As a representative for flow in pulses connected with bangs (flow type a) the flow of yeast cells (5.4 mg/ml) suspended in water at -25 °C is shown (Fig. 3). After forwarding the piston at a speed of 0.92 cm/s for 14–15 s a pressure slightly above 200 MPa was

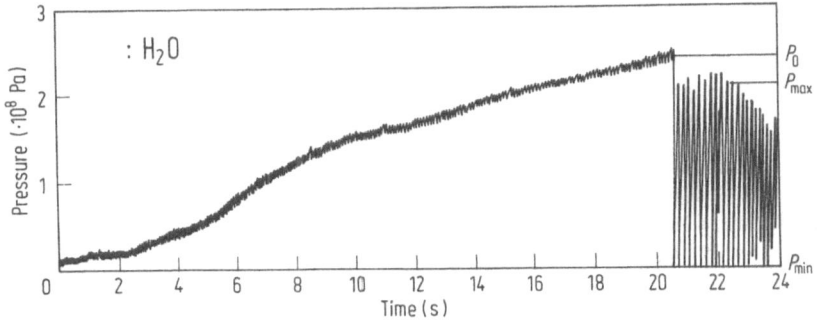

Figure 3. Pressure recording from freeze-pressing of deionized water. Flow with bangs = type a. (From Biotechnol. Bioeng. *9*, 267–269, 1967)

reached when the material under pressure started to flow through the hole of the disk. The flow which was explosive and accompanied by a loud bang caused a drop in pressure to nearly zero. The velocity of the material leaving the orifice of the disk was measured to be higher than 200 m/s. After 0.15 s the pressure had risen to nearly the original 200 MPa and the process was repeated such that bangs with 0.15 s intervals sounded. Most of the time of a pulse (0.15 s) was used for rise of pressure, whereas the fall in pressure was nearly instantaneous. This type of flow (type a) was seen when *S. cerevisiae* (5.4 mg/ml) was suspended in water or different salt solutions (0.10–0.15 M KCl, $CaCl_2$ and NaCl) and pressed at —15, —25 and —35 °C. The disintegration was 53–80%. A slightly greater disintegration with the same flow pattern was seen with 2.0 M NaCl at —35 °C (84%). In general, greater disintegration was seen at —25 °C and —35 °C in the presence of salts. The pressures required for initiation of flow (P_0) in water agreed with the phase diagram of water. Presence of salts lowered the pressures (P_0). The pressures required for flow after 2 s (P_{max}) were lower than P_0 and rather independent of moderate concentrations of salts (≤ 0.15 M). Thus salts may facilitate phase changes in water. Once phase changes have occurred, further changes seem to be facilitated. The facilitation by salts was probably caused by eutectic melting under pressure. A greater effect was noticed for NaCl (eutectic point, e.p. $= -21.8$ °C) and for CaCl (e.p. $= -54.9$ °C) than for KCl (e.p. $= -11.1$ °C).

Suspension in Different Concentrations of Gelatin (78)

The presence of 20% gelatin increased the disintegration, more so at —25 and —35 than at —15 °C. Likewise 5 and 10% gelatin increased the disintegration at —25 and —35 °C, whereas it was decreased at —15 °C. About 80% disintegration was seen after pressing once in 5–20% gelatin at —25 and —35 °C.

When 20% gelatin was pressed, the flow was smoother with no bangs at least not after the initiation of the flow (flow type c, Fig. 5). The pressure initially dropped 70–150 MPa the fall depending on temperature and initial pressure. Then there was a smooth flow under slightly decreasing pressure. The pressure oscillations were less than 30 MPa. Concentrations of gelatin of 5 and 10% behaved in ways intermediate to water and 20% gelatin (flow type b; Fig. 4).

Presence of Both Salts and Gelatin (78)

Whereas the disintegration in 0.10 M NaCl was greater than in water, addition of 0.10 M NaCl to 5–20% gelatin reduced the disintegration. The pressures required for initiation of flow (P_0) were usually lower in the presence of salts, whereas the successive pressures (P_{max}, P_{min}) showed less difference similar to the effect of salts in water. Concentrations of 10 and 20% gelatin always yielded a smooth flow type (type c; Fig 5).

Freeze-pressing of Various Concentrations of Yeast [79]

At —25 °C the flow during freeze-pressing became smoother and the disruption was greater with increasing concentrations of yeast cells, at least with dry

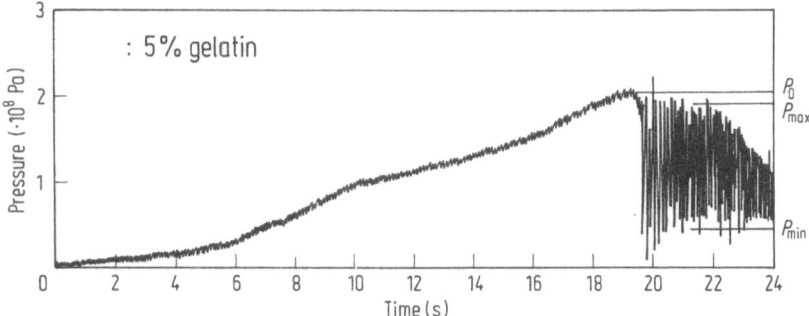

Figure 4. Pressure recording from freeze-pressing of 5% gelatin dissolved in deionized water. Flow with moderate bangs and a tendency to smooth flow = type b. (From Biotechnol. Bioeng. *18*, 449–463, 1976)

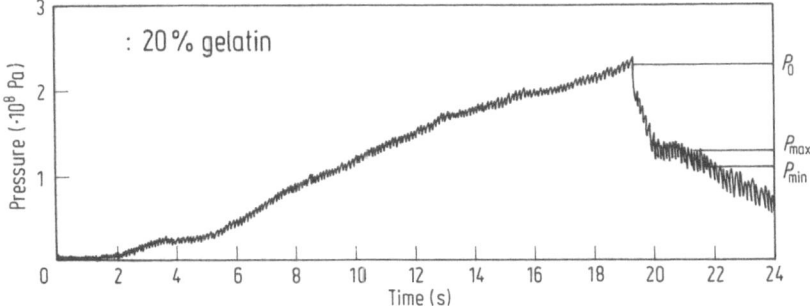

Figure 5. Pressure recording from freeze-pressing of 20% gelatin dissolved in deionized water. Smooth flow = type c. (From Biotechnol. Bioeng. *18*, 449–463, 1976)

weights $\geqq 270$ mg/ml (Table 11). The increase in disruption was under certain circumstances concurrent with a decrease of the pressure required for flow, both the initial (P_0) and the subsequent (P_{max}). The pressing of the highest concentration of yeast cells occurred with a smooth flow (type c, Fig. 5).

The smoothing effect on the flow by gelatin and high concentrations of cells was probably mediated by an increase of the viscosity, such that after melting the material was extruded through the orifice of the pressure chamber too slowly to escape the advancing piston. The retardation of the process was also seen in conductivity measurements, as the time for recrystallization increased with increasing concentrations of gelatin and cells from 0.07–0.1 s in water and salt solutions to 1–4 s in 20% gelatin and *S. cerevisiae* (0.270 g/g) [74, 76].

Interpretation of the Physical Data [74–79]

Freeze-pressing is intimately connected to changes in the crystal structure of water at sub-zero temperatures under pressure. Phase transition is initiated at the orifice in the X-press disk, where the frozen sample in the X-press is unsupported. The crystal defects may spread into the sample as the compression continues. At temperatures between 0 and $-22\,°C$, transition from ice I to liquid (melting) without any appreciable compression of the sample precedes the flow.

At temperatures between $-22\,°C$ and $-35\,°C$ the following transformations may precede flow, (a) from ice I to ice III, (b) from ice I to supercooled liquid and ice III, (c) from stabilized ice III to ice V and even (d) from stabilized ice V to ice VI. The phase transitions were deduced from measurements of changes in specific volume of the samples at compression. For salts, in general (a) occured, whereas for deionized water and samples containing biological material (a–c) appeared. Case (d) was observed only for 20% (w/w) gelatin. These results as well as the increase in conductivity under pressure before flow of the material indicate that phase transition into the liquid state occurs even at temperatures far below the triple point of liquid—ice I—ice III (Fig. 1) at $-22\,°C$ and 210 MPa. Flow starts near the phase boundaries between ice I and liquid and between ice I and ice III. The phase-boundary between ice I and liquid may be extended into the region of ice III and even ice II for deionized water and concentrated yeast cells which is a kind of supercooling. The phase transitions are initiated at the orifice and propagate into the sample on further compression. Usually flow and decrease of pressure starts before all the sample

Table 11. Disruption of *S. cerevisiae* at different concentrations of cells, temperatures, orifices, and velocities of flow through the orifices in the X-press.

Conc.	Temp.	Orifice diam.	Average velocity	Pressure (10^8 Pa)				Bangs at flow[a]	Intact cells
				initial	after 2s				
(mg/ml)	(°C)	(mm)	(m/s)		max	min	mean		(%)
270	−15	0.1	250	1.8	1.7	1.7	=	0	20
5.4	−25	2.5	0.0003	2.3	2.3	0	1.15	x	34
5.4	−25	1	7.1	2.3	1.7	0.4	1.04	(x)	37
270	−25	2.5	0.0003	1.7	1.2	1.1	1.2	0	26
270	−25	2.5	1.13	1.9	1.4	0.8	1.1	(0)	27
270	−25	1	7.1	2.1	1.4	1.1	1.2	(0)	24
270	−25	0.1	335	2.5	2.7	2.7	=	0	8
5.4	−35	2.5	0.0003	2.8			1.4	x/0	39
270	−35	2.5	0.0003	2.9	1.3	1.2	1.3	0	22
270	−35	2.5	1.13	2.4	1.2	1.0	1.1	0	26
				2.2	2.2	0.1	1.1	x	44
270	−35	1	7.1	2.4	1.8	1.4	1.6	0	6
				2.6	2.2	0.9	1.5	(x)	14
270	−35	1	14	3.9	1.9	1.7	1.8	0	6
270	−35	0.1	413	3.9	3.8	3.8	=	0	5
270	−45	2.5	1.13	3.0	2.0	0.1	1.1	x	47
270	−45	2.5	7.1	3.8	1.7	1.4	1.5	0	16

[a] x = bangs throughout the flow; 0 = no bangs.

has transformed. The velocity of phase transition is greater in deionized water and salt solutions than in the presence of biological material. When the sample has started to flow biological material like gelatin and yeast cells aid in the retention of the liquid state probably partly by physical binding of water, partly by the higher viscosity which results in slower flow and smaller drop of pressure. Rapid phase transformations, which occur in high concentrations of water allow rapid melting and recrystallization and promote explosive, pulse-like discharges of the sample. The slower phase transformations and smoother flow in the presence of biological material yields greater disintegration (Table 11). The raised conductivity during the flow indicates that the ice is melted such that the abrasive effect of ice is less important than recently proposed [122].

Recent Developments of Laboratory-scale Freeze-pressing (X-press)

The observation that the disintegration was increased when high concentrations of biological material were pressed at temperatures below the triple point at higher speed than was possible with the original manual hydraulic jack, has been exploited in a recent combination of equipment (Fig. 6). An inexpensive cooling bath (which may also be used for other purposes freeze-drying), a light-weight hydraulic press and a leight-weight oil-pump is easily handled in most laboratories. The disintegration achieved compares favourably with the original, more laborious method (Table 11).

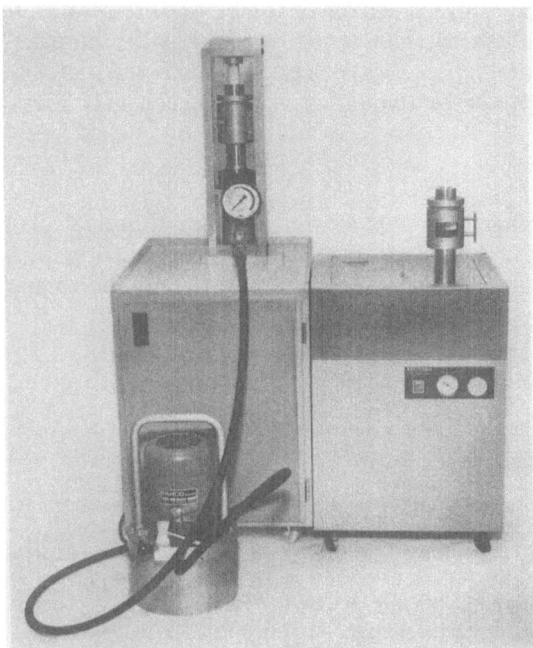

Figure 6. Laboratory equipment for freeze-pressing cooling bath (freezer, right), hydraulic press (top left), and electrical oil-pump (bottom left). A 25 ml X-press is shown both on the top of the freezer and adapted in the hydraulic press. (From Biotechnol. Bioeng. *18*, 449–463, 1976)

Large-scale Disintegration of Micro-organisms by Freeze-pressing

For the disintegration of larger quantities of micro-organisms and other biological material a press operating semicontinuously has been built. The material to be freeze-pressed is frozen such that cylindrical rods are formed which fit into the pressure chamber. A piston driven by a hydraulic pump forces the material through an orifice. At a sample temperature of $-35\ °C$ and a press temperature of $-20\ °C$ about 90% disruption was achieved when undiluted baker's yeast (*Saccharomyces cerevisiae*, 270 mg/g) was pressed through the orifice of the pressure chamber in a smooth flow [80]. Approximately 10 kg of material was freeze-pressed per hour. The energy demand was about 1 kWh per kg at a power efficiency of 20%. In a more recently built press ca 20 kg per hour may be handled. There should be no major obstacle to increasing the capacity of the process.

Disruption of Mammalian Cells and Tissues by Extrusion under Controlled Pressure

Freeze-pressing of most mammalian cells produces such extensive homogenization that the subcellular organization such as cell membranes, nuclei and mitochondria disintegrate. To preserve these organelles pressure extrusion of non-frozen cells and tissues at lower pressures have been used [34, 130]. Pressure extrusion at 3.5 MPa (35 kp cm^{-2}) of rabbit polymorphonuclear leucocytes disrupted 80–90% of the cells [130]. Leucocyte granules (lysosomes), usually considered to be extremely fragile, were abundantly present in the disintegrated material and separated from soluble components by zonal centrifugation [130]. After extrusion once at pressures of 6–13 MPa, chicken erythrocytes were completely disrupted and the nuclei remained intact. In contrast, extrusion twice or three times at 6 MPa damaged the nuclei (unpublished). The results not only offer straight-forward methods to prepare mammalian organelles but also support the concept that focused disruption preserves the integrity of intracellular units.

Acknowledgements

The mechanical skill of Messrs Lage and Jan Lovén and the technical skill of Mr Bertil Larsson is gratefully acknowledged. The close cooperation with Dr K. E. Magnusson and the comments and suggestions given by colleagues through the years have been highly appreciated.

References

1. Åkesson B, Lundvik L (1978) Simultaneous purification and some properties of aspartate: tRNA ligase and seven other amino-acid: tRNA ligases from Escherichia coli. Eur J Biochem 83:29–36
2. Allenmark S, Servenius B (1978) Characterization of bacterial L-(−)-tyrosine decarboxylase by isoelectric focusing and gel chromatography. J Chromatography 153:239–245
3. Augenstein DC, Thrasher K, Sinskey AJ (1974) Optimization in the recovery of a labile intracellular enzyme. Biotechnol Bioeng 16:1433–1447
4. Baharsefat M, Esterabadi H, Manhouri H, Yamini B (1967) Mycoplasma agalactiae II. Immuno-electrophoresis patterns of M. agalactiae antigen. Arch Inst Razi 19:91–95

5. Baker R, Ridell M, Lind A, Ouchterlony Ö (1979) Immunodiffusion studies of various structural preparations from mycobacterial cells. Int Arch Allergy 59:328–336
6. Belin L (1972) Immunological analyses of birch pollen antigens, with special reference to the allergenic components. Int Arch Allergy 42:300–322
7. Belin L (1980) Prevalence of symptoms and immuno-response in relation to exposure to infected humidifiers. Europ J Respir Dis, Suppl 107. 61:155–162
8. Belin L (1980) Clinical and immunological data on "wood-trimmer's disease" in Sweden. Europ J Respir Dis, Suppl 107. 61: 169–176
9. Benz F, Liersch M, Nüesch J, Triechler H (1971) Methionine metabolism and cephalosporin C synthesis in Cephalosporium acremonium. Eur J Biochem 20:81–88
10. Berg J-O (1981) Cellular localization of glycoside hydrolases in Bacteroides fragilis. Curr Microb 5:13–17
11. Berntsson E, Broholm K-A, Kaijser B (1978) Serological diagnosis of pneumococcal disease with enzyme-linked immunosorbent assay (ELISA). Scand J Infect Dis 10:177–181
12. Billaudelle H, Edebo L, Hammarsten E, Hedén C-G, Malmgren B, Palmstierna H (1960) Studies on the chemical and immunological structure of Bordetella pertussis. Acta Path Microbiol Scand 50: 208–224
13. Bolin T, Sjödahl R, Sundqvist T, Tagesson C (1981) Passage of molecules through the wall of the gastointestinal tract. Scand J Gastroenterol 16:897–901
14. Bowyer D, King JP (1977) Methods for the rapid separation and estimation of the major lipids of arteries and other tissues by thin-layer chromatography on small plates followed by microchemical assays. J Chromat 143:473–490
15. Bowyer D, Davies PF (1978) Effect of concentration of perfusing free fatty acid on arterial lipid synthesis in perfused normal and atherosclerotic rabbit aortas. Atherosclerosis 31:409–419
16. Branefors P (1979) Serological study of somatic antigens from Haemophilus influenzae and two related species. Int Arch Allergy 59:143–149
17. Bridgman PW (1912) Water in the liquid and five solid forms under pressure. Proc Amer Acad Arts Sci 47:441–458
18. Bridgman PW (1970) The Physics of High Pressure. Dover Publ, New York
19. Broadbent DA, Cartwright NJ (1971) Bacterial attack on phenolic ethers. Resolution of a Nocardia O-demethylase and purification of a cytochrome P 450 component. Microbios 4:7–12
20. Brookman JSG (1973) An extreme pressure pump for continuous cell disintegration. Biotechnol Bioeng 15:693–705
21. Brookman JSG (1974) Mechanism of cell disintegration in a high pressure homogenizer. Biotechnol Bioeng 16:371–383
22. Brown NC, Canellakeis ZN, Lundin B, Reichard P, Thelander L (1969) Ribonucleoside diphosphate reductase. Purification of the two subunits, proteins B1 and B2. Eur J Biochem 9:561–573
23. Cartwright NJ, Holdom KS, Broadbent DA (1971) Bacterial attack on phenolic ethers. Dealkylation of higher ethers and further observation on O-demethylases. Microbios 10:113–130
24. Cartwright NJ, Smith ARW (1967) Bacterial attack on phenolic ethers. An enzyme system demethylating vanillic acid. Biochem J 102:826–841
25. Cederberg E, Neujahr HY (1970) Distribution of β-carotene in subcellular fractions of Blakeslea trispora. Experientia 26:366–367
26. Dobrogosz WJ (1981) Enzymatic activity. In: Gerhardt P et al. (eds) Manual of Methods for General Bacteriology. American Society for Microbiology, Washington DC 20006, Chapter 18, ISBN 0-914826-30-1
27. Domagk GF, Chilla R, Domschke W, Engel H-J, Sörensen N (1969) Darstellung und Eigenschaften kristallisierter Glucose-6-phosphat-Dehydrogenase aus Candida utilis. Hoppe-Seyler's Z Physiol Chem 350:626–634
28. Doulah MS, Hammond TH, Brookman JSG (1975) A hydrodynamic mechanism for the disintegration of Saccharomyces cerevisiae in an industrial homogenizer. Biotechnol Bioeng 17:845–858
29. Edebo L (1960) A new press for the disruption of micro-organisms and other cells. J Biochem Microbiol Technol Eng 2:453–479
30. Edebo L (1961) Disintegration by freeze-pressing. I. Effects on bacteria. Acta Path Microbiol Scand 52:300–320

31. Edebo L (1961) Disintegration by freeze-pressing. II. Effects on fungi and phages. Acta Path Microbiol Scand 52:361–371
32. Edebo L (1966) Disintegration by freeze-pressing. III. A method for disintegration of large quantities of material. Biotechnol Bioeng 8:461–463
33. Edebo L (1967) Disintegration by freeze-pressing. IV. A modified apparatus for batch treatment. Biotechnol Bioeng 9:267–269
34. Edebo L (1969) Disintegration of cells. In: Perlman D (ed) Fermentation Advances. Academic Press New York pp 249–271
35. Edebo L, Holme T (1961) Disruption of bacteriophages by freezepressing. Abstr Intern Biophys Congr, Stockholm, p 287
36. Edebo L, Magnusson K-E (1973) Disintegration of cells and protein recovery. Pure Appl Chem 36:325–338
37. Engler CR, Robinson CW (1980) Disruption of Candida utilis cells in high pressure flow devices. Biotechnol Bioeng 23:765–780
38. Erge D, Maier W, Gröger D (1973) Untersuchungen über die enzymatische Umwandlung von Chanoclavin I. Biochem Physiol Pflanzen 164:234–247
39. Eriksson NE, Ahlstedt S, Belin L (1976) Diagnosis of reaginic allergy with house dust, animal dander and pollen antigens in adult patients. Int Arch Allergy 52:335–346
40. Eriksson S, Sjöberg B-M, Hahne S, Karlström O (1977) Ribonucleoside diphosphate reductase from Escherichia coli. J Biol Chem 252:6132–6138
41. Fjällbrant B (1969) Cervical mucus penetration by human spermatozoa treated with anti-spermatozoal antibodies from rabbit and man. Acta Obst Gynec Scand 48:71–84
42. Follows M, Hetherington PJ, Dunnill P, Lilly MD (1971) Release of enzymes from baker's yeast by disruption in an industrial homogenizer. Biotechnol Bioeng 13:549–560
43. Fossum K (1970) Proteolytic enzymes and biological inhibitors. IV. Bacterial proteinase inhibitors and their effect upon enzymes of various origin. Acta Path Microbiol Scand 78 B:755–759
44. Grabar P (1953) Biological actions of ultrasonic waves. Adv Biol Med Physics 3:191–246
45. Grov A, Helgeland SM (1971) Immunochemical characterization of Staphylococcus epidermidis and Micrococcus cell walls. Acta Path Microbiol Scand 79B:812–818
46. Grov A, Rude S (1967) Immunochemical characterization of Staphylococcus aureus cell walls. Acta Path Microbiol Scand 71:409–416
47. Hejgaard J (1970) Determination of α-acyl-α-hydroxy acids in biological fluids by head-space gas chromatography. Anal Biochem 37:368–377
48. Helgeland SM, Grov A (1971) Immunochemical characterization of staphylococcal and micrococcal mucopeptides. Acta Path Microbiol Scand 79B:819–826
49. Helting T, Ögren S, Lindahl U, Pertoft H, Laurent T (1972) Glycosaminoglycan synthesis in mouse mastocytoma. Biochem J 126:587–592
50. Henke H (1972) Untersuchungen an Tryptophan-Biosynthese Enzymen aus Coprinus und anderen Basidiomyceten. Thesis, Eidgenössische Technische Hochschule, Zürich, Diss. Nr. 4825. ISBN 3 260 03236 3
51. Hetherington PJ, Follows M, Dunnill P, Lilly MD (1971) Release of protein from baker's yeast (Saccharomyces cerevisiae) by disruption in an industrial homogenizer. Trans Inst Chem Engrs 49:142–148
52. Hofstad T (1967) An anaerobic oral filamentous organism possibly related to Leptotrichia buccalis. 2. Composition of cell walls. Acta Path Microbiol Scand 70:461–468
53. Hofstad T, Sveen K, Dahlén G (1977) Chemical composition, serological reactivity and endotoxicity of lipopolysaccharides extracted in different ways from Bacteroides fragilis, Bacteroides melaninogenicus and Bacteroides oralis. Acta Path Microbiol Scand 85B:262–270
54. Holme T, Edebo L (1965) Studies of Salmonella antigens by the agar gel precipitin test. Acta Path Microbiol Scand 65:287–294
55. Holmgren A, Kallis G-B, Nordström B (1981) A mutant thioredoxin from Escherichia coli tsnC 7007 that is nonfunctional as subunit of phage T7 DNA polymerase. J Biol Chem 256: 3118–3124
56. Holmgren A, Ohlsson I, Grankvist M-L (1978) Thioredoxin from Escherichia coli. Radioimmunological and enzymatic determination in wild type cells and mutants defective in phage T7 DNA replication. J Biol Chem 253: 430–436
57. Hopgood MF, Walter DJ (1969) Succinic acid production by rumen bacteria. III. Enzymatic studies on the formation of succinate by Ruminococcus flavefaciens. Aust J Biol Sci 22:1413–1424

58. Hostalek Z, Tinterova M, Jechova V, Blumauerova M, Suchy J, Vanek Z (1969) Regulation of biosynthesis of secondary metabolites. I. Biosynthesis of chlortetracycline and tricarboxylic acid cycle acitivity. Biotechnol Bioeng 11:539–548
59. Hughes DE (1951) A press for disrupting bacteria and other microorganisms. Brit J Exp Pathol 32:97–109
60. Hughes DE, Wimpenny JWT, Lloyd D (1971) The disintegration of microorganisms. In: Norris JR, Ribbons DW (eds) Methods in Microbiology vol V B. Academic Press London New York, pp 1–54
61. Jansson L, Lindahl U (1970) Evidence for the existence of a multichain proteoglycan of heparan sulfate. Biochem J 117:699–702
62. Kelemen MV, Sharpe JEE (1979) Controlled cell disruption: A comparison of the forces required to disrupt different micro-organisms. J Cell Sci 35:431–441
63. Koch H, Kiefer J (1972) RNA-synthesis in irradiated diploid yeast at 50 per cent survival level. Int J Radiat Biol 21:223–234
64. Kristoffersen T, Hofstad T (1970) Chemical composition of lipopolysaccharide endotoxins from human oral fusobacteria. Arch Oral Biol 15:909–916
65. Krogh HK, Tönder O (1968) Adherence of erythrocytes to stratum corneum of skin tissue sections. Int Arch Allergy 34:170–180
66. Krogh HK, Tönder O (1976) Comparison of human stratum corneum, callus and psoriatic scales by means of serological methods. Int Arch Allergy 53:434–440
67. Lagerkvist U, Rymo L, Waldenström J (1966) Structure and function of transfer RNA. II. Enzyme-substrate complexes with valyl ribonucleic acid synthetase from yeast. J Biol Chem 241:5391–5400
68. Lagerkvist U, Waldenström J (1964) Structure and function of transfer RNA. I. Species specificity of transfer RNA from E. coli and yeast. J Mol Biol 8:28–37
69. Lagerkvist U, Waldenström J (1967) Purification and some properties of valyl ribonucleic acid synthetase from yeast. J Biol Chem 242:3021–3025
70. Lambert PA, Smith ARW (1976) Antimicrobial action of dodecyldiethanolamine: Activation of ribonuclease in Escherichia coli. Microbios 17:35–49
71. Lindahl T, Ljungquist S, Siegert W, Nyberg B, Sperens B (1977) DNA N-glycosidases. Properties of uracil-DNA glycosidase from Escherichia coli. J Biol Chem 252:3286–3294
72. Lindahl U (1970) Attempted isolation of a heparin proteoglycan from bovine liver capsule. Biochem J 116:27–34
73. Löfkvist T, Sjöquist J (1963) Purification of staphylococcal antigens. With special reference to antigen A (Jensen). Int Arch Allergy 23:289–305
74. Magnusson K-E (1975) Physical studies on the freeze-pressing of biological material. Linköping University Medical Dissertation No. 28. Linköping University, Linköping, Sweden
75. Magnusson K-E (1977) The continuation of the phase-boundary between ice I and liquid into region of ice III and II and its relation to freeze-pressing of biological material. Cryobiology 14:68–77
76. Magnusson K-E (1977) A physical description of freeze-pressing of biological material with the X-press. Cryobiology 14:78–86
77. Magnusson K-E, Edebo L (1974) Estimation of the disruption in freezepressed Saccharomyces cerevisiae by an electronic particle counter. Biotechnol Bioeng 16:1273–1282
78. Magnusson K-E, Edebo L (1976) Influence of salts and gelatin on disintegration of Saccharomyces cerevisiae by freeze-pressing. Biotechnol Bioeng 18:449–463
79. Magnusson K-E, Edebo L (1976) Influence of cell concentration, temperature, and press performance on flow characteristics and disintegration in the freeze-pressing of Saccharomyces cerevisiae with the X-press. Biotechnol Bioeng 18:865–883
80. Magnusson K-E, Edebo L (1976) Large-scale disintegration of microorganisms by freeze-pressing. Biotechnol Bioeng 18:975–986
81. Maier W, Erge D, Gröger D (1972) Über Aktivierungsreaktionen bei Claviceps. Biochem Physiol Pflanzen 163:432–442
82. Marquardt J, Holm SE, Lycke E (1965) Immunoprecipitating factors of vaccinia virus. Virology 27:170–178
83. Mathieson JM, Pearce JM (1977) The isolation of minimally degraded hyaluronate from rat skin. Biochem J 161:419–424
84. Mattern P (1968) Etat actual et résultats des techniques immunologiques utilisées à l'Institut

Pasteur de Dakar pour le diagnostic et l'étude de la trypanosomiase humaine africaine. Bull WHO 38:1–8

85. Milde E-J (1971) Reactivity between rheumatoid factor and rheumatoid tissue. Ann Rheum Dis 30:84–90

86. Millbank JW, Neale MJ (1969) An improved French pressure cell. Biotechnol Bioeng 11:711–718

87. Milner HW, Lawrence NS, French CS (1950) Colloidal dispersion of chloroplast material. Science 111:633–634

88. Mitra SK, Lustig F, Åkesson B, Lagerkvist U, Strid L (1977) Codon-anticodon recognition in the valine codon family. J Biol Chem 252: 471–478

89. Mosqueira FG, Higgins JJ, Dunill P, Lilly MD (1981) Characteristics of mechanically disrupted baker's yeast in relation to its separation in industrial centrifuges. Biotechnol Bioeng 23:335–343

90. Myrvold HE, Brandberg Å (1977) Microembolism in experimental septic shock. Eur Surg Res 9: 34–47

91. Myrvold HE, Lewis D (1977) Platelets, fibrinogen and pulmonary haemodynamics in early experimental septic shock. Circ Shock 4: 201–209

92. Neujahr HY, Varga JM (1970) Degradation of phenols by intact cells and cell-free preparations of Trichosporon cutaneum. Eur J Biochem 13:37–44

93. Nord C-E, Linder L, Wadström T, Lindberg AA (1973) Formation of glycoside hydrolases by oral Streptococci. Arch Oral Biol. 18:391–402

94. Nordstoga K, Fjölstad M (1967) The generalized Shwartzman reaction and Haemophilus infections in pigs. Path Vet 4:245–253

95. Norstoga K, Fjölstad M (1970) Porcine salmonellosis. II. Production of the generalized Shwartzman reaction by intravenous injections of disintegrated cells of Salmonella cholerae-suis. Acta Vet Scand 11:370–379

96. Norlin M, Navalkar R (1966) Immunoprecipitinogenic spectra of culture filtrates of "anonymous" mycobacteria compared to those of bacillary cells. Bull Int Un Tuberc 38:52–56

97. Ögren S, Lindahl U (1971) Degradation of heparin in mouse mastocytoma tissue. Biochem J 125:1119–1129

98. Parish WE (1971) Studies on vasculitis. II. Some properties of complexes formed of antibacterial antibodies from persons with or without vasculitis. Clinical Allergy 1:111–121

99. Parish WE (1971) Studies on vasculitis. III. Decreased formation of antibody to M protein, group A polysaccharide and to some exotoxins, in persons with cutaneous vasculitis after streptococcal infection. Clinical Allergy 1:295–309

100. Parish WE (1971) Studies on vasculitis. IV. The low incidence of antibacterial anaphylactic antibodies in the sera of persons with cutaneous vasculitis following bacterial infection. Clinical Allergy 1:433–446

101. Parish WE, Rook AJ, Champion RH (1965) A study of auto-allergy in generalized eczema. Brit J Derm 77:479–526

102. Pazoutova S, Votruba J, Rehacek J (1981) A mathematical model of growth and alkaloid production in the submerged culture of Claviceps purpurea. Biotechnol Bioeng 23:2837–2849

103. Pouliot P, Viens P, Targett GAT (1974) Lymphocyte transformation and mouse cell-mediated immune response (CMI) to Trypanosoma musculi infection. IRCS 2:1567

104. Rabow L (1979) Lactate dehydrogenase (LD), hydroxybutyrate dehydrogenase (HBD), and LD-isoenzymes in brain tissue. Acta Neurochir 36:61–70

105. Rabow L, Kristensson K (1977) Changes in lactate dehydrogenase isoenzyme patterns in patients with tumours of the central nervous system. Acta Neurochir 36:71–81

106. Reichard P, Baldesten A, Rutberg L (1961) Formation of deoxycytidine phosphates in extracts from Escherichia coli. J Biol Chem 236:1150–1157

107. Riazuddin S, Lindahl T (1978) Properties of 3-methyladenine-DNA glycosylase from Escherichia coli. Biochemistry 17:2110–2118

108. Ribi E, Perrine T, List R, Brown W, Goode G (1959) Use of pressure cell to prepare cell walls from mycobacteria. Proc Soc Exptl Biol Med 100:647–649

109. Ridell M (1975) Taxonomic study of Nocardia farcinica using serological and physiological characters. Int J Syst Bact 25:124–132

110. Ridell M (1981) Immunodiffusion studies of some Nocardia strains. J Gen Microbiol 123: 69–74

111. Ridell M, Goodfellow M, Minnikin DE, Minnikin SM, Hutchinson IG (1982) Classification

of Mycobacterium farcinogenes and Mycobacterium senegalense by immunodiffusion and thin-layer chromatography of longchain components. J Gen Microbiol 128:1299–1307

112. Ridell M, Norlin M (1973) Serological study of Nocardia by using mycobacterial precipitation reference systems. J Bact 113:1–7

113. Rogers HJ, Perkins HR, Ward JB (1980) Microbial cell walls and membranes. Chapman & Hall, London. ISBN 0-412-12030-5

114. Rymo L, Lagerkvist U, Wonacott A (1970) Crystallization of lysyl transfer ribonucleic acid synthetase from yeast. J Biol Chem 245:4308–4316

115. Rymo L, Lundvik L, Lagerkvist U (1972) Subunit structure and binding properties of three amino acid transfer ribonucleic acid ligases. J Biol Chem 247:3888–3897

116. Sahlin M, Gräslund A, Ehrenberg A, Sjöberg B-M (1982) Structure of the tyrosyl radical in bacteriophage T4-induced ribonucleotide reductase. J Biol Chem 257:366–369

117. Samuelsson T, Lundvik L (1978) Purification and some properties of asparagine, lysine, serine, and valine: tRNA ligases from Bacillus stearothermophilus. J Biol Chem 253:7033–7039

118. Schmauder H-P, Gerullis Ch, Gröger D (1976) Wirkung von 5-Fluorindol und 5-Fluor-,D,L-tryptophan auf Wachstum und Alkaloidsynthese in Claviceps. Biochem Physiol Pflanzen 170: 201–210

119. Schmauder H-P, Gröger D (1976) Vergleichende Untersuchungen zur Tryptophan-Biosynthese bei Claviceps. Biochem Physiol Pflanzen 169:201–205

120. Schmauder H-P, Gröger D (1976) Zur Charakterisierung der Anthranilat-Synthetase bei Claviceps. Biochem Physiol Pflanzen 169:471–486

121. Schmauder H-P, Gröger D (1973) Untersuchungen zur Aromatenbiosynthese und Alkaloidbildung in Claviceps-Arten. Biochem Physiol Pflanzen 164:41–57

122. Schnaitman CA (1981) Fractionation. In: Gerhardt P et al. (eds) Manual of Methods for General Bacteriology. American Society for Microbiology, Washington DC 20006, Chapter 18. ISBN 0-914826-30-1

123. Schwartz JH (1967) Initiation of protein synthesis under the direction of Tobacco Mosaic Virus RNA in cell-free extracts of Escherichia coli. J Mol Biol 30:309–322

124. Schwartz JH, Reeves JY, Broome JD (1966) Two L-asparaginases from E. coli and their action against tumours. Proc Nat Acad Sci 56:1516–1519

125. Standford JL, Beck A (1969) Bacteriological and serological studies of fast growing mycobacteria identified as Mycobacterium friedmannii. J Gen Microbiol 58:99–106

126. Standford JL, Gunthorpe WJ (1969) Serological and bacteriological investigation of Mycobacterium ranae (fortuitum). J Bact 98: 375–383

127. Sund M-L, Linder L (1980) Regulation of β-fructofuranosidase (invertase) in Streptococcus mitis. J Gen Microbiol 118:85–94

128. Sund M-L, Linder L, Andersson C (1978) Isoelectric focusing studies in the β-fructofuranosidases and α-glycosidases of Streptococcus mitis. J Gen Microbiol 106:337–342

129. Sutherland IW (1978) Separation and purification of bacterial antigens. In: Weir DM (ed) Handbook of Experimental Immunology. 3rd ed. Blackwell Scientific Publications, Oxford ISBN 0 632 00096 1

130. Tagesson C, Stendahl O, Magnusson K-E, Edebo L (1973) Disintegration of single cells in suspension. Isolation of rabbit polymorphonuclear leucocyte granules. Acta Path Microbiol Scand 81B:464–472

131. Tanner W, Jung P, Behrens NH (1971) Dolicholmonophosphates: mannosyl acceptors in a particulate in vitro system of S. cerevisiae. FEBS Letters 16:245–248

132. Thelander L (1967) Thioredoxin reductase. Characterization of a homogenous preparation from Escherichia coli B. J Biol Chem 242: 852–859

133. Turner MW, McGregor IA (1969) Studies on the immunology of human malaria. I. Preliminary characterization of antigens in Plasmodium falciparum infections. Clin Exp Immunol 5:1–16

134. Varga JM, Neujahr HY (1970) Purification and properties of catechol 1,2-oxygenase from Trichosporon cutaneum. Eur J Biochem 12:427–434

135. Wålinder O (1969) Protein-bound acid-labile phosphate. Isolation of 1-^{32}P-phosphohistidine and 3-^{32}P-phosphohistidine from some mammalian and microbial cell extracts incubated with adenosine triphosphate-^{32}P*. J Biol Chem 244:1065–1069

136. Wallinder I-B, Neujahr HY (1971) Cell wall and peptidoglycan from Lactobacillus fermenti. J Bact 105:918–926

137. Watt PJ, Ward ME, Glynn AA (1971) A comparison of serological tests for the diagnosis of gonorrhoea. Brit J Vener Dis 47:448–451
138. Virkar PD, Narendranathan TJ, Hoare M, Dunnill P (1981) Studies on the effect of shear on globular proteins: Extension to high shear fields and to pumps. Biotechnol Bioeng 23:425–429
139. Yoshida A, Hedén C-G, Cedergren B, Edebo L (1961) A method for the preparation of undigested bacterial cell walls. J Biochem Microbiol Technol Eng 3:151–159
140. Zetelaki K (1969) Disruption of mycelia for enzymes. Process Biochem 4 (DEC): 19–27

Disintegration of Microorganisms in a 20 l Industrial Bead Mill

Horst Schütte, Karl Heinz Kroner, Helmut Hustedt and Maria-Regina Kula

Summary

The suitability of a 20 l high speed industrial bead mill has been examined for the disintegration of microorganism in large-scale. The release of enzymes and proteins from suspensions of several yeast and bacterial species was followed during the disruption procedure. Certain operating variables for the disintegration were investigated, in particular the effect of agitator speed, the flow rate, and the bead diameter. The heat production and the power consumption were also measured during these experiments. Disintegration of bacteria is more difficult in comparison with yeast due to the small size, nevertheless enzymes could be solubilized on high yield by treatment in the mill but with lower capacity than yeast.

Introduction

Intracellular enzymes are of growing interest for future developments in enzyme technology, e.g. coenzyme dependent processes. The first step in their isolation is a disruption of the outer envelope of the microbial cell and a scale-up of this procedure becomes necessary. At present, mechanical methods for disintegration seem to be applicable for large scale processes only. High-pressure industrial homogenizers (Hetherington 1971, Follows 1971, Brookman 1974, Doulah 1975) or high speed industrial bead mills (Rehacek 1969, Currie 1972, Marffy and Kula 1974, Mogren 1974, Rehacek and Schaefer 1977, Limon-Lason 1979) are currently preferred for this purpose. Here we report upon our continuing investigations on the release of protein and enzymes from different microorganisms in a commercially available agitator mill (Netzsch LME 20) (Schütte et al. in press) and give information for the effective utilization in detail.

Materials and Methods

Microorganisms

Commercial baker's yeast (*Saccharomyces cerevisiae*) was obtained from Deutsche Hefewerke GmbH. (Northeim, Germany), brewer's yeast (*Saccharomyces carlsbergensis*) was a gift of the Hofbrauhaus Wolters AG. (Braunschweig, Germany), *E. coli* MRE 600 was purchased from E. Merck (Darmstadt, Germany). All other microorganisms used were cultivated in the pilot plant of the GBF. Cells were stored frozen at −20 °C until use except the packaged lots of baker's yeast which were kept at 4 °C and used within two weeks after delivery.

The Agitator Mill:

The Netzsch LME 20-mill (Netzsch Maschinenfabrik Selb, Germany) is a variable speed homogenizer with a horizontally positioned grinding chamber (Fig. 1). Technical details of the model LME 20 are summarized in Table 1.

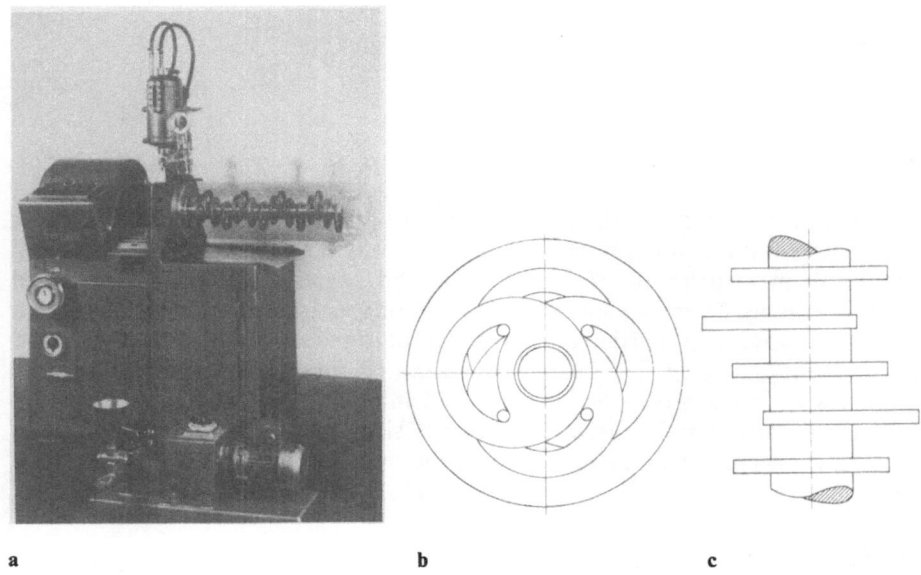

a   b   c

Figure 1. Netzsch LME 20-mill **a** general view of mill; the arrangement of agitator discs on the drive shaft viewed **b** end-on, **c** from the side

Table 1. Technical Data of the Netzsch Mill Type LME 20

Internal working volume of the grinding tank	22,7 l
Power rating of drive motor	17.5 KW
Variable speed range of agitator	700–1450 rpm
	(5.2–10.6 m/sec)
Flow rate capacity	50–500 l/h
Discs mounted on the agitator shaft	16
Slot width of the microseparator	0.2 mm

The discs are mounted eccentrically on the central shaft. The griding chamber is equipped with a cooling jacket which effectively prevents an undesirable temperature rise in the suspension during the disruption process. The seal and bearing assembly has a separate cooling circuit. With a given slot width of 0.2 mm the manufacturer recommends that the machine could be operated with beads of minimal size of 0.45 mm diameter. Lead free glass beads were obtained from the Netzsch company.

Analytical Procedure

Samples of the cell homogenates were centrifuged at 45000 xg for 1 hour. The supernatant was assayed for enzyme activity and protein concentrations using standard methods (Schütte et al. in press).

Normalization of enzyme activity and degree of disintegration

Commercial baker's yeast was used as a relatively defined and cheap enzyme source. This choice enabled us also to compare our data with the published performance of other homogenizer. To normalize the enzyme activities of different batches of yeast a disintegration was run to completion with a representative sample of each batch in the Dyno-Mill Type KDL. 100% values for the different bacteria were determined in a similar manner in the Dyno-Mill or using the Manton-Gaulin Homogenizer Type 15 M-8TA, respectively.

Disintegrator Operation

Suspensions from baker's yeast, brewer's yeast or *Candida boidinii* were prepared in a 200 l jacketed tank using a propeller stirrer (Ekato Type EMK 40). 50 mM potassium phosphate (pH 7.8) containing 0.05% (v/v) 2-mercaptoethanol was used as a buffer. The suspension was pumped to the mill with a variable speed Netzsch-Mohno pump Type 2 NL 20 A. The yeast suspension was precooled to 5–10 °C and passed through the homogenizer in a single pass mode. To find the steady-state conditions, samples from the outlet were taken in 5 minute intervalls. The disintegrator was cooled with aqueous ethylene glycol solution at −10 °C. The flow rate through the cooling jacket was 700 l/h and through the seal assembly 300 l/h. The temperature of the homogenate and of the cooling medium was measured at the inlets and outlets from the mill. The power input was measured by means of an amperometer. Bacteria were processed in a similar manner but here buffers were used according to the requirements of the particular enzyme.

Table 2. Parameters of the recycling experiments in the Dyno-Mill and in the Netzsch-Mill

	Dyno Mill KDL	Netzsch-Mill LME 20
grinding tank	640 ml	22,700 ml
load volume of beads	544 ml (85%)	19,295 ml
bead diameter	0.55–0.85 mm	0.55–0.85 mm
flow rate	5,0 l/h	177.4 l/h
speed of rotation	2,000 rpm (6.7 m/sec)	920 rpm
	3,000 rpm (10 m/sec)	1,380 rpm
yeast suspension	40%	40%
buffer	50 mM potassium phosphate, pH 7.8	
volume of suspension	2,000 ml	71,000 ml

Disruption Kinetics

Conditions for recycling experiments with the Neztsch LME 20 mill were selected by linear extrapolation from similar experiments in the 0.6 l Dyno-Mill Type KDL as described by Marffy and Kula (1974). The increase of protein concentration and enzymatic activity in the supernatant was followed as a function of time. Parameters of the recycling experiments are shown in Table 2.

Results and Discussion

Cell disruption in bead mill homogenizers was found to be a first order reaction as reported by several authors (Currie 1972, Marffy and Kula 1974, Limon-Lason 1979). If the disintegration process in the Netzsch LME-20 mill is a first order reaction it should be represented by the equation

$$\log [R_m/(R_m - R)] = K \cdot t$$

where R_m is the experimentally observed maximum protein concentration, R is the protein released and K is the first order rate constant. In recycling experiments we followed the increase of protein concentration and glucose-6-phosphate dehydrogenase and α-glucosidase activity as a function of time. For an average impeller tip speed of 6.7 m/sec or 10 m/sec, a first order kinetic for enzyme and protein solubilization can be observed with the Netzsch-mill as well with in the Dyno-mill. The results of the recycling experiments are given in Table 3.

Table 3. Comparison of the K-values of the Dyno-Mill and the Netzsch-Mill under identical operational parameters

K-value 10^4 K l s^{-1}	Dyno-Mill KDL		Netzsch-Mill LME 20	
	6.7 m s^{-1}	10 m s^{-1}	6.7 m s^{-1}	10 m s^{-1}
Protein	2.82	2.94	2.00	2.18
G-6-P-DH	2.85	3.92	2.20	2.20
α-glucosidase	3.61	4.18	2.98	2.97

Cell disintegration in the Netzsch-mill can be described as a first order process (Fig. 2). The rate constant, however, did not increase as expected with increasing rotational speed in the Netzsch mill over the range tested, while the K-values in the Dyno-mill under the same conditions increased. Disintegration in a bead mill is a rather complex process which is influenced by a number of parameters. Only some of these parameters can actually be influenced by selecting or changing operating variables. In prior experiments (Currie 1972, Marffy and Kula 1974) the rate constant was found to depend on the agitator speed as well as on the flow rate of suspension, the microorganism concentration, the bead size, the load volume of beads, the density of beads and the disruption temperature. Fixed parameters such as the geometry of the grinding chamber and the stirrer design are important for the mixing characteristics of a mill and will influence the

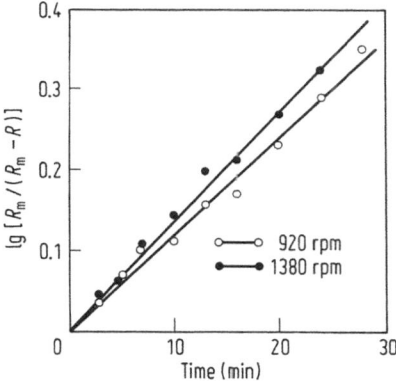

Figure 2. Protein release during recycling experiments in the Netzsch LME 20-mill

residence time distribution which in turn will influence the rate constant (Limon-Lason 1979, Rehacek and Schaefer 1977). Experiments to clarify this point for the Netzsch mill are in progress.

Protein Release

The release of soluble protein was measured by varying the speed of rotation in the range of 700 to 1450 rpm (Fig. 3). The protein concentration in solution increased with increasing rotational speed up to approximately 1100 rpm corresponding to an agitator tip speed of 8 m/sec. At higher rotational speed up to 1450 rpm, protein yield in a single pass doesn't increase noticably. We used three sizes of glass beads with a diameter of 0.45–0.65 mm, 0.55–0.85 mm and 1 mm and changed the bead load volume in the range of 70–85%. With glass

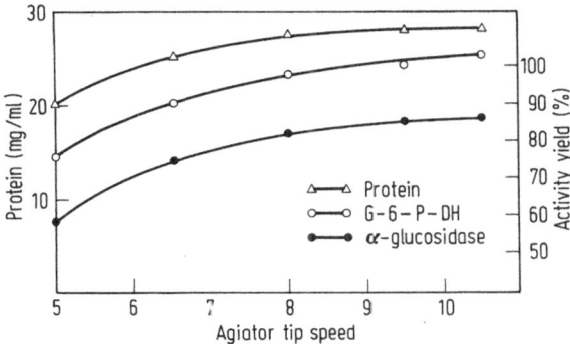

Figure 3. Effect of agitator speed on protein and enzyme release from baker's yeast.
Constant parameters: bead diameter: 0.55–0.85 mm
                    bead load volume: 80%
                    cell suspension: 40% (w/v)
                    flow rate: 100 l/h
                    tip speed: [m/sec]
protein △————△; G-6-P-DH ○————○; α-glucosidase ●————●

beads of a diameter of 0.55–0.85 mm and 80% bead load volume, the highest release of soluble protein was obtained (Fig. 3). 80% loading volume of glass beads was selected for further experiments.

During disruption at 1100 rpm a temperature rise of the yeast suspension of 2 °C was observed and a power of 5.5 KW was needed in comparison to 7 °C temperature increase and 7.5 KW during operation with 1450 rpm. With a bead load volume of 90% a temperature rise of 14.5 °C was measured and the yield of soluble protein already decreased.

Enzyme Release

We studied the effect of agitator speed on the release of soluble cytoplasmic enzymes such as glucose-6-phosphate dehydrogenase from baker's yeast (Fig. 3) and brewer's yeast and formate dehydrogenase from *Candida boidinii*. Here also the enzyme activities increase in the soluble fraction with increasing rotational speed up to 1100 rpm. For the different yeast species a degree of disintegration of approximately 90% was observed by pumping in a single pass mode through the agitator. These results show that the mill is a very effective disintegrator for yeast cells.

Effect of yeast concentration in the cell suspension on the release of enzymes

The influence of the concentration of microorganisms in the slurry on enzyme release was checked using 30, 40 and 50% cell suspension. We could show that with a rotational speed of 700 rpm maximal enzyme release is observed at 40% concentration. In contrast, at higher speed (1100 rpm) a concentration of cells between 30–50% is apparently not as important to achieve a high degree of disintegration (Mogren 1974, Schütte et al. in press).

Effect of bead diameter on the release of enzymes

In the course of our experiments with baker's yeast we analyzed two enzymes, glucose-6-phosphate dehydrogenase which is a cytoplasmic enzyme and

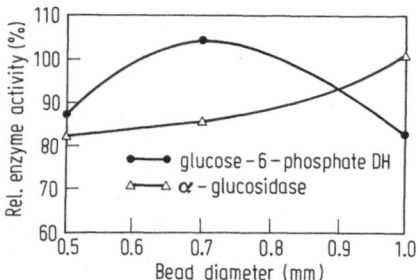

Figure 4. Effect of glass bead diameter on the release of α-glucosidase and glucose-6-phosphate dehydrogenase from baker's yeast.
Constant parameters: agitator tip speed: 10 m/sec. (1400 rpm)
                         bead load volume: 80%
                         cell suspension: 40%
                         flow rate: 100 l/h

α-glucosidase which is predominantly located in the periplasmic space. Three different sizes of lead-free glass beads were used with a diameter of 0.45–0.58 mm, 0.55–0.85 mm and 1 mm. The results are summarized in Fig. 4. Glucose-6-phosphate is solubilized best with 0.55–0.85 mm glass beads, but α-glucosidase with 1 mm beads. This observation must be related to the fact that complete disintegration of the cell is not necessary to solubilize α-glucosidase. The results show that the location of the desired enzyme in the cell is important for the choice of the optimal bead diameter.

Effect of the flow rate on the disintegration of baker's yeast

It was found that the enzyme activity solubilized decreased only slightly with increasing flow rate and processing at 360 l/h resulted only in a 10% decrease in activity yield in comparison with a flow rate of 40 l/h. Experiments with the variation of the flow rate point out that the degree of disintegration is not a simple function of the residence time.

Wet milling was found to be a very effective process for the disruption of several yeast species e.g. *Saccharomyces cerevisiae*, *Saccharomyces carlsbergensis*, and *Candida boidinii* used for the production of enzymes. Very high yields of enzyme activity can be solubilized by processing approximately 50 kg of yeast under optimized conditions (Table 4). By accepting a slightly lower yield of activity, up to 200 kg of baker's yeast could be disrupted per hour.

Table 4. Release of different enzymes from various yeasts using the Netzsch LME 20 mill under optimized conditions

Microorganism	Enzyme	Activity solubilized after one pass through the mill %	Throughput kgh^{-1}
Saccharomyces carlsbergensis	Glucose-6-phosphate dehydrogenase	86	60
	α-glucosidase	100	50
Saccharomyces cerevisiae	Glucose-6-phosphate dehydrogenase	100	70
	α-glucosidase	100	50
Candida boidinii	Formate dehydrogenase	95	40
	Isopropanol dehydrogenase	89	40

Disintegration of Bacteria

Investigation of the disruption of bacteria in the industrial bead mill was limited by the availability and higher cost of large quantities of bacteria (>20 kg) needed for the experiments. Therefore, we mainly studied the influence of bead diameter and flow rate on the disintegration process which we considered most interesting. To study the effect of the size of glass beads, we employed beads with a diameter of 0.45–0.58 mm and 0.55–0.85 mm. The flow rate of bacterial suspension was varied in the range of 50 to 150 l/h. Bacterial cells are approximately only 1/10 the size of a yeast cell which will hamper their disintegration in glass bead mills. Using the smaller bead size, a higher disintegra-

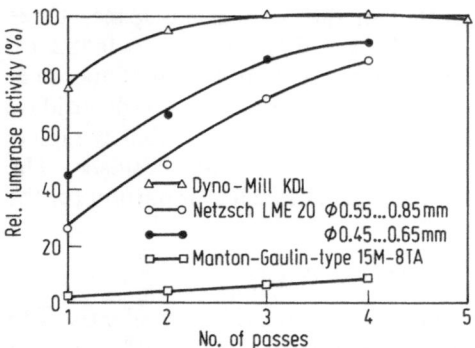

Figure 5. Disruption of Brevibacterium ammoniagenes using various types of homogenizers.

Dyno-Mill Type KDL	△————△	
Netzsch LME 20-Mill	○————○ ; beads 0.55–0.85 mm	
	●————● ; beads 0.45–0.58 mm	
Manton-Gaulin-Type 15M-8TA	□————□ ; 550bar	

tion efficiency is to be expected because the number of possible impacts during operation will increase about threefold for the same load volume. The kinetic energy for the smaller bead size decreases of course in comparison with the larger bead size. For the release of fumarase from *Brevibacterium ammoniagenes* the small glass beads were found preferable (Fig. 5). In this special case wet milling of *Brevibacterium ammoniagenes* is by far more effective than treatment in a high pressure homogenizer. Using a Manton-Gaulin-Homogenizer in four passes with pressure of 560 bar, only 10 % of fumarase activity was solubilized. The disruption efficiency decreases drastically when the flow rate is changed from 50 l/h to 100 l/h. Similar results were obtained when *Bacillus sphaericus* was disintegrated in the mill. Also the release of leucine dehydrogenase decreases with increasing flow rate of the cell suspension but it will increase if the agitator tip speed is increased and the small bead size is used. The results demonstrate that small glass beads with a diameter of 0.45–0.58 mm rate are optimal for the disruption of bacteria in the Netzsch LME 20 mill. However, under such conditions the mill is operated at the limit of its performance and the micro-separator may be easily damaged which may lead to crushing of the beads and a loss of the bead charge with the product stream. To prevent this, only well sieved glass beads of a diameter of 0.5–0.65 mm should be used for disintegration of bacteria. The influence of the flow rate is shown in some detail for the solubilization of penicillin acylase from *Escherichia coli* (Fig. 6). The same activity yield was observed after three passes through the agitator mill with a flow rate of 100 l/h as with a flow rate of 50 l/h in a single pass. Table 5 summarizes the activity yield for several intracellular enzymes from bacteria obtained in the industrial bead mill during the first steps of a large scale enzyme isolation. Approximately 10–20 kg of wet bacterial cells could be processed per hour with high enzyme yield. These data show that for scale-up of a disruption process the Netzsch LME 20 mill —

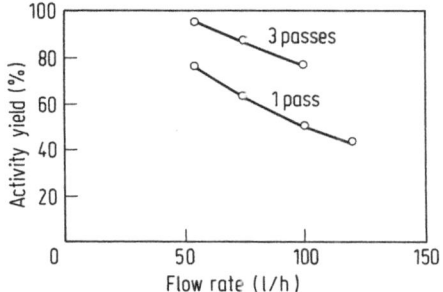

Figure 6. Effekt of flow rate on the release of penicillin acylase from E. coli

Constant parameters: agitator tip speed: 8.5 m/sec. (1200 rpm)
                    glass bead diameter: 0.55–0.85 mm
                    bead load volume: 85%
                    cell suspension: 40%

Table 5. Release of different enzymes from various bacteria using the Netzsch LME 20 mill

Microorganism	Enzyme	Activity solubilized	Number of passes	Throughput kgh^{-1}
Bacillus cereus	Leucine dehydrogenase	84	2	16
Bacillus sphaericus	Leucine dehydrogenase	80	3	15
Brevibacterium ammoniagenes	Fumarase	85	3	12
Escherichia coli MRE 600	Isoleucyl-tRNA synthetase	95	2	20
Escherichia coli	Penicillin acylase	95	3	7
Lactobacillus confusus	Lactate dehydrogenase	92	2	20

originally designed for paint production — can also be successfully employed for disintegration of bacteria. Table 6 shows the preferred operating conditions which we elucidated during our studies with the Netzsch LME 20 mill. Since the mode of action of the mill is very different from the mode of action of the high pressure homogenizer, the two machines are useful alternatives for the difficult task of solubilization of intracellular enzymes.

Table 6. Preferred operating conditions for the disintegration of yeast and bacteria using the Netzsch LME 20 mill.

a) yeast:
    agitator speed   : 8 m/sec. (1100 rpm)
    flow rate   : 100 l/h
    yeast concentration   : 40% (w/v)
    bead size   : 0.55–0.85 mm[a]
    load volume of beads : 80%
b) bacteria:
    agitator speed   : 10 m/sec. (1400 rpm)
    flow rate   : 50 l/h
    bacteria concentration : 40% (w/v)
    bead size   : 0.5–0.65 mm
    load volume of beads : 85%

[a] for enzymes located in the periplasmic space: 1 mm

Acknowledgements

The authors are indepted to Dr. W. Hummel and Mr. W. Hahn for the supply of various microorganisms. We sincerely thank Mrs. M. Jeckeln, Mrs. A. Schulz, Mr. J. Gottschlich, Mr. R. Kraume-Flügel and Mr. W. Stach for skilful technical assistance and Dipl.-Ing. W. Wania and the staff of the pilot plant of the GBF for their help during cultivation and harvest of the microorganisms.

References

Brookman JSG (1974) Mechanism of cell disintegration in a high pressure homogenizer. Biotechnol Bioeng 16:371–383

Currie JA, Dunnill P, Lilly MD (1972) Release of protein from baker's yeast (Saccharomyces cerevisiae) by disruption in an industrial agitator mill. Biotechnol. Bioeng. 14:725–736

Doulah MS, Hammond TH, Brookman JSG (1975) A hydrodynamic mechanism for the disintegration of Saccharomyces cerevisiae in an industrial homogenizer. Biotechnol Bioeng 17: 845–858

Follows M, Hetherington PJ, Dunnill P, Lilly MD (1971) Release of enzymes from baker's yeast by disruption in an industrial homogenizer. Biotechnol. Bioeng. 13:549–560

Hetherington PJ, Follows M, Dunnill P, Lilly MD (1971) Release of protein from baker's yeast (Saccharomyces cerevisiae) by disruption in an industrial homogenizer. Trans. Inst. Chem. Eng. 49:142–148

Limon-Lason J, Hoare M, Orsborn CB, Doyle DJ, Dunnill P (1979) Reactor properties of a high-speed bead mill for microbial cell rupture. Biotechnol. Bioeng. 21: 745–774

Marffy F, Kula M-R (1974) Enzyme yields from cells of brewer's yeast disrupted by treatment in a horizontal disintegrator. Biotechnol. Bioeng. 16: 623–634

Mogren H, Lindblom M, Hedenskog G (1974) Mechanical disintegration of microorganisms in an industrial homogenizer. Biotechnol. Bioeng. 16: 261–274

Rehacek J, Beran K, Bicik V (1969) Disintegration of microorganisms and preparation of yeast cell walls in a new type of disintegrator. Appl Microbiol 17: 462–466

Rehacek J, Schaefer J (1977) Disintegration of microorganisms in an industrial horizontal mill of novel design. Biotechnol Bioeng 19: 1523–1534

Schütte H, Kroner KH, Hustedt H, Kula M-R: Experiences with a 20 l industrial bead mill for the disruption of microorganisms. Enzyme and Microbial Technology (in press)

III. Enzyme Purification

Enzymatic Activity Can be Recovered from Solvent-Denatured Catalase

Paul F. Schubert and R. K. Finn

The correlation of solvent precipitation data with conformational changes is reviewed briefly. For catalase, additional information about the conformation can be gained from shifts in the Soret spectrum upon treating with dithionite. Here we report observations for *crude* catalase and compare these with previous work using *crystalline* catalase; some reinterpretation of earlier conclusions is warranted. Furthermore, this knowledge of the conformational state allows design of processes so as to optimize the yield during purification. For example, catalase which had been denatured during an alcohol precipitation may be partially renatured by redissolving in water and then precipitating with ammonium sulfate.

Although enzymes have been introduced in a large number of technical applications in recent years, their tendency to denature and lose catalytic activity has always been a major disadvantage. This disadvantage is particularly apparent when enzyme replacement is a significant cost factor in the operation of an industrial process. As a result, much research has been focused on the mechanism of protein folding and denaturation [1, 2, 3]. However, these studies often do not address the special needs of enzyme technology.

For example, since denaturation has been viewed as irreversible, little effort has been devoted to the development of simple methods for renaturing enzymes. Also, laboratory studies with enzymes have traditionally used highly purified or crystalline material. The possibility that the dominant conformation encountered in a technical application may be different from that of the crystalline state has been largely ignored. The work reported here is directed to these applied needs. It is hoped this report will both contribute to the understanding of enzyme processes and also complement fundamental protein structure and function studies.

I. Renaturation of "Irreversibly" Denatured Catalase

The mechanism for the denaturation of enzymes has been extensively studied but is still not fully understood. The loss of catalytic activity is due to changes in the conformation of the protein. A variety of denaturing agents such as heat or organic solvents may bring about the unfolding of the protein from its native conformation. Of particular interest here is the distinction between reversible denaturation and irreversible denaturation. Stated briefly, an enzyme which has been reversibly denatured will regain its catalytic properties if the denaturing agent is removed. On the other hand, an irreversibly denatured enzyme does not refold spontaneously to its native conformation, not even if the denaturing agent is removed.

This traditional distinction between reversible and irreversible denaturation has had a significant influence on the technological development of enzyme processes. Such processes have often included extraordinary measures to stabilize the enzyme, but the enzyme is nevertheless discarded when its catalytic activity has decreased. A similar philosophy regarding enzymes has developed with respect to their separation and purification. If catalytic activity is lost due to harsh conditions, the loss is regarded as permanent.

A broader view of denaturation is now becoming evident. Thus, a number of studies involving renaturation have been reported which emphasize mechanisms of denaturation and the kinetics of reversibly denatured proteins. Typically, denaturants such as urea, guanadine hydrochloride, detergents, heat and organic solvents have been used in recent works [3–10]. However, there have been few studies that address the important technical problem of the reactivation of "irreversibly" denatured enzymes. A recent example of such reactivation was provided by Martinek et al. [11, 12]. In their work, immobilized trypsin and chymotrypsin were thermally denatured; the denaturation was termed irreversible since the catalytic activity was not recovered upon cooling. However, when the S—S bonds of the enzyme were reduced and subsequently reoxidized with glutathione, much of the catalytic activity of the enzyme was restored. This "renaturation cycle" could be repeated several times with the same immobilized enzyme.

The denaturation of crude catalase during precipitation by organic solvents provides a familiar example of an "irreversibly" denatured protein. However, an elegant method to renature catalase to an active form may be devised. The method follows from a careful study of the kinds of conformational changes that can occur during solvent precipitation.

The solvent precipitation scheme proposed in an earlier paper [13] considered not only conformational transitions that occur in solution but also considered the phase equilibria between soluble and precipitated protein.

$$NC \rightarrow ID \leftrightharpoons RD \qquad \text{(solution)}$$
$$\quad \updownarrow \qquad \updownarrow \qquad \updownarrow$$
$$NC_p \quad ID_p \quad RD_p \qquad \text{(precipitated)}$$

NC   Native conformation
ID   Irreversibly denatured
RD   Reversibly denatured

The combined effects of temperature, chain length of solvent and concentration of solvent must all be taken into account when devising a scheme to purify enzymes by precipitation. The dominant conformation in the precipitated solid may be influenced to some degree by controlling the solvent concentration at which the precipitation occurs. This influence on the conformational state is possible since the equilibrium in solution is reached in several minutes while the phase equilibrium (full precipitation) requires at least one hour. In fact, it was shown [13] that efficient recoveries of active catalase could be obtained even at high concentrations of ethanol if due attention was given to the kinetics and to the equilibria shown above.

The role of ethanol concentration on conformational transitions has been previously demonstrated in a work by Parodi et al. [14]. The conformation of lysozyme in an aqueous ethanol solution was monitored by spectral methods. At intermediate alcohol concentrations, an α-helical conformation of the enzyme became the dominant form. When the ethanol concentration was increased further, a random conformation replaced the α-helical form as the major conformation. The reversiblility of the transitions was not determined so that these results cannot be directly compared to the precipitation work we have undertaken with catalase.

The general approach to renature any inactive, solvent precipitated enzyme would be to first redissolve the enzyme in buffer and then to modify the solution conditions to bring about a conformational change from the previously precipitated "irreversibly" denatured form to a conformation that can spontaneously refold to the catalytically active form. One procedure would be to conduct this precipitation under conditions of high solvent concentration. This would take direct advantage of the results presented above.

Another approach to renaturation is based on reports [15–18] saying that a high concentration of inorganic salts may induce a change in an enzyme to a conformation which in some cases may be indistinguishable from the thermally denatured form. Although this work was not done on the catalase-ammonium sulfate system, it provided the basis for the following experiments.

Crude beef liver catalase (Nutritional Biochemicals Corp., Cleveland, Ohio) that had been initially precipitated with an organic solvent, purified and then lyophilized to a powder, was used. This material was first redissolved in buffer and then the enzyme was reprecipitated at 4 °C from a series of samples in which different concentrations of ammonium sulfate were used. For each treated sample, the precipitate was separated from the supernatant by centrifugation (14,000 × g for 10 minutes) at 4 °C. The supernatant was carefully decanted and the precipitate was dissolved again in buffer and assayed for catalase activity and protein content by the methods described previously [13]. The results are shown in Table 1.

Table 1. Precipitation of Catalase (Crude) at High Ammonium Sulfate Concentrations

pH 6.0	4 °C	3.4 mg/ml protein	
g salt/ml soln		Run 1	Run 2
		% of initial total activity	
.50		94	100
.55		110	100
.60		110	—
.67		102	130
.70		130	
sat.			170

At moderate concentrations of ammonium sulfate, quantitative recovery of the active form was achieved. However, at high salt concentrations additional catalytically active enzyme was obtained in both experiments. The renaturation is a real effect and further work is required to relate it to distinct denatured conformations.

This example of renaturation takes advantage of conformational transitions that do not require S—S bond oxidations. Most importantly, it should serve to emphasize that renaturation has been neglected in the consideration of enzyme process technology.

II. Differences Between Crude and Crystalline Conformations of Catalase Observed by Soret Band Absorption

Some better understanding of different conformations of the protein that may occur during the course of enzyme purification is also important if these procedures are to become less empirical. Unfortunately, the analytical tools most useful for these studies can seldom be applied to crude mixtures. The heme proteins, however, afford an unusual opportunity to use absorption in the Soret band of the visible spectrum as illustrated below.

Preparation of crude catalase: Crude catalase was purified directly from fresh beef liver. Enzyme from 160 g of homogenized tissue was extracted into 200 ml of pH 6.0, 0.1 M phosphate buffer. After an initial centrifugation to clarify the solution, a first fraction of protein was precipitated out by the addition of 25 gm of ammonium sulfate to each 100 ml of the clarified liquid. This first precipitate was removed by centrifugation and an additional 40 gm of ammonium sulfate was added to the supernatant to precipitate the catalase. The fraction recovered, although crude, was sufficient for analysis of the Soret band region. Crude catalase may also be obtained commercially (Nutritional Biochemicals Corp., Cleveland, Ohio). As described above, this material had been initially precipitated with an organic solvent, purified and lyophilized to a powder. Highly purified catalase was also obtained from this vendor.

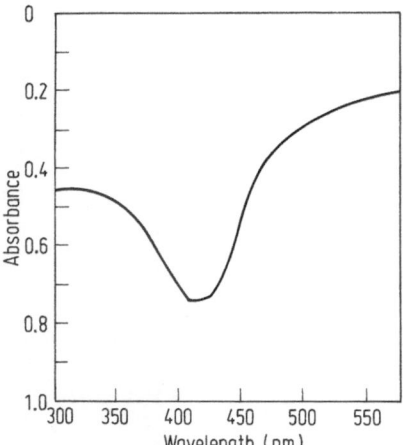

Figure 1. Soret region spectrum or commercial purified catalase

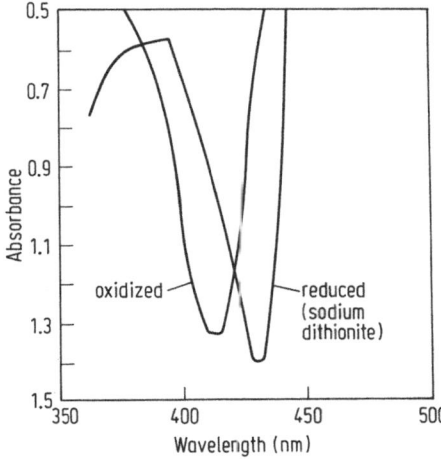

Figure 2. Soret region spectrum of crude catalase, ammonium sulfate precipitated

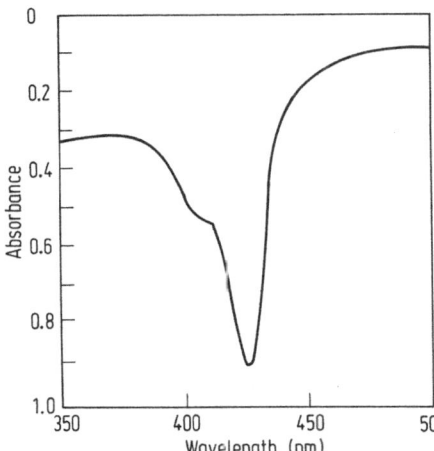

Figure 3. Soret region spectrum of crude catalase (organic solvent precipitated and lyophilized)

Soret region absorption: Figs. 1, 2, and 3 show the Soret region absorption for the visible spectrum of highly purified catalase and the samples of crude catalase described above. These spectra are significantly different in several respects. The peak in the Soret region for the highly purified catalase occurred at 409 nm; the location of this peak was not affected by the addition of sodium dithionite. The catalase obtained from the ammonium sulfate precipitation had a maximum Soret band absorption at 412 nm. After the addition of sodium dithionite, the iron in the heme group was reduced and the adsorption peak shifted to 430 nm. The solvent precipitated fraction was apparently in the reduced form; the maximum Soret absorption was at 428 nm.

The results are relevant not only to the solvent purification problem but, when coupled with fundamental studies using crystalline catalase, provide significant new insight into the conformations of catalase. Yang and Samejima [19] followed the acid denaturation of the crystalline form of this enzyme. The helical content of the native (crystalline) protein was reported to be 50% compared to 20–25% for the acid denatured conformation. Tanford and Lovrien [20] used sedimentation techniques to compare crystalline catalase with lyophilized enzyme prepared from the same material. The lyophilized fraction was shown to contain one-half and one-quarter size fragments of the original protein. The valuable extension of this work by Deisseroth and Dounce [21] is of particular interest here. The crystalline enzyme showed a single sedimentation peak. The heme group absorbed at 409 nm in the Soret region and was not reducible by sodium dithionite. On the other hand, a lyophilized fraction prepared from the same crystalline material showed two sedimentation peaks. Component I was composed of one-half molecules and was catalytically inactive. Component II was identified to be a mixture of the original crystalline form together with a conformationally modified molecule; the relative composition of the mixture depended upon the conditions of lyophilization. Unlike the crystalline enzyme, however, the heme groups of both Components I and II were reduced by sodium dithionite; this reduction caused a shift in the Soret band absorption from 409 nm to 428 nm. The iron was reoxidized immediately by the addition of hydrogen peroxide. The two components could be distinguished from each other by absorption differences in the 550–600 nm region. It is emphasized that all of this work was done using *crystalline* material.

Figure 1, the spectrum of the commercially purchased highly purified catalase, matches the spectrum obtained from the crystalline material of Deisseroth and Dounce. In Fig. 3 the spectrum obtained from crude, solvent-precipitated catalase matches that of Component II described above. Thus, the shoulder at 408 nm would be attributed to the "native" (crystalline) conformation and the peak at 428 nm attributed to the "conformationally modified" reduced form. However, this matching spectrum was obtained from material which had never been crystallized.

This comparison (particularly the effect of sodium dithionite on the heme group) confirms a difference in the conformation of *crystallized* catalase to that of the enzyme in its *true native state*. Surprisingly, this fact seems to have been overlooked in the denaturation studies, all of which have used the crystalline material. It now appears that the "denaturation" induced by lyophilization of crystalline material should be more accurately viewed as not only partial dissociation into subunits but also change to a conformation resembling the original state and not as change to a new denatured form.

III. Summary Statement

These two examples, 1) reactivation of "irreversibly denatured" enzyme and 2) demonstration of a key difference of the crystalline form from the crude native enzyme, serve to illustrate the bridge that must be built between science and the technology required for efficient enzyme purification. There is much to be learned about relatively impure enzyme preparations; their characterization and their relation to purified preparations requires further study by biotechnologists.

References

1. Anfinsen CB and Scheraga HA (1975) Adv. Protein Chem *29*, 205
2. Tanford C (1968) Adv. Protein Chem *23*, 122
3. Tanford C (1970) Adv. Protein Chem *24*, 2
4. McCoy LJr, Wong KP (1981) Biochemistry *20*, 3062
5. Pasta P, Carrea G, Longli R and Antonini E (1980) Biochem Biophys Acta *616*, 143
6. Porter D, Cardenas J (1980) Biochemistry *19*, 3447
7. Kime MJ, Ratcliffe RG, Moore P and Williams R (1980) Eur J Biochemistry *110*, 493
8. Lacks S, Springhorn S, and Rosenthal A (1979) Anal Biochem *100*, 357
9. Clarke S (1981) Biochim Biophys Acta, *670*, 195
10. Orsini G and Goldberg ME (1978) J Biol Chem *253*, 3453
11. Martinek K, Mozhaev VV, Smirnov MD, Berezin IV (1980) Biotechnol Bioeng *22*, 247
12. Martinek K, (1980) Biochim Biophys Acta *615*, 426
13. Schubert PF and Finn RK (1981) Biotechnol Bioeng *23*, 1245
14. Parodi R, Bianchi E and Ciferi A (1973) J Biol Chem *248*, 4047
15. Bigelow CC (1964) J Mol Biol *8*, 696
16. von Hippel P and Wong K (1964) Science *145*, 577
17. von Hippel P, and Wong K (1965) J Biol Chem *240*, 2909
18. Oobatake M, Takahashi S (1979) J Biochem (Tokyo) *86*, 65
19. Yang JT, and Samejima T (1963) J Biol Chem *238*, 3262
20. Tanford C and Lovrien R (1962) J Am Chem Soc, *84*, 1892
21. Deisseroth A and Dounce AL (1967) Arch Biochem Biophys, *120*, 671

Extractive Purification of Enzymes

Helmut Hustedt, Karl Heinz Kroner, Horst Schütte and Maria-Regina Kula

Summary

A short review of the current status of enzyme isolation and purification by
liquid-liquid extraction in aqueous two-phase systems is given. The method
can be used mainly for the separation of the desired enzyme or also protein
in general from cell debris, contaminating proteins including interfering activities,
nucleic acids and polysaccharides. Working on large scale phase separation is
performed by commercially available separators or in settling tanks. So far, more
than 20 enzymes were separated this way from broken cells of procaryotic and
eucaryotic microorganisms. Enzyme yields were usually in the range of
90–100 % and the purification factor related to removal of contaminating protein
varied between 1 and 8. A number of those enzymes were further purified by
subsequent partition steps. As a special example, the extractive purification of
aspartase from E. coli is discussed here.

Furthermore, a general scheme for extractive enzyme purification is derived
taking into account several parameters important for the process design. In the
last chapter continuous enzyme isolation by liquid-liquid extraction is discussed.

Introduction

Extracellular enzymes have been used in several industries on a large scale for many
years. Because of their much higher potential today intracellular enzymes are also
of increasing importance as technical catalysts as well as in the medical field for
diagnostic and therapeutic purposes. This fact, as well as the fast developments
in the production of biological active proteins from microorganism by the
recombinant DNA-technology require the elaboration of adequate separation
and purification technologies for such substances which are suitable for large
scale work. For some years we have investigated the use of aqueous two-
phase systems for the isolation and purification of enzymes and biological active
proteins by liquid-liquid extraction (Kula et al. 1977, Kroner et al. 1978 and
1982a, Hustedt et al. 1978 and 1980). These systems, for example PEG/dextran
or PEG/potassium phosphate systems, have a high water content in the range of
75–90 %, which makes them suitable for the gentle partition of biopolymers
(Albertsson 1971).

The partition behaviour of proteins in aqueous two-phase systems can be influenced by the kind of polymers, the molecular weight of the polymers, the
length of the tie-line, the concentration and kind of ions in the system, the pH-

value and the temperature (Albertsson 1971). The effect of these parameters is normally increasing with increasing molecular weight of the partitioned substance. The Nernst Theory of the independence of the partition coefficient on the concentration holds also for aqueous phase systems for a fairly wide range of concentration. The properties of aqueous phase systems and the partition behaviour of biopolymers especially in PEG/dextran systems was investigated in detail by P. A. Albertsson and co-workers (Albertsson 1971).

Separation Possibilities of Aqueous Phase Systems and Extraction Process Design

In our laboratories so far polyethylene glycol/dextran and polyethylene glycol/salt two-phase systems were used for protein extraction; the salts most frequently used were potassium phosphate, sodium phosphate and ammonium sulfate (Kula et al. 1982). The purified and fractionated dextran fractions initially used could be replaced successfully by the much cheaper crude dextran (Kroner et al. 1982 b). The polymers are non-toxic and biodegradable and their use should not lead to problems with waste water treatment. But there is no reason that other phase forming polymers could not be used successfully for protein extraction.

Aqueous two-phase systems can be used for the separation of intracellular proteins from cell debris as well as for further purification. The removal of cell debris which is normally the first step in a purification procedure after cell disintegration is discussed in the next chapter. Further purification may include the separation of nucleic acids, polysaccharides, interfering activities and contaminating protein. According to their more hydrophilic nature, nucleic acids and polysaccharides partition normally in polyethylene glycol/salt systems in favor of the hydrophilic salt phase and can be removed this way from the desired protein if this is shifted into the polyethylene glycol rich phase (Kroner et al. 1982a, unpublished results of the authors). For the removal of interfering enzymatic activities as well as for the removal of other contaminating protein no general rules can be given; suitable separation conditions must be found by experimentation as which is the faster or the simpler to obtain the desired protein. Purification factors obtained by a single step partition are usually in the range between 2 and 5 depending on the special circumstances.

Single-, step liquid-liquid extraction is normally not suited for protein purification to a purity of 90 % or more, but is for the first steps in an isolation procedure when large volumes or large amounts of protein have to be handled. The purity obtained this way should normally be sufficient for the use of the enzymes as technical catalysts. Otherwise, liquid-liquid extraction can be followed by conventional purification methods such as chromatography or multiple step countercurrent extraction can be tried.

Furthermore, the removal of the polymers has to be taken into account which means in practice the removal of the polyethylene glycol. This is usually done by shifting the protein into the salt phase which contains in equilibrium with the PEG-rich upper phase only a small amount of dissolved PEG (Albertsson 1971, Hustedt et al. 1978, unpublished results of the authors). In general, this residual amount of PEG, if necessary, can easily be removed by ultrafiltration. Another possibility of polyethylene glycol removal is the binding of the protein to an ion exchanger after adjustment of pH and ionic strength. In some cases it may also be possible to enhance the polyethylene glycol concentration performing a precipitation by this polymer. However, in this way one may obtain precipitates which are difficult to separate from the liquid. Ultrafiltration of polyethylene glycol rich phases is in principle also possible after sufficient dilution; but this results in large volumes and in addition the separation of the polyethylene glycol with high average molecular weight may be difficult. The shifting of the protein to the salt phase is usually the method of choice. So far every protein investigated, including more hydrophilic species such as β-interferon (Menge et al. unpublished results) could be shifted into a polyethylene glycol poor salt phase.

Taking also into account that the components constituting the phase system should be optimally utilized to avoid unnecessary costs for polymers and chemicals, the foregoing considerations lead to the following scheme for extraction procedures (Fig. 1): For the first partition step conditions should be selected that the cell debris (and contaminating protein, if possible) partition to the bottom phase while the desired protein, in the figure called "product", is in the top phase with high yield (90–100%) since phase systems PEG/crude dextran as well as PEG/salt systems can be used at this stage. After phase separation, salt is added to the product containing PEG-phase now creating

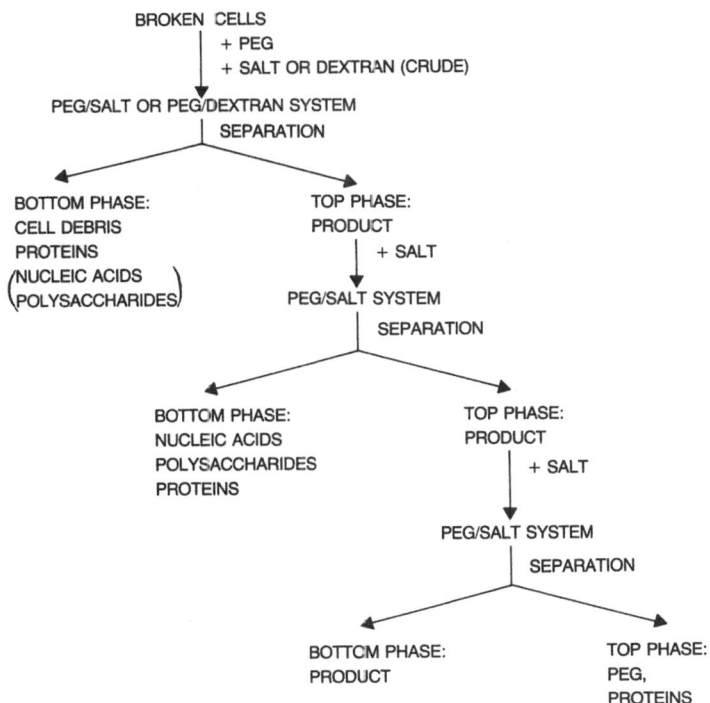

Figure 1. General scheme of enzyme isolation by subsequent partition steps

a PEG/salt system. At this step, conditions should be chosen that the desired protein partitions again into the PEG-phase whereas nucleic acids, polysaccharides and a lot of contaminating proteins are extracted into the bottom phase. After phase separation salt is again added to the top phase of the second partition step forming a new polyethylene glycol/salt system. Here conditions should be adjusted in such a way that the desired protein is in the bottom phase while a further portion of contaminating protein should be left in the top phase. Special interfering activities should be removed also by the three steps. Sometimes it is already possible to remove at the first step nucleic acids, polysaccharides and a lot of contaminating protein together with the cell debris with the bottom phase. In such cases it may be sufficient to shift the protein at the second partition step to the salt phase thus limiting separation to at a two-stage process.

After ultrafiltration and concentration of the final salt phase, the enzyme is usually sufficiently pure for the use as a catalyst. In other cases, e.g. when isolating proteins for medical use, further purification can be done by chromatographic methods. The purification of proteins to high purity by continuous multistage countercurrent extraction also seems to be possible on large scale (Hustedt at al. 1980, Kula et al. 1982), but must still be investigated in more detail.

Examples of Extractive Protein Isolation

Extraction of Enzymes from disrupted microbial cells

During the past years a large number of enzymes have been extracted in our laboratories or pilot plant from broken cells of procaryotic and eucaryotic microorganisms using aqueous phase systems. Examples are listed in Table 1. In most cases the yield is more than 90 %, and in many cases substantial portions of contaminating proteins are also removed together with cell debris as can be seen from the purification factor which varies from 1 to 8. In all cases the cell debris was removed with the bottom phase; separation of the cell debris with the top phase is also possible (Hustedt et al. 1981) but such conditions are less favourable for phase separation on a large scale as compared with the removal of the cell debris in the bottom phase.

At the first partition step phase, separation is generally performed by commercially available disc-stack separators (Kroner et al. 1978, Kula et al. 1981, Kroner et al. 1982a). Usually the bottom phase is sufficiently fluid to permit the use of liquid-liquid separating equipment. If the viscosity of the bottom phase is very high thus leading to a high flow resistance, nozzle separators can be successfully used for phase separation (Kroner et al. 1982a, Kroner et al. 1982b). When using such machines, the flow rate must be sufficiently high to avoid loss of the product containing top phase through the nozzles which occurs when the flow resistance is too low (Kroner et al. 1982a). This can be seen from Fig. 2 which shows the purity of top and bottom phase in relationship to the throughput for the separation of a PEG/phosphate system including 20 % disrupted wet cells of the yeast *Candida boidinii*: the top phase is pure independent of the flow rate in the investigated range, whereas the purity of the bottom phase increases with

Table 1. Extraction of Enzymes from Broken Cells

Enzyme	Organism	Kind of Phase System	Yield (%)	Purification Factor
α-glucosidase	Saccharomyces cerevisiae	PEG/salt	95	3.2
Glucose-6-phosphate dehydrogenase		PEG/salt	91	1.8
Hexokinase		PEG/salt	92	1.6
Glucose isomerase	Streptomyces species	PEG/salt	86	2.5
Pullulanase	Klebsiella	PEG/dextran	91	2
Phosphorylase	pneumoniae	PEG/dextran	85	1
Isoleucyl tRNA synthetase	Escherichia coli	PEG/salt	93	2.3
Fumarase		PEG/salt	93	3.4
Aspartase		PEG/salt	96	6.6
Penicillin acylase		PEG/salt	90	8.2
Leucine dehydrogenase	Bacillus sphaericus	PEG/dextran	98	2.4
Glucose dehydrogenase	Bacillus species	PEG/salt	95	2.3
β-glucosidase	Lactobacillus cellobiosus	PEG/salt	98	2.4
Lactate dehydrogenase	Lactobacillus species	PEG/salt	95	1.5
Leucine dehydrogenase	Bacillus cereus	PEG/salt	98	1.3
Fumarase	Brevibacterium ammoniagenes	PEG/salt	83	7.5
Formaldehyde dehydrogenase	Candida boidinii	PEG/dextran	94	n.d.
Formate dehydrogenase		PEG/dextran	91	n.d.
Formate dehydrogenase		PEG/salt	94	1.5
Isopropanol dehydrogenase		PEG/salt	98	2.6

increasing throughput minimizing the loss of product at an optimal flow rate of around 190 l/h. It should also be pointed out here that before the removal of the cell debris a heat treatment or an acid treatment of the cell homogenate can be carried out. The precipitate formed thereby can then be removed together with the cell debris.

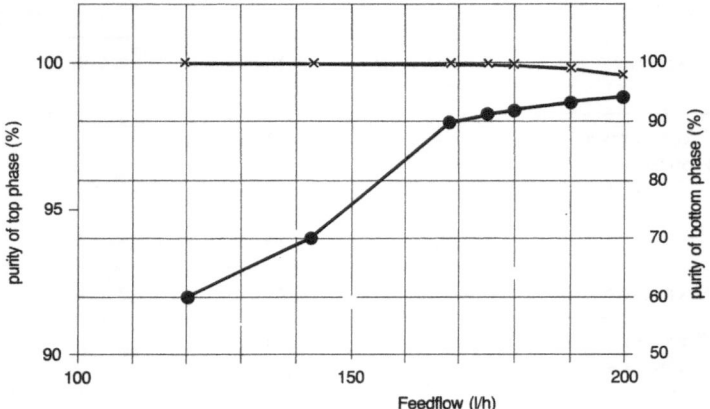

Figure 2. Relationship between phase purity and feed flow rate for the nozzle-separator (YEB 1344 (α-Laval). System conditions: 18 % PEG 400, 7 % PEG 1550, 8 % potassium phosphate, 20 % disrupted cells of Candida boidinii (×) top phase, (○) bottom phase. (From Kroner et al. 1982a)

Examples of Enzyme Purification by Several Subsequent Partition Steps

In the past years a number of intracellular proteins have been partially purified by our group using several subsequent partition steps. Most of these are listed in Table 2. The final purification factors varied between 2 and 33. The outstanding high purification of β-interferon is a special case since it was not extracted from microbial cells, but from the supernatant of human fibroblast cultures.

In this connection it should be emphasized that many of the processes reported in Table 2 are not optimized with regard to the removal of contaminating proteins because the removal of interfering activities or other unwanted side products of microbial growth was often of major importance. The final yield was in all cases 70 % or more. Such high yields are generally characteristic for the extractive protein purification though the processes are carried out at room temperature. The high yields are due in part to the stabilizing effects of the polymers and the short residence times.

In Fig. 3 the purification of aspartase from E. coli by three partition steps is outlined as an example. After cell disruption in a glass bead mill or a Manton-Gaulin homogenizer, PEG and phosphate are added forming a PEG/phosphate system. After mixing, the phases are separated by a liquid-liquid separator. Cell debris, a lot of contaminating protein including more than 99 % of interfering fumarase activity and most of the polysaccharides are removed with the bottom phase, while the aspartase is extracted in the top phase with a yield of around 95 %. At the second step a salt solution is added to this phase making up a new PEG/salt system which can be separated overnight in a settling tank. At this stage the aspartase is in the top phase whereas other proteins including the residual fumarase and residual amounts of polysaccharides are removed with the bottom phase. At the third step, salt solution and a small amount of PEG 10,000 are again added to the enzyme-containing PEG-rich phase of the second step; now the conditions are adjusted shifting the aspartase to the salt phase whereas a further part of the contaminating protein and PEG are left

Table 2. Examples of Proteins Partially Purified by Several Subsequent Partition Steps[a] in the Department "Hochmolekulare Naturstoffe", GBF, D-3300 Braunschweig

Enzyme	Organism	No. of Partition Steps	Overall Purification Factor	Overall Yield %	Remarks
Fumarase	Brevibacterium ammoniagenes	2	22	70	removal of polysaccharides
Aspartase	Escherichia coli	3	18	82	removal of interfering fumarase
Pullulanase	Klebsiella pneumoniae	4	6.3	70	70% pure enzyme, removal of α-amylase and proteases
α-1,4-glucan phosphorylase	Klebsiella pneumoniae	2	2.5	81	removal of glucosyl transferase
Glucose dehydrogenase	Bacillus species	3	33	83	removal of nucleic acids and polysaccharides
Leucine dehydrogenase	Bacillus sphaericus	2	3.1	87	
Formate dehydrogenase	Candida boidinii	3	4.2	78	70% pure enzyme, removal of polysaccharides and nucleic acids
Aspartate β-decarboxylase	Pseudomonas dacunhae	3	6	78	removal of polysaccharides and nucleic acids
Leucine dehydrogenase	Bacillus cereus	2	2.4	89	
d-Lactate dehydrogenase	Lactobacillus confusus	2	1.9	91	
β-Interferon	Human fibroblasts	2	> 350	75	

[a] The first partition step always aimed to remove cell debris (with the exception of interferon)

in the top phase. The system can be separated again in a settling tank. The overall purification is 18-fold and the final yield 82%. So far extractive purification procedures have been carried out starting with up to 150 kg of wet cell mass. The results for this scale can be normally calculated with high precision from experiments in 10 ml-scale. This is shown in Table 3 in which the yield of formate dehydrogenase obtained in a 10 ml scale is compared with the yield at the process scale. The values are highly consistent. These results are, above all, independent of the partition coefficient and of the protein concentration and process scale. Phase equilibrium is easily attained due to the unusual low interfacial tension which is in the range of 10^{-2} to 10^{-3} mN/m for PEG/dextran systems and in the range of 10^{-1} to 10^{-2} mN/m for PEG/salt systems (Albertsson 1971). Scale-up is further facilitated by the commercial availability of the necessary equipment and machinery due to the high standard of extraction technology in the chemical industry.

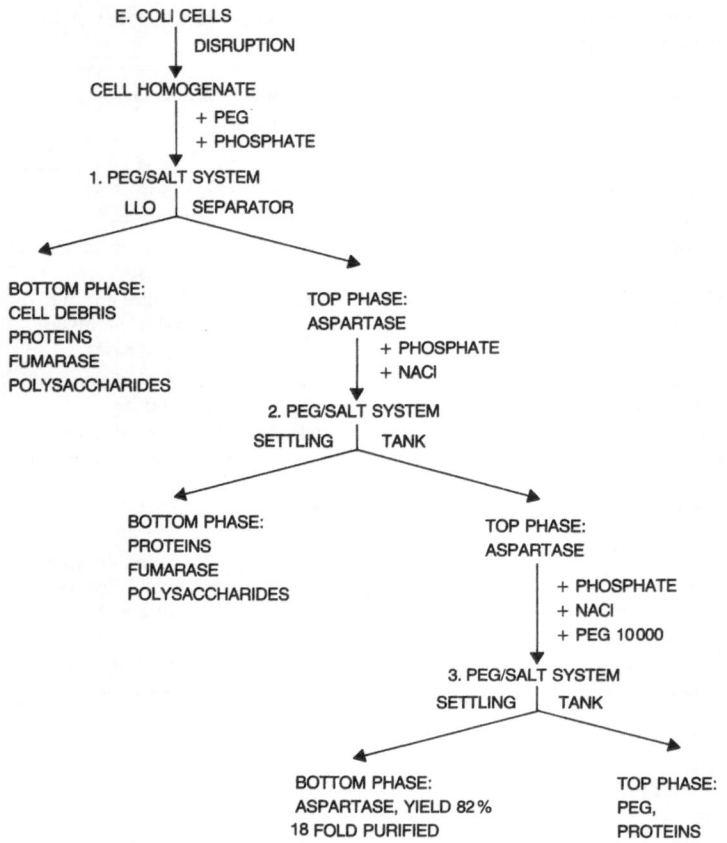

Figure 3. Flow diagram of the extractive purification of aspartase from E. coli

Continuous Processing

As was pointed out by Dunnill and Lilly, continuous protein purification may offer the following advantages (Dunnill and Lilly 1972):
— a smaller equipment can be used
— the faster throughput during critical stages reduces deactivation
— continuous fermentation can be fully utilized.
  As disadvantages they claim:
— the reliability of machines must be higher
— sophisticated process monitoring and control is necessary
— contamination and strain mutation can decrease the advantages of long-time operation.

  Further important drawbacks may be the low flexibility as well as the fact that normally for each process a special plant is necessary. However, these drawbacks do not apply to the application of liquid/liquid extraction for continuous processing since all extractive purification processes are generally performed according to the same scheme as was pointed out in chapter 2. For this reason

Table 3. Performance of scale-up for the extractive purification of formate dehydrogenase (data taken from Kroner et al. 1982 a)

Step No.	Yield of FDH in 10 ml scale (%)	Yield of FDH in process scale (%)	Scale-up factor
1	95	94 (250 l)	25 000
2	84	80 (350 l)	35 000
3	93	93 (386 l)	38 600
4	100	100 (233 l)	23 300
Final yield	74	70	$2-4 \times 10^4$

as well as for the reason that the extraction technology requires only mixing of liquids and phase separation operations, this technology seems to be specially suited for continuous processing. Phase separation by a separator is always a continuous process and also phase separation in a settling tank can be continuously performed without difficulties. Continuous mixing or dispersion of the liquids forming the phase system can also easily be performed reaching phase equilibrium due to the low interfacial tension or small droplet size, respectively. For example in preliminary experiments we used static mixers of the type SMV from the Sulzer company and experienced no essential difficulties neither at the mixing of the first system containing the cell debris nor at the following steps. Of course, more intensive mixing was necessary at the first step due to the higher viscosity.

Fig. 4 shows as an example the flow diagram of the continuous isolation of a bacterial enzyme: The cell suspension is pumped through a glass bead mill for cell disruption. If the flows through a heat exchanger where the temperature is adjusted to around 20 °C. From there the suspension is pumped together with the corresponding amount of a PEG/salt dispersion through a mixing unit consisting of three consecutive static mixers. The length of each mixer was 5 cm, the internal diameter was 0.9 cm and the hydraulic diameter around 0.2 cm.

The total flow at this stage was ~400 ml/min. After passing the mixer the dispersion flows into the first separator (a gyrotester from α-Laval was used), where phase separation occurs. The overflow containing the product is pumped together with a corresponding amount of salt solution through a second "mixing unit", for which a simple tube coil (tube length 100 cm, internal diameter 0.4 cm) was used. From there the new dispersion flows into the second separator (also an α-Laval Gyrotester) where again phase separation is performed and the α-Laval enzyme is now extracted in the bottom phase. At the first stage cell debris, many contaminating proteins and polysaccharides are removed. At the second stage, PEG and further contaminating proteins are left in the top phase. The flow should have been laminar at both stages since the Reynold-numbers were in both cases lower than 20. However, a small opposite effect was caused by the pulsation of the pumps.

The procedure was carried out for a period of around 5 h. During this time no changes of the process data such as product concentrations in the phases, phase ratio and phase purity were observed. The final yield was the same as in the batch process.

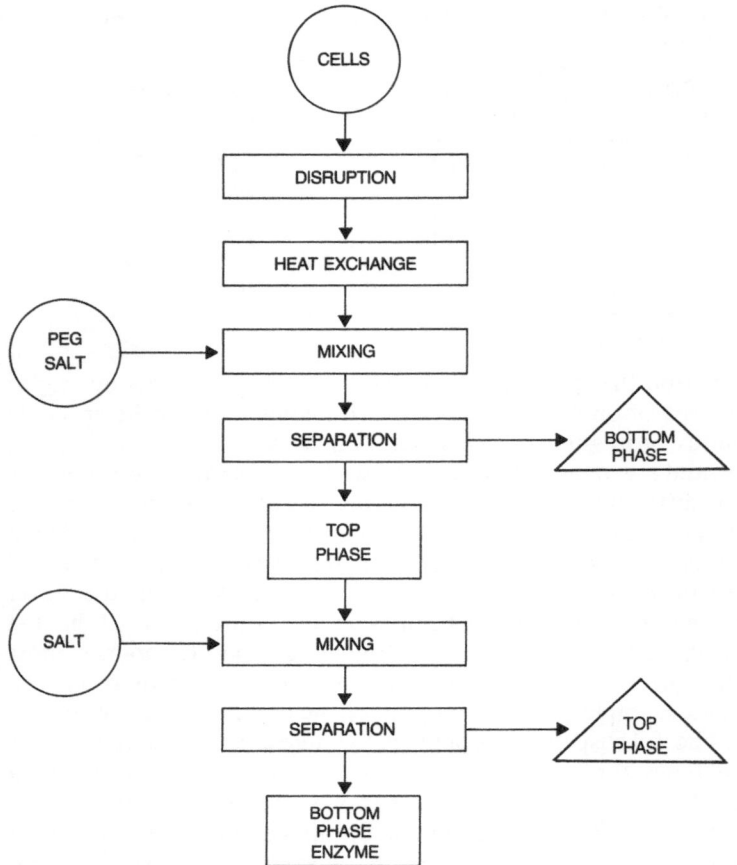

Figure 4. Flow diagram of continuous enzyme isolation by two subsequent partition steps

There seem to be no problems to also carry out continuous 3-stage processes according to the same scheme. Furthermore we see no problems for the scale-up of such procedures. Because of the small residence times and the other advantages already mentioned, continuous processing may be more favorable than batch procedures under certain circumstances.

Concluding Remarks

The extractive purification of enzymes and biologically active proteins appears to be an efficient technique especially suited to large-scale isolation processes. During the whole process the protein is kept in solution and problems inherent to solid-liquid separation are avoided as well as yield reductions due to pre-cipitation steps. Liquid-liquid separation is performed using conventional equipment and much knowledge from other fields of chemical engineering can be applied. Reductions of process time, labor costs, investment costs and energy requirements as well as increased yields can more than compensate for the costs of the chemicals necessary for constituting the phase systems. No problems with scale-up were experienced so far. The process has been carried out in our pilot plant starting with up to 150 kg of wet cells or one kg of enzyme protein, respectively.

There are several aspects for the further development of the method which have not yet been sufficiently studied: continuous processing, multistage operation, improvements in the selectivity of extraction, recycling of the compounds constituting the phase system etc. So far two extraction processes are being used for commercial enzyme production on an industrial scale. We assume that this number will increase in the near future.

References

Albertsson PA (1971) Partiton of Cell Particles and Macromolecules, Wiley, New York

Dunnill P and Lilly MD (1972) Continuous Enzyme Isolation, in: Gaden EL (ed) Biotechn Bioeng Symposium Nr 3, Wiley, New York, p 97–113

Hustedt H, Kroner KH, Stach W and Kula M-R (1978) Procedure for the simultaneous large-scale isolation of pullulanan and 1,4-α-Glucan phosphorylase from Klebsiella pneumoniae involving liquid-liquid separations, Biotechn Bioeng 20, 1989–2005

Hustedt H, Kroner KH and Kula M-R (1980) Enzyme Purification by liquid-liquid extraction, in: Weetall HH and Royer GP (eds), Enzyme Engineering Vol 5, Plenum Publ Corp, New York, p 45–47

Hustedt H, Kroner KH and Kula MR (1981), Large-scale isolation of fumarase by partition, Poster presented at the 1 Eng Foundation Conf on Advances in Fermentation Products Recovery Process Technology, Banff, Cañada, 1981

Kroner KH, Hustedt H, Granda S and Kula M-R (1978) Technical aspects of separation using aqueous two-phase systems in enzyme isolation processes, Biotechn Bioeng 20, 1967–1988

Kroner KH, Schütte H, Stach W and Kula M-R (1982a) Scale-up of formate dehydrogenase by partition, Journ Chem Techn Biotechn 32, 130–137

Kroner KH, Hustedt H and Kula M-R (1982b) Evaluation of crude dextran as phase forming polymer for the extraction of enzymes in aqueous two-phase systems in large-scale, Biotech Bioeng 24, 1015–1045

Kula M-R, Kroner KH, Hustedt H, Granda S and Stach W (1977) Verfahren zur Abtrennung von Enzymen, German Patent Nr 26 39 129, US-Patent 4.144.130

Kula M-R, Kroner KH, Hustedt H and Schütte H (1981) Technical Aspects of Extractive Enzyme Purification, in: A Constantinides, WR Vieth and K Venkatasubramanian (eds), Biochemical Engineering II, Ann NY Acad Sci 369, 341–354 New York Academy of Science, New York

Kula M-R, Kroner KH and Hustedt H (1982) Purification of enzymes by liquid-liquid extraction; in: A Fiechter (ed), Advances in Biochemical Engineering, Vol 24, Springer-Verlag, Berlin, Heidelberg, New York, p 73–118

IV. Soluble Enzymes: Application

Enzymic Developments in the Production of Maltose and Glucose

William M. Fogarty and Catherine T. Kelly

Introduction

Major advances in starch technology and starch processing have taken place in the last two decades as a result of the development of new amylolytic enzymes (Fogarty and Kelly, 1979, 1980; Norman, 1979). As a result, a range of products with specific compositions and physical properties may be produced. The various commercially used enzymes and their applications are summarized in Table 1. Thermostable α-amylases e.g. those from *Bacillus amyloliquefaciens* (Borgia and Campbell, 1978) and *Bacillus licheniformis* (Saito, 1973; Madsen et al., 1973; Morgan and Priest, 1981) are used in liquefaction processes which with the aid of heat convert insoluble starch granules into partial hydrolysates with lower viscosities. In these processes the hydrolytic action is concluded when the degree of polymerisation is about 10–12. Due to their thermostabilities and hydrolytic action patterns thermostable α-amylases (E.C. 3.2.1.1, α-1,4 glucan 4-glucanohydrolase, endo-amylase) are particularly suited as thinning agents in liquefaction processes but in order to obtain hydrolysates with low molecular weight carbohydrates suitable for fermentation and additional uses, other microbial amylolytic enzymes must be used.

Thermolabile amylolytic enzymes of fungal origin are used as saccharifying agents following liquefaction of starch (Allen and Spradlin, 1974; Barfoed, 1976, Norman, 1979; Fogarty, 1981). Their major attribute is their ability to produce products of low molecular weight which are not obtainable with the thermostable α-amylases. The enzymes most widely used in this context are fungal α-amylase for production of maltose syrups (Allen and Spradlin, 1974; Barfoed, 1976; Underkofler et al., 1965) and amyloglucosidase for production of dextrose (Underkofler, 1968).

This presentation will summarize current developments in a specific area of starch enzymology. Firstly it will review work on microbial β-amylases and other enzymes capable of producing maltose in high yield and secondly it will examine data which throw new light on the nature of some microbial α-glucosidases.

Production of High-Maltose Syrups

Large quantities of high-maltose syrups ($>80\%$) are not produced in Europe or in the U.S.A. but there is an increasing demand for this product in Japan (Maeda and Tsao, 1979). It is expected that world demand for high-maltose syrups will expand considerably in the next decade. The most important charac-

W. M. Fogarty and C. T. Kelly

Table 1. Functions and Applications of Microbial Enzymes in Starch Processing Technology

Enzyme	Type	Source	Function	Substrate	End-Products
α-Amylase	Thermostable	*Bacillus amyloliquefaciens*	Liquefaction	Starch	G_6 + α-limit dextrins
		Bacillus lichenformis	Liquefaction	Starch	G_2, G_3, G_5 + α-limit dextrins
	Thermolabile	*Aspergillus* spp.	Saccharification	Liquefied (thinned) Starch	G_2, G_3 + α-limit dextrins
Amyloglucosidase	Thermolabile	*Aspergillus* spp. *Rhizopus* spp.	Saccharification	Liquefied (thinned) Starch	G_1
Amyloglucosidase + α-amylase	Thermolabile	*Aspergillus* spp.	Saccharification	Liquefied (thinned) Starch	G_1, G_2, G_3[a]

[a] The composition of the end-product may be varied by adjusting the ratio of the two enzymes

teristics of high-maltose syrups are low viscosity in solution, low hygroscopicity, resistance to crystallisation, mild, i.e. low sweetness, good heat stability and lack of colour formation.

Thermolabile α-amylases are used as saccharifying enzymes following lique-faction of starch. These enzymes are not sufficiently thermostable to be used in liquefaction processes and they hydrolyse substrate to a greater degree than the liquefaction enzymes. The product therefore contains a higher proportion of maltose and thus it is in the manufacture of maltose syrups that the thermolabile enzymes find major application (Allen and Spradlin, 1974; Barfoed, 1976). The enzyme used commercially is that obtained from strains of *Aspergillus oryzae* (Table 2). It has optimum values for temperature and pH at 55 °C and 4.7, respectively. These characteristics contrast markedly with those of the thermo-stable enzymes used in liquefaction. The main end-products formed with the fungal enzyme are maltose and maltotriose and although large levels of maltotriose are formed initially, the level of maltose ultimately reaches a maximum at about 50%.

Table 2. Properties of Commercially Used α-Amylases
Thermolabile Endo-Acting α-Amylase

Source:	Fungi e.g.
	Aspergillus oryzae
Function:	Saccharification
	Preparation of high-maltose and high conversion syrups from liquefied starch
Properties:	Temp. opt.: 55 °C
	pH opt.: 4.7
	Specificity: α-1,4 D-glucosidic bonds

Thermostable Endo-Acting Amylases

Source:	Bacteria
Function:	Viscosity reduction of starch concentrates. Preparation of dextrins
Types:	1. *Bacillus amyloliquefaciens* enzyme
	Temp. opt.: 70 °C
	pH opt.: 5.7 (at 40 °C)
	Specificity: α-1,4 D-glucosidic bonds
	2. *Bacillus licheniformis* enzyme
	Temp. opt.: 92 °C
	pH opt.: 5.7 (at 40 °C)
	Specificity: α-1,4 D-glucosidic bonds

New Maltose Producing Amylases

In the case of α-amylases two types of end-product profiles exist, e.g. in the case of thermolabile enzymes it is the di- and tri-saccharides, maltose and maltotriose and in the case of thermostable enzymes the degree of polymerisation of the product is larger. Recently, a number of α-amylases have been characterised which exhibit highly interesting properties including their end-product profiles which appear very attractive from a commercial viewpoint.

An α-amylase found in the culture broth of *Streptomyces praecox* (Wako et al., 1978) is unique in its action pattern, in that maltose is the sole end-product in the final hydrolysate. Unlike β-amylase which hydrolyses alternate α-1,4 glucosidic bonds by an exo-mechanism of substrate attack, this enzyme produces a variety of oligosaccharides in the early stages of the hydrolytic process. Furthermore, the enzyme produces maltose exclusively from maltotriose without production of glucose (Suganuma et al., 1980).

Wako et al. (1978) observed that in the hydrolysis of starch by this enzyme:
(1) glucose was not produced in any significant amount during the hydrolytic reaction;
(2) maltooligosaccharides (G_2–G_5) were produced initially and G_4 (maltotetraose) accumulated in a relatively large amount;
(3) at the final stage, G_2 (maltose) was the sole product in the reaction mixture.

These results can be explained (Table 3) on the basis of the conversion route $2G_3 \rightarrow 3G_2$ either by condensation or transglycosylation mechanisms. The larger saccharides, G_5 or G_6 are hydrolyzed rapidly to give G_2, G_3 and G_4. Maltotriose is degraded into G_2 via G_4 as a transfer product. Thus at the final reaction stage maltose is the sole product.

Table 3. Degradation of Starch by α-Amylase from *Streptomyces praecox*

Intermediate products
G_2, G_3, G_4, G_5, G_6
Final product
G_2

Mechanism
$2G_3 \rightarrow 3G_2$
$G_4 \rightarrow 2G_2$
$G_5 \rightarrow G_2 + G_3$
$G_6 \rightarrow G_2 + G_4$

G_2, maltose; G_3, maltotriose etc. etc.

An α-amylase of *Thermoactinomyces vulgaris* (Shimizu et al., 1978) also produces maltose from starch in yields of about 74% and the other main product is glucose. It has activity optima of pH 5.0 and 70 °C and is stabilized by calcium (Table 4). Another unique feature of this enzyme is its ability to hydrolyse pullulan to panose, a property not previously detected in α-amylases. End-product analyses of the degradation of starch and pullulan are outlined in Table 5.

The α-amylase of *T. vulgaris* was purified to homogeneity (Sakano et al., 1982). Both pullulan-hydrolysing and starch-hydrolysing activities were inhibited by p-chloromercuribenzoate (p-CMB), maltotriitol, panitol, isopanitol and microbial α-amylase inhibitors to about the same extent. K_m, V_{max} and K_i values

Table 4. Properties of *Thermoactinomyces vulgaris and Streptomyces hygroscopicus* α-Amylases

	T. vulgaris	S. hygroscopicus
pH optimum	5.0	5.0–6.0
Temp. optimum (°C)	70	50–55
Molecular weight	ND	48 000
Inhibited by EDTA	ND	—
Effect of Calcium	Stabilized	Stabilized

ND = not determined

Table 5. End-Product Analysis of the Action of α-Amylase of *Thermoactinomyces vulgaris* on Starch and Pullulan

Composition of hydrolysate (%)					
Substrate	Glucose	Maltose	Maltotriose	Panose	Higher Saccharides
Starch	11.8	74.1	0.1	3.3	10.7
Pullulan	0.7	1.5	—	96.5	0.4

Adapted from Shimizu et al. (1978)
Assay conditions: substrate, 2.0% (w/v) in 0.05 M acetate buffer, pH 4.5, containing 5 mM CaCl$_2$, 60 °C, 48 hours.

supported the view that the hydrolytic action of the enzyme on the two substrates was due to a single catalytic site. Furthermore, it was suggested that this α-amylase could attack α-1,6-D-glucosidic linkages in hydrolysates of pullulan as well as α-1,4-D-glucosidic linkages in starch and pullulan.

Hidaka and Adachi (1980) described an α-amylase from *Streptomyces hygroscopicus* that converts starch to maltose in 75% yield. The enzyme had a pH optimum of 5.0–6.0, a temperature optimum of 50–55 °C and a molecular weight of 48,000 (Table 4). Maltotriose and maltotetraose are the principal primary products formed. Maltotetraose is then hydrolysed directly to maltose. Conversion of maltotriose to maltose is achieved by either (1) a condensation-hydrolytic process and/or (2) a synthetic step followed by a hydrolytic reaction (Fig. 1).

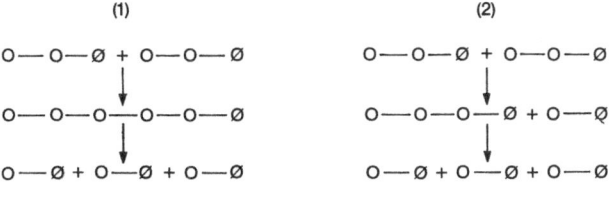

(1)

O—O—Ø + O—O—Ø

O—O—O—O—O—Ø

O—Ø + O—Ø + O—Ø

1. Condesation – hydrolysis

(2)

O—O—Ø + O—O—Ø

O—O—O—Ø + O—Ø

O—Ø + O—Ø + O—Ø

2. Transglycosylation and hydrolysis

Figure 1. Breakdown of maltotriose by Streptomyces hygroscopicus α-amylase

It would appear therefore that the mechanism of maltose production by *S. hygroscopicus* α-amylase takes place as follows:
(1) Hydrolysis of α-1,4-glucosidic linkages in an endofashion producing malto-triose and maltotetraose in the early stages of reaction;
(2) hydrolysis of maltotetraose to maltose;
(3) breakdown of maltotriose by condensation and/or transglycosylation mechanisms followed by hydrolysis of the synthesized products, resulting in production of maltose in excess over glucose.

A summary of the distribution of end-products of the three systems described is outlined in Table 6. Because of their pH and temperature characteristics the enzymes of *T. vulgaris* and *S. hygroscopicus* would be more suited to commercial processes than that of *S. praecox*. A low viscosity corn syrup can easily be produced using *S. hydroscopicus* α-amylase (Hidaka and Adachi, 1980). It has low viscosity, low browning tendency and does not crystallize. Since the oligosaccharide composition of the syrup is similar to that from malt, application of the syrup in the brewing industry is expected.

Table 6. End-Products Produced from Starch by *Streptomyces praeox*, *Thermoactinomyces vulgaris* and *Streptomyces hygroscopicus* α-Amylases

Enzyme	End-Product Composition (%)
S. praeox	Maltose (>80)
T. vulgaris	Maltose (74), Glucose (12), Higher saccharides (11)
S. hygroscopicus	Maltose (75), Glucose (11), Maltotriose (6), Higher saccharides (8)

Bacillus polymyxa β-Amylase

β-Amylase (EC 3.2.1.2 α-1,4 glucan maltohydrolase, saccharogenic amylase) is an important enzyme in the brewing and distilling industries. It is present in both germinated and ungerminated cereals and other plants e.g. sweet potato, soya bean and sorghum (Manners, 1962). Its occurrence as an extracellular enzyme in microorganisms was first established in our laboratories (Bourke and Fogarty, 1975; Fogarty and Griffin, 1973; Griffin and Fogarty, 1973a, b, c; W. M. Fogarty and M. T. O'Reilly, unpublished observations) and detected subsequently in other bacteria (Higashihara and Okada, 1974; Shinke et al., 1974; 1979; Takasaki, 1976b; Thomas et al., 1980; Napier, 1977; Murao et al., 1979). The enzyme acts in an exo-fashion from the non-reducing chain ends of amylose, amylopectin or glycogen and hydrolyses alternate α-1,4-glucosidic bonds producing the β-anomeric form of maltose.

β-Amylase is unable to hydrolyse or by-pass α-1,6 glucosidic linkages in amylopectin and glycogen. Degradation of these branched structures is therefore incomplete. Action of β-amylase on amylopectin results in 50%–60% conversion to maltose with formation of a β-limit dextrin which is essentially the parent polymer with the outer chains trimmed down close to the outermost branch points.

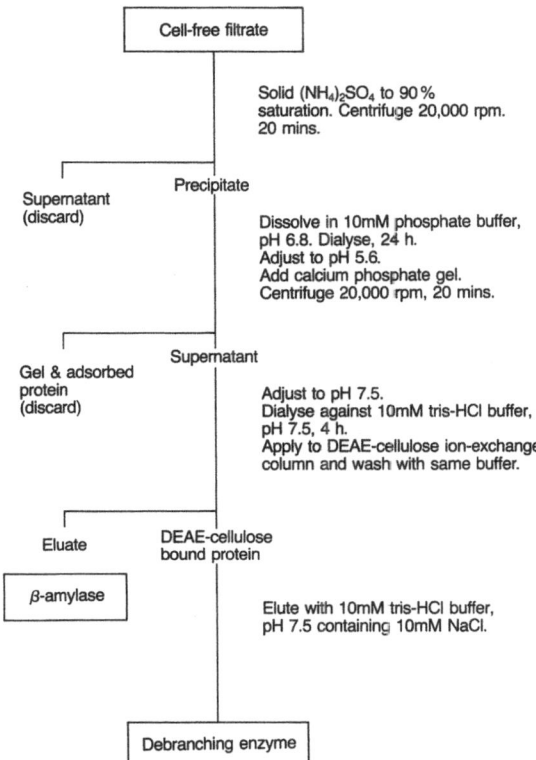

Figure 2. Purification of the starch degrading system of *Bacillus polymyxa*

Investigations in our laboratories resolved the *Bacillus polymyxa* system into two components, a debranching enzyme specific for α-1,6 linkages of amylopectin and an α-1,4 glucan maltohydrolase (β-amylase) (Griffin and Fogarty, 1973b, c). The resolution of the two components was effected by salt precipitation, adsorption and ion-exchange chromatography on DEAE-cellulose (Fig. 2). Quantitative conversion of starch to maltose is thus obtained by virtue of the uniqueness of this system i.e. by virtue of the nature of the two enzymes. The β-amylase was competitively inhibited by the Schardinger dextrins, cyclohexaamylose and cyclo-heptaamylose (α- and β-cyclodextrins). Inhibition by Schardinger dextrins is also a property shared by some plant β-amylases. This functional specificity was exploited as a means of obtaining the enzyme in a very highly purified state. The ligand, cyclohexaamylose was bound to a spacer arm (Fig. 3), in this case a bifunctional epoxide — 1,4-bis (2,3-epoxy propoxy) — butane, which had been linked to an insoluble support — Sepharose 6B. Addition of a partially purified

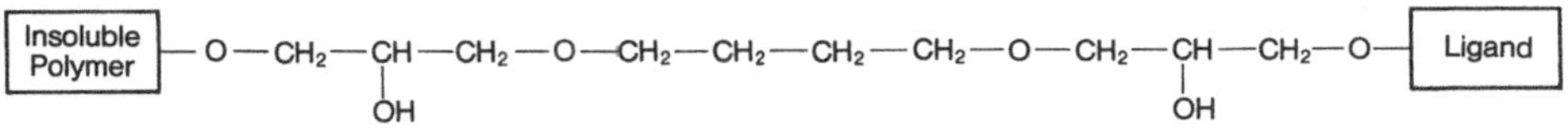

Insoluble Polymer — O — CH$_2$ — CH — CH$_2$ — O — CH$_2$ — CH$_2$ — CH$_2$ — CH$_2$ — O — CH$_2$ — CH — CH$_2$ — O — Ligand
                                         | |
                                        OH   OH

Figure 3. Support system used in the purification of α-amylase by affinity chromatography

enzyme preparation to a column of this materials (Fig. 4) led to the binding of the enzyme to the competitive inhibitor. Inert protein and other activities were unaffected and passed through the column. The β-amylase was disassociated from the bound inhibitor by eluting the column with a soluble form of the inhibitor. The specificity of the system is such that it is highly successful even with crude enzyme, containing a number of impurities.

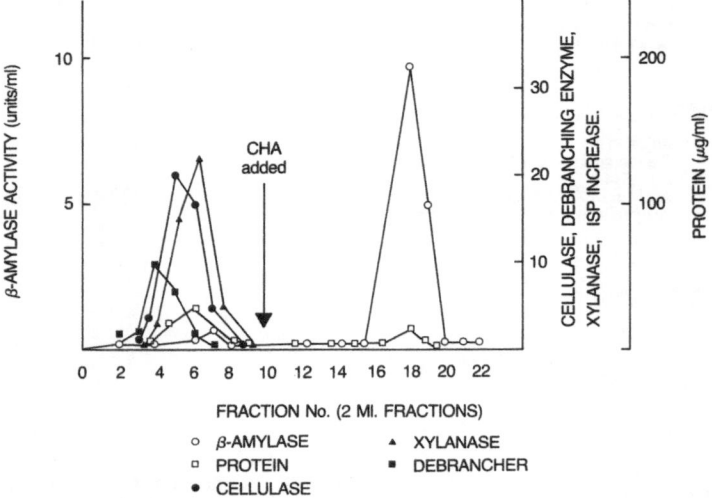

Figure 4. Purification of β-amylase by affinity chromatography on Sepharose 6B-cyclohexaamylose complex (CHA). The purified enzyme was eluted from the column using a soluble form of the competitive inhibitor

The production of high maltose syrups is regarded as an area of predictable growth. It is inevitable that with the development of β-amylase technology will come the expansion of the commercial exploitation of starch-debranching enzymes on a larger commercial scale than heretofore, not only for use in association with β-amylase but also in conjunction with amyloglucosidase in the production of high dextrose syrups.

Other Microbial β-Amylases

Since the original discovery of β-amylase in *Bacillus polymyxa*, its presence in a number of other species is now well established (Table 7). The pH optima of these enzymes tend to be higher than the corresponding plant enzymes and in most cases they are inhibited by sulphydryl reagents e.g. p-CMB.

The β-amylase of *Bacillus circulans* (Napier, 1977; Fogarty and Nash, 1982) exhibits some characteristics that distinguish it from other microbial β-amylases. The enzyme is active at higher temperatures and is much less susceptible to p-CMB than other β-amylases and this activity is not dependent on the absence

Table 7. Properties of some Microbial β-Amylases

Organism	Molecular Weight	Optimum pH	Optimum Temp. (0 °C)	Reference
Bacillus cereus				
var. *mycoides*	35,000	7.0	50	Takasaki (1976 b)
Bacillus circulans	53,000–63,000	6.5–7.5	60	Napier (1977)
		7.0–7.5	60	Fogarty and Nash (1982)
Bacillus megaterium				
NCIB 9323	35,000	6.0	50	
NCIB 9376	32,000	6.5	50	W. M. Fogarty and M. T. O'Reilly (unpublished observations)
Bacillus polymyxa	59,000	6.8	37	Bourke and Fogarty (1975); Fogarty and Griffin (1975)
	44,000	7.5	45	Murao et al. (1979)
Bacillus sp.				
IMD 198	58,000	6.8	55	Bourke and Fogarty (1979)
Bacillus sp.				
BQ 10	160,000	6.0–7.0	45–55	
	80,000			Shinke et al. (1975a, b)
Pseudomonas sp.				
BQ 6	37,000	6.5–7.5	45–55	Shinke et al. (1975c)

of sulphydryl inhibitors such as heavy metals. The enzyme has a K_m for starch of 0.71 mg/ml and an isoelectric point of pH 4.7 and does not degrade maltotriose (Fogarty and Nash, 1982).

It is important in producing syrups in excess of 80% maltose, that if the β-amylase being used does not degrade maltotriose that formation of intermediates of uneven chain length (DP5, DP7, etc.) are prevented as these eventually lead to the formation of the trimer. This problem may be alleviated by using a β-amylase that degrades maltotriose, as for example that of *Bacillus* sp. IMD 198 (Bourke and Fogarty, 1979) or if the enzyme lacks this property by using a liquefied starch with a low DE as starting material (Takasaki, 1976a, Yamanobe and Takasaki, 1979).

Developments in α-Glucosidases

α-Glucosidases have potential applications in fundamental research, clinical laboratories and industry and such usefulness is reflected in the current interest being displayed in these enzymes.

Although α-glucosidases occur widely in the microbial kingdom this presentation will refer only to bacterial enzymes. α-Glucosidases hydrolyse terminal, non-reducing α-1,4 linked glucose residues in different substrates, releasing α-D-glucose. Their specificity is mainly toward the glucosidic residue i.e. the α-D-glucopyranosyl radical. Many bacterial α-glucosidases display highest activity towards maltose and in many instances are true maltases (Table 8). In some cases the specificity is so broad that it also permits hydrolysis of α-1,6 linkages and aryl and alkyl-α-D-glucosides (Table 9).

Table 8. Substrate Specificities of some Bacillus α-Glucosidases

Substrate	Bacillus amylolyticus	Bacillus megaterium NCIB 9376	Bacillus megaterium NCIB 9323	Bacillus sp. 7196
	Relative Activities			
Maltose	100	100	100	100
Maltotriose	61	18	24	22
Panose	7	0	0	0
Melibiose	10	0	0	0
p-Nitrophenyl-α-D-glucoside	0	0	0	166
Phenyl-α-D-glucoside	0	0	5	2.5

None of the enzymes displayed activity towards isomaltose, sucrose, methyl-α-D-glucoside or starch.

Table 9. Primary and other Substrate Specificities of some Bacterial α-Glucosidases

Enzyme Source	Primary Substrate	α-1,6- activity	aryl- activity	alkyl- activity	Reference
Bacillus amylolyticus	Maltose	—	—	—	Kelly et al. (1980)
B. amyloliquefaciens	Sucrose	+	+	trace	Urlaub and Wöber (1978)
B. brevis	Maltose	+	—	—	McWethy and Hartman (1979)
B. cereus	Maltose	trace	trace	—	Yamasaki and Suzuki (1974)
B. firmus	Maltose	—	+	—	W. M. Fogarty and M. Moriarty, unpublished data
B. megaterium NCIB 9376	Maltose	—	—	—	
B. megaterium NCIB 9323	Maltose	—	+	—	
Bacillus sp. 7196	p-NPG[a]	—	+	—	W. M. Fogarty and M. T. O'Reilly, unpublished data.
Bacillus sp. KP1035	Maltose	—	—	—	Suzuki et al. (1978)
B. subtilis P-11	Maltose	+	—	—	Wang and Hartman (1976)
Pseudomonas amyloderamosa	Maltose	+	+	trace	Amemura et al. (1974)
Pseudomonas fluorescens	Maltose	—	+	—	Guffanti and Corpe (1976)

[a]p-NPG = p-nitrophenyl-α-D-glucoside

A novel α-glucosidase in *B. amyloliquefaciens* (Urlaub and Wöber, 1978) surprisingly displayed highest activity towards sucrose and p-nitrophenyl-α-D-glucoside (Table 10). In addition it hydrolysed maltose, maltotriose, isomaltose and isomaltotriose. No other α-glucosidases have been reported with such high activity towards sucrose. The application of an enzyme in any form is dependent upon a knowledge of its precise and fundamental properties. For this reason we undertook to look further at the *B. amyloliquefaciens* system. Cells of *B.*

Table 10. *B. amyloliquefaciens* α-Glucosidase Activity with Various Substrates

Substrate	Relative enzyme activity (%)
Maltose	40
Maltotriose	32
Isomaltose	30
Isomaltotriose	18
Soluble starch	—
Amylose	—
Phytoglycogen	—
Sucrose	140
p-Nitrophenyl-α-D-glucoside	100
p-Nitrophenyl-β-D-glucoside	—
Methyl-α-D-glucoside	Trace

Urlaub and Wöber (1978)

amyloliquefaciens ATCC 23350 grown for 26 hours in a medium containing trehalose as carbon source were harvested, disrupted and the nucleic acids precipitated by streptomycin sulphate. The dialyzed intracellular contents were applied to DEAE-Bio Gel-A ion-exchange resin and eluted with 0.01 M sodium phosphate buffer pH 6.0 containing 0.1 M NaCl. This provided two active peaks which were subsequently identified as a maltase and an α-glucosidase. The overall purification scheme is summarised in Table 11. When the organism was grown in a similar medium but with sucrose as carbon source a third peak with high activity was obtained. This was shown to be a typical invertase and the overall purification scheme for this enzyme is outlined in Table 12.

Table 11. Overall Purification of the Maltase and α-Glucosidase of *B. amyloliquefaciens* ATCC 23,350

Fraction	Volume (units/mg)	Protein (mg)	Maltase (units)	Specific activity (units/mg)	Purification fold	% Yield	Protein (mg)	α-Glucosidase (units)	Specific activity (units/mg)	Purification fold	% Yield
			Maltase					α-Glucosidase			
Intracellular contents	62	32×10^3	4.3	13×10^{-5}	1	100	32×10^3	689	0.021	1	100
Streptomycin sulphate	70	807.5	4.2	6.1×10^{-3}	46.9	98	807.5	682	0.84	40	98.9
DEAE-Bio Gel-A ion exchange	15	8.01	3.83	0.478	3,676.9	89	2.03	507	249.7	11,890.4	73.5

Table 12. Overall Purification of the Invertase of *B. amyloliquefaciens* ATCC 23,350

Treatment	Volume (ml)	Protein (mg)	Enzyme activity (units)	Specific activity (units/mg)	Purifi- cation fold	% Recovery
Intracellular contents	80	1.6×10^3	86	0.053	1	100
Streptomycin sulphate	98	285.3	81.7	0.286	5.39	95
DEAE-Bio Gel-A ion exchange	32	8.3	21.1	.2.54	47.9	61.3

Examination of the properties of the three systems showed that they were three distinct entities (Table 13). The α-glucosidase showed activity only towards p-nitrophenyl-α-D-glucoside, (p-NPG), isomaltose and isomaltotriose whereas the maltase had highest activity towards maltose, sucrose and maltotriose and the invertase was only active on sucrose and raffinose (Table 14). Thus the substrate specificities of the three enzymes are also quite distinct.

Table 13. Some Properties of *B. amyloliquefaciens* ATCC 23350 Enzymes

	α-Glucosidase	Maltase	Invertase
pH Optimum	4.9	6.5	6.5
Temp. Optimum (°C)	40	30	40
Molecular Weight	26,000	72,000	30,000
Isoelectric Point	4.6	4.7	4.2

Table 14. Substrate Specificities of the three Enzymes of *B. amyloliquefaciens* ATCC 23350

Substrate (1%, w/v)	% Relative activity
α-Glucosidase	
p-Nitrophenyl-α-D-glucoside[a]	100
Isomaltose	24
Isomaltotriose	13
Maltase	
Maltose	100
Sucrose	84
Maltotriose	32
Melezitose	10
Invertase	
Sucrose	100
Raffinose	22

[a] Assayed at 0.5% (w/v)

The following substrates were investigated with each enzyme: p-nitrophenyl-α-D-glucoside, p-nitrophenyl-β-D-glucoside, α-methyl-D-glucoside, isomaltose, isomaltotriose, panose, maltose, maltotriose, sucrose, trehalose, gentiobiose, turanose, melibiose, melezitose, raffinose and starch. Activity was detected only with those substrates listed in the table.

The components of another strain of *B. amyloliquefaciens* — ATCC 23844 — were similarly resolved into three entities, and α-glucosidase, a maltase and an invertase. Again the properties were quite distinct in each case and likewise the substrate specificities.

That the enzymes were separate components was further illustrated by induction experiments. Thus α-glucosidase could be selectively induced by glucose, fructose, sucrose and lactose and no maltase was produced. Maltase was induced by the presence of maltose and raffinose. Sucrose and raffinose induced highest yields of invertase.

In summary this investigation showed that the intracellular contents of *B. amyloliquefaciens* ATCC 23844 and the type strain *B. amyloliquefaciens* ATCC 23350 showed a much higher specificity for sucrose than that reported for the α-glucosidase of *B. amyloliquefaciens* was a single enzyme with a broad substrate Wöber (1978). On examination of the effect of carbon inducers, it was found here that the *B. amyloliquefaciens* strains produced an invertase induced by sucrose. The invertase which was produced separately from the α-glucosidase accounted for part of the sucrose hydrolysis. Furthermore, it had been reported that the α-glucosidase of *B. amyloliquefaciens* was a single enzyme with a broad substrate specificity. Our results show that this deduction is incorrect. In this study we established the existence of three enzymes, an α-glucosidase, a maltase and an invertase.

The system reported by Urlaub and Wöber showed a broad substrate specificity (Table 10) and our studies show that the activities reported can be attributed to one or other of our individual enzymes. Thus, the activity towards p-NPG, isomaltose and isomaltotriose is specifically related to the α-glucosidase and the maltose and maltotriose activities in particular and some of the sucrose activity is related to the maltase enzyme. The bulk of the sucrose activity is related to the invertase. The distinctive and separate nature of the invertase and maltase is further confirmed by the lack of or the presence of specificities for melezitose or raffinose (Kulp, 1975). It can be concluded therefore from our studies that earlier work was unsuccessful in separating the different enzyme activities in the intracellular contents of *B. amyloliquefaciens*.

These results make a strong case for more rigorous resolution before establishment of fundamental properties of enzyme systems. We are currently examining other, similar bacterial systems and it would appear that in those with broad substrate specificities, high resolution techniques indicate the presence of distinct enzymic entities. These will be reported in due course. Such observations are helpful in overcoming the current ambiguities relating to the true nature of α-glucosidases.

Acknowledgement

The authors wish to thank Drs. Griffin, Bourke and Kadam and Ms. Ann Nash for their contributions to the development of this work.

References

Allen WG and Spradlin JE (1974). Amylases and their properties. Brew Dig 49(7):48–50, 52–53, 65

Amemura A, Sugimoto T and Harada T (1974). Characterization of intracellular α-glucosidase of *Pseudomonas* SB15. J Ferment Technol 52:778–780

Barfoed HC (1976). Enzymes in starch processing. Cereal Foods World 21:588–604

Borgia PT, Campbell LL (1978). α-Amylase from five strains of *Bacillus amyloliquefaciens*: Evidence for identical primary structures. J Bacteriol 134:389–393

Bourke EJ, Fogarty WM (1975). The β-amylase of *Bacillus polymyxa*. Proc Soc Gen Micro 2(4):80–81

Bourke EJ, Fogarty WM (1979). The β-amylase of *Bacillus subtilis* IMD 198. Soc Gen Microbiol Quart 6(4):153

Fogarty WM (1981). Some recent developments in starch-degrading enzymes. The Institute of Brewing (Scottish Section), Aviemore Conference May, 1982

Fogarty WM, Griffin PJ (1973). Studies on the production of an extracellular amylase of *Bacillus polymyxa* in batch culture. J Appl Chem Biotechnol 23:166–167

Fogarty WM, Griffin PJ (1975). Purification and properties of β-amylase produced by *Bacillus polymyxa*. J Appl Chem Biotechnol 25:229–238

Fogarty WM, Kelly CT (1979). Starch-degrading enzymes of microbial origin. In Bull AH (ed) Progress in Industrial Microbiology Vol 15 Elsevier Publ Company, Amsterdam p 87–150

Fogarty WM, Kelly CT (1980). Amylases, amyloglucosidases and related glucanases. In: Rose AH (ed) Microbial Enzymes and Bioconversions: Economic Microbiology Vol 5. Academic Press, London p 115–170

Fogarty WM, Nash AM (1982). The β-amylase of *Bacillus circulans*. J Chem Technol Biotechnol, in preparation

Griffin PJ, Fogarty WM (1973a). Production of an amylolytic enzyme by *Bacillus polymyxa* in batch cultures. J Appl Chem Biotechnol 23:301–308

Griffin PJ, Fogarty WM (1973b). Preliminary observations on the starch-degrading system elaborated by *Bacillus polymyxa*. Biochem Soc Trans 1:397–400

Griffin PJ, Fogarty WM (1973c). Further studies on the amylolytic system elaborated by *Bacillus polymyxa*. Biochem Soc Trans 1:1097–1100

Guffanti AA, Corpe WA (1976). Partial purification and characterization of α-glucosidase from *Pseudomonas fluorescens*. Arch Microbiol 107:269–276

Hidaka H, Adachi T (1980). Studies on the α-amylase from *Streptomyces hygroscopicus* SF-1084. In: Marshall JJ (ed). Mechanisms of saccharide polymerization and depolymerization. Academic Press, New York p 101–118

Higashihara M, Okada S (1974). The β-amylase of *Bacillus megaterium* strain No 32. Agric Biol Chem 38:1023–1029

Kelly CT, Heffernan ME and Fogarty WM (1980). A novel α-glucosidase produced by *Bacillus amylolyticus*. Biotechnol Lett 2:351–356

Kulp K (1975) Carbohydrases In: Reed G (ed). Enzymes in Food Processing. Academic Press, New York p 53–122

Madsen GB, Norman BE and Slott S (1973). A new heat stable bacterial amylase and its use in high temperature liquefaction. Die Stärke 25(9):304–308

Maeda H, Tsao GT (1979). Maltose production. Process Biochem 14(7):2–5, 27

Manners DJ (1962). Enzymic synthesis and degradation of starch and glycogen. Adv Carbohyd Chem 17:371–430

McWethy SJ, Hartman PA (1979). Extracellular maltase of *Bacillus brevis*. Appl Environ Microbiol 37:1096–1102

Morgan FJ, Priest FG (1981). Characterization of a thermostable α-amylase from *Bacillus licheniformis* NCIB 6346. J Appl Bacteriol 50:107–114

Murao S, Ohyama K and Arai M (1979). β-Amylases from *Bacillus polymyxa* No 72. Agric Biol Chem 43:719–726

Napier EJ (1977) United States Patent 4,011,136

Norman BE (1979). The application of polysaccharide degrading enzymes in the starch industry. In: Berkeley RCW, Gooday GW and Ellwood DC (eds). Microbial Polysaccharides and Polysaccharides, Academic Press, London p 339–376

Saito N (1973). Thermophilic extracellular α-amylase from *Bacillus licheniformis*. Arch Biochem Biophys 155:290–298

Sakano Y, Hiraiwa S-i, Fukushima J and Kobayashi T (1982). Enzymatic properties and action patterns of *Thermoactinomyces vulgaris* α-amylase. Agric Biol Chem 46:1121–1129

Shinke R, Nishira H and Mugibayashi N (1974). Isolation of β-amylase producing microorganisms. Agric Biol Chem 38:665–666

Shinke R, Kunimi Y and Nishira H (1975a). Isolation and characterization of β-amylase producing microorganisms. J Ferment Technol 53:687–692

Shinke R, Kunimi Y and Nishira H (1975b). Production and some properties of β-amylase of *Bacillus* sp. BQ10. J Ferment Technol 53:693–697

Shinke R, Kunimi Y and Nishira H (1975c). Production and some enzymatic properties of β-amylase of *Pseudomonas* sp BQ 6. J Ferment Technol 53:698–702

Shinke R, Aoki K, Nishira H and Yuki S (1979). Isolation of a rifampicin-resistant asporogenous mutant from *Bacillus cereus* and its high β-amylase productivity. J Ferment Technol 57:53–55

Shimizu M, Kanno M, Tamura M and Suekane M (1978). Purification and some properties of a novel α-amylase produced by a strain of *Thermoactinomyces vulgaris*. Agric Biol Chem 42:1681–1688

Suganuma T, Mizukami T, Moori K, Ohnishi M and Hiromi K (1980). Studies of the action pattern of an α-amylase from *Streptomyces praecox* NA-273. J Biochem 88:131–138

Suzuki Y, Ikemoto T and Abe S (1978). Purification and some properties of maltases I and II from thermophilic *Bacillus* sp KP1035. J Ferment Technol 56:8–14

Takasaki Y (1976a). Studies on amylases effective for production of maltose Part 1. Production and utilization of β-amylase and pullulanase from *Bacillus cereus* var *mycoides*. Agric Biol Chem 40:1515 –1522

Takasaki Y (1976b). Studies on amylases from *Bacillus* effective for production of maltose. Part 11. Purification and enzymic properties of β-amylase and pullulanase from *Bacillus cereus* var *mycoides*. Agric Biol Chem 40:1523–1530

Thomas M, Priest FG and Stark JR (1980). Characterization of an extracellular β-amylase from *Bacillus megaterium sensu stricto*. J Gen Micro 118:67–72

Underkofler LA, Denault LJ and Hou EF (1965). Enzymes in the starch industry. Die Stärke 17:179–183

Underkofler LA (1968). Development of a commercial enzyme process: glucoamylase. Adv Chem Ser 5:343–358

Urlaub H, Wöber G (1978). α-Glucosidase, a membrane bound enzyme of α-glucan metabolism in *Bacillus amyloliquefaciens*. Purification and partial characterization. Biochim Biophys Acta 522: 161–173

Wako K, Takahashi C, Hashimoto S and Kaneda J (1978). Studies on maltotriose and maltose forming amylases from *Streptomyces*. J Jap Soc Starch Sci 25:155–161

Wang L-H, Hartman PA (1976). Purification and some properties of an extracellular maltase from *Bacillus subtilis*. Appl Environ Microbiol 31:108–118

Yamanobe T, Takasaki Y (1979). Production of maltose from starch of various origins by β-amylase and pullulanase of *Bacillus cereus* var *mycoides*. Nippon Nogei Kagaku Kaishi 53:77–80

Yamasaki Y, Suzuki Y (1974). Purification and properties of α-glucosidase from *Bacillus cereus*. Agric Biol Chem 38:443–453

Enzymatic Polymerization of Lignin

Kaj Forss

Summary

The formation of lignin in coniferous wood by enzymatic polymerisation is discussed on the basis of published experimental findings and current theories concerning the composition and structure of lignin.

Introduction

Together with cellulose and the other plant polysaccharides, commonly called hemicelluloses, lignin occurs in vascular plants in amounts ranging between 20 and 35 per cent. After cellulose, lignin is the most abundant organic substance on earth. Part of the lignin is located in the cell walls, where, in all probability, it is bound to some of the hemicellulose components. Another part is located in the middle lamella between the cells, where the lignin "glues" the fibres together.

It is known that lignin is composed of the three phenylpropane units shown in Fig. 1, linked together by carbon-carbon and ether bonds.

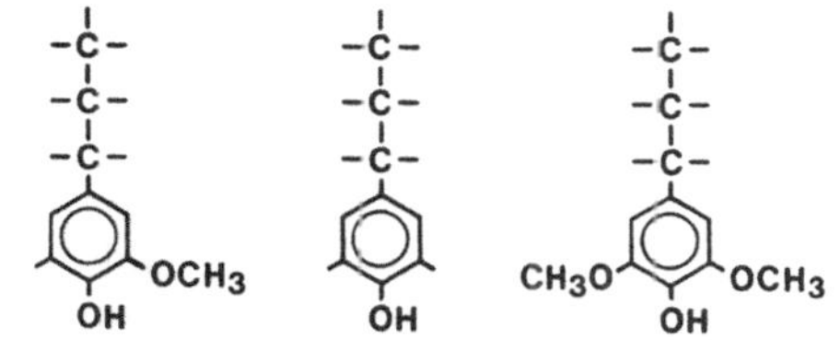

GUAIACYL UNIT   p-HYDROXYPHENYL-PROPANE UNIT   SYRINGYL UNIT

Figure 1. The phenylpropane units constituting the building blocks of lignin

These units are present in different proportions in the lignins of grasses, deciduous wood and coniferous wood. Despite more than a century of research, the structure of these lignins is not yet completely clear. This presentation deals with the enzymatic polymerization of lignin, with particular emphasis on coniferous lignin, which for industrial reasons is of the greatest interest. In chemical wood pulping the main objective is to separate the wood fibres from each other by converting the lignin into water-soluble derivatives by means of some suitable reactions.

Formation of Lignin from its Precursors

From an enzymatic point of view, lignin chemistry goes back to 1875 when Tiemann [24] determined the constitution of the glucoside coniferin (Fig. 2).

CH$_2$OH
|
CH
‖
CH

OCH$_3$
OC$_6$H$_{11}$O$_5$ **Figure 2.** Coniferin

At the end of the century Tiemann and Mendelsohn [25] as well as Klason [20] expressed the hypothesis that lignin is related to coniferin and coniferyl alcohol. In 1908 two French scientists Cousin and Hérissey [2] treated iso-eugenol with the juice from the mushroom *Russula delica* and obtained a crystalline product. The structure of this product remained unsolved until 1933, when Erdtman [3] identified it as a phenylcoumaran formed by enzyme-catalysed oxidative coupling (Fig. 3).

CH$_3$ H$_3$C   CH = CH – CH$_3$
| |
CH   HC —
‖ |
CH   HC —— O
                          OCH$_3$

2   – 2(H$^+$ + e)
        OCH$_3$ ————►   OCH$_3$
OH   OH

Figure 3. Enzyme-catalysed oxidative coupling of iso-eugenol

At that time, it had already been supposed that lignin is made up of ether bonded phenylpropane units, and in the same year Freudenberg and Sohns [14] had proposed the presence of phenylcoumaran structures in the lignin molecule. Erdtman thought that if Freudenbergs concept of lignin could be shown to be correct, lignin might possibly be formed via oxidative polymerization of a propyl guaiacol oxidised in the side chain. This view was adopted and further developed by Freudenberg in an extensive series of studies concerning the biosynthesis of

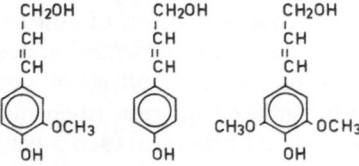

CH$_2$OH   CH$_2$OH   CH$_2$OH
| | |
CH   CH   CH
‖   ‖   ‖
CH   CH   CH

       OCH$_3$ CH$_3$O   OCH$_3$
OH   OH   OH

CONIFERYL   p–COUMARYL   SINAPYL   **Figure 4.** The primary precursors of lignin according to
ALCOHOL   ALCOHOL   ALCOHOL   Freudenberg

lignin [13]. He regarded the three p-hydroxycinnamic alcohols, trans-coniferyl, trans-p-coumaryl and trans-sinapyl alcohol (Fig. 4) which are present in wood as their glucosides, as the primary precursors of lignin.

Oxidative Coupling of Phenols by Phenol Oxidases

Freudenberg succeeded in showing that coniferin diffuses from the cambium, which is the thin living part of the tree between the wood and the bark, into newly formed wood cells, where it is hydrolysed under the influence of beta glucosidases. By treating an aqueous solution of coniferyl alcohol with atmospheric oxygen in the presence of juice squeezed from the mushroom *Psalliota campestris*, Freudenberg obtained an insoluble polymer, "DHP"-Dehydrierungspolymerisat which he regarded as identical with lignin.

According to Freudenberg the first step in the polymerization is the formation of mesomeric radicals from the three precursors under influence of a phenol oxidase as shown in Fig. 5.

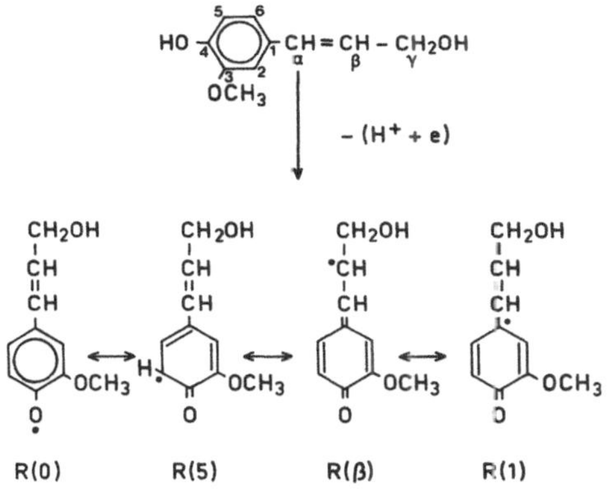

Figure 5. Mesomeric radical formed during the dehydrogenation of coniferyl alcohol

A closer examination of this first oxidative step in lignin polymerization reveals three important features. Firstly, only phenylpropane units with free phenolic hydroxy groups can form radicals, as the actual process is the removal of one electron from the phenoxy anion. Secondly, only phenyl propane units with conjugated double bonds in the side chain can react with the unpaired electron in the beta position. Thirdly, the radicals are quinoidic.

The next step in the polymerization is the coupling in pairs of the radicals in their various mesomeric forms. After rearrangement with or without addition of water, the quinoid intermediates are transformed into dilignols. If carbohydrate is added instead of water, a lignin carbohydrate bond is formed. The three most important dilignols are shown in Fig. 6.

The stereochemistry of the formation of the dilignol ($\pm$) pinoresinol is very interesting, as Harkin [17] has indicated. The hydrogens on the beta carbon forming the new beta-beta bond must be in the cis orientation otherwise the two tetrahydrofuran rings cannot close (Fig. 7). Although pinoresinol contains

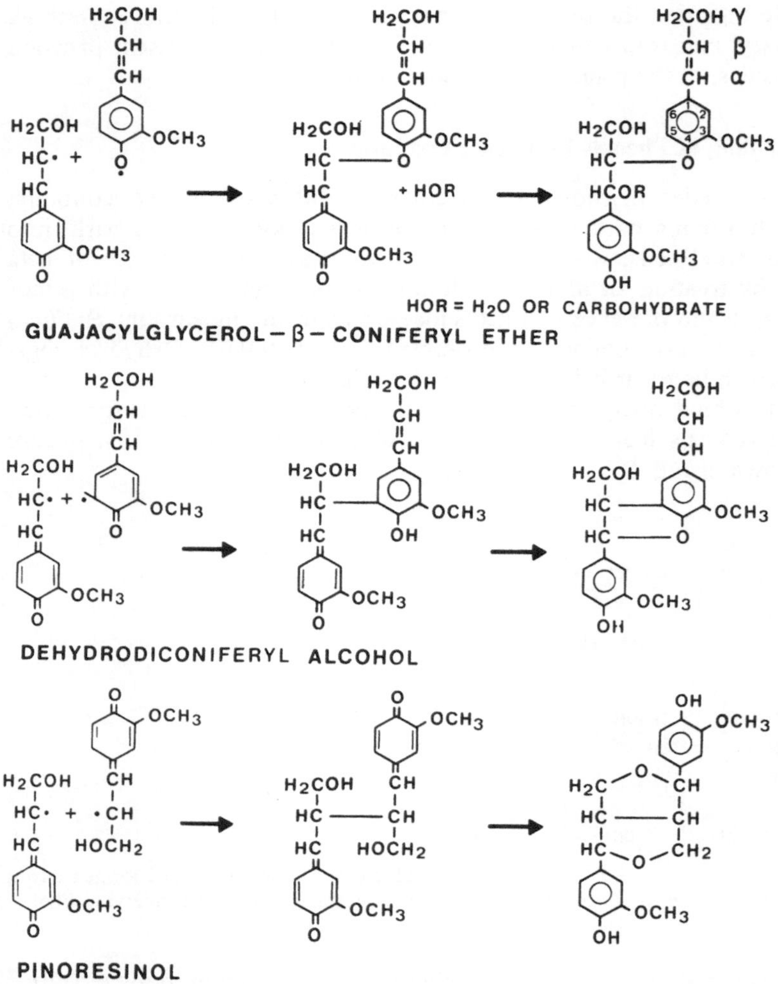

GUAJACYLGLYCEROL — β — CONIFERYL ETHER

DEHYDRODICONIFERYL ALCOHOL

PINORESINOL

Figure 6. Dilignols formed by dehydrogenation of coniferyl alcohol

four asymmetric carbon atoms and theoretically can exist in 16 optically active isomers, only 6 are possible. Of these 4 isomers, (±) pinoresinol and (±) epipinoresinol, have been detected in enzymatic oxidation of coniferyl alcohol. Thus, the pinoresinol encountered during lignification is a racemic mixture of + and − forms and consequently optically inactive. The same holds true for all the other higher lignols that contain asymmetric carbon atoms, for lignin itself and for all the lignin degradation products ever obtained.

As can be seen from Fig. 6, the dilignols formed possess free phenolic hydroxy groups and are consequently still able to form radicals, but none of the radicals formed can react with the unpaired electron in the beta position.

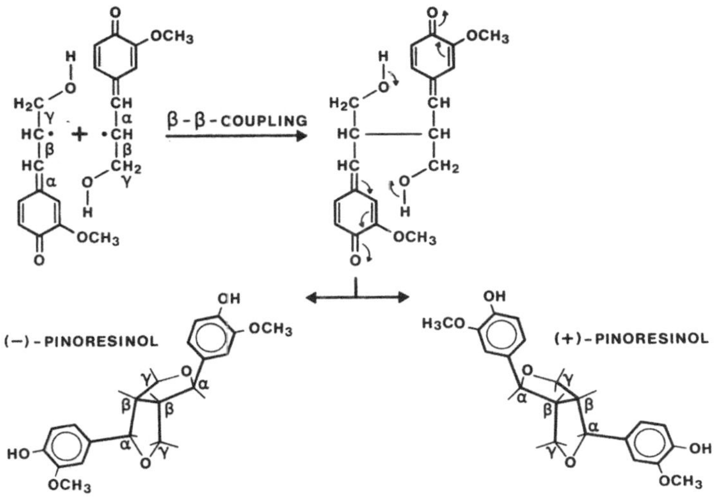

Figure 7. The stereochemistry of pinoresinol formation

With this restriction the polymerization can proceed beyond the dilignol stage through oxidative coupling but also through addition of phenols to quinone-methides. Such a reaction is exemplified in Fig. 8.

Without going into the other numerous suggested reactions, we can divide the final polymerization of the lignin macromolecule into "bulk polymerization" and "end-wise-polymerization".

In bulk polymerization, dilignols, oligolignols and polylignols react with each other. Fig. 8 shows the formation of a hexalignol via the bulk polymerization of one molecule of dehydrodiconiferyl alcohol with two molecules of the quinone-methide, which by addition of water gives the guaiacyl-glycerol-β-coniferyl ether

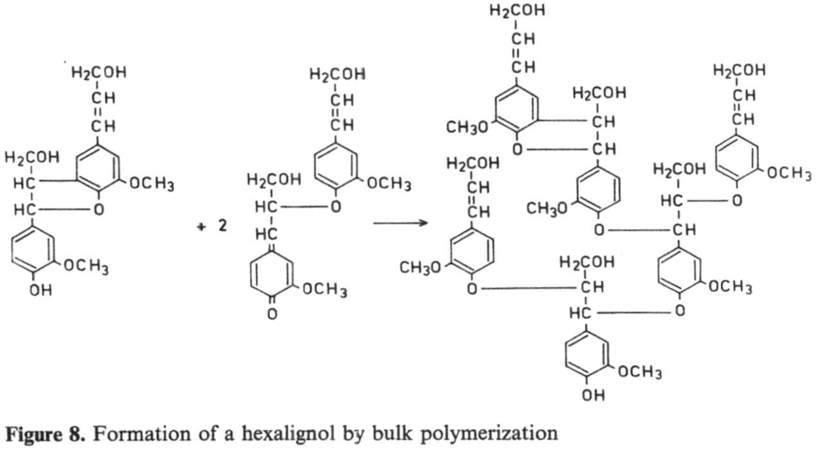

Figure 8. Formation of a hexalignol by bulk polymerization

in Fig. 6. The polymerization occurs only between the units possessing a free phenolic hydroxy group or a quinonemethide structure.

Of the three dilignols shown in Fig. 6, two have unsaturated coniferyl alcohol side chains which would be preserved in bulk polymerization. Lignin would consequently contain a large number of double bonds. This is in conflict with experimental findings which indicate that lignin contains few, if any double bonds.

In order to explain the low number of unsaturated side chains in lignin the bulk polymerization model has been supplemented by an end-wise polymerization model according to which the beta radical of a monomeric lignin precursor is coupled with a radical of a higher lignol. A example of end-wise polymerization is shown in Fig. 9.

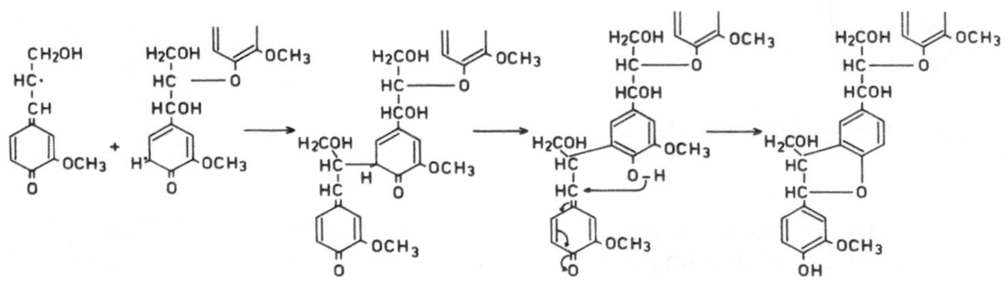

Figure 9. End-wise polymerization

Because the monomeric precursor with its unpaired electron in the beta position reacts with the polymeric end-group with its unpaired electron either in position 0–4 or C–5, the end-wise polymerisation results in a linear, unbranched polymer. Furthermore, the phenolic hydroxy groups disappear, either as the result of coupling of radicals or through ring closure to form phenylcoumaran structures. The phenolic hydroxy groups are therefore found only in the end groups. Both the linearity of the final molecule and the extensive disappearance of phenolic hydroxy groups conflict with experimental findings.

Current opinion appears to be that lignin is built up by a combination of both the bulk and the end-wise polymerization models. Bulk polymerization gives rise to branching of the model, but does not explain the complete, or almost complete, absence of unsaturated side chains. The end-wise polymerization model explains the absence of double bonds, but not the compact structure of the lignin molecule, nor its relatively high content of phenolic hydroxy groups.

Since the end of the 1940s various authors have summarized their knowledge and understanding of coniferous lignin in the form of proposals for lignin formulation. The formulations that have won most regard were put forward by Adler [1] and Freudenberg. Freudenberg's latest formulation [12] is from 1965 and is shown in Fig. 10.

The C_9 units form a branched linear structure held together primarily by aryl ether and carbon-carbon bonds. In accordance with the theory of polymerization

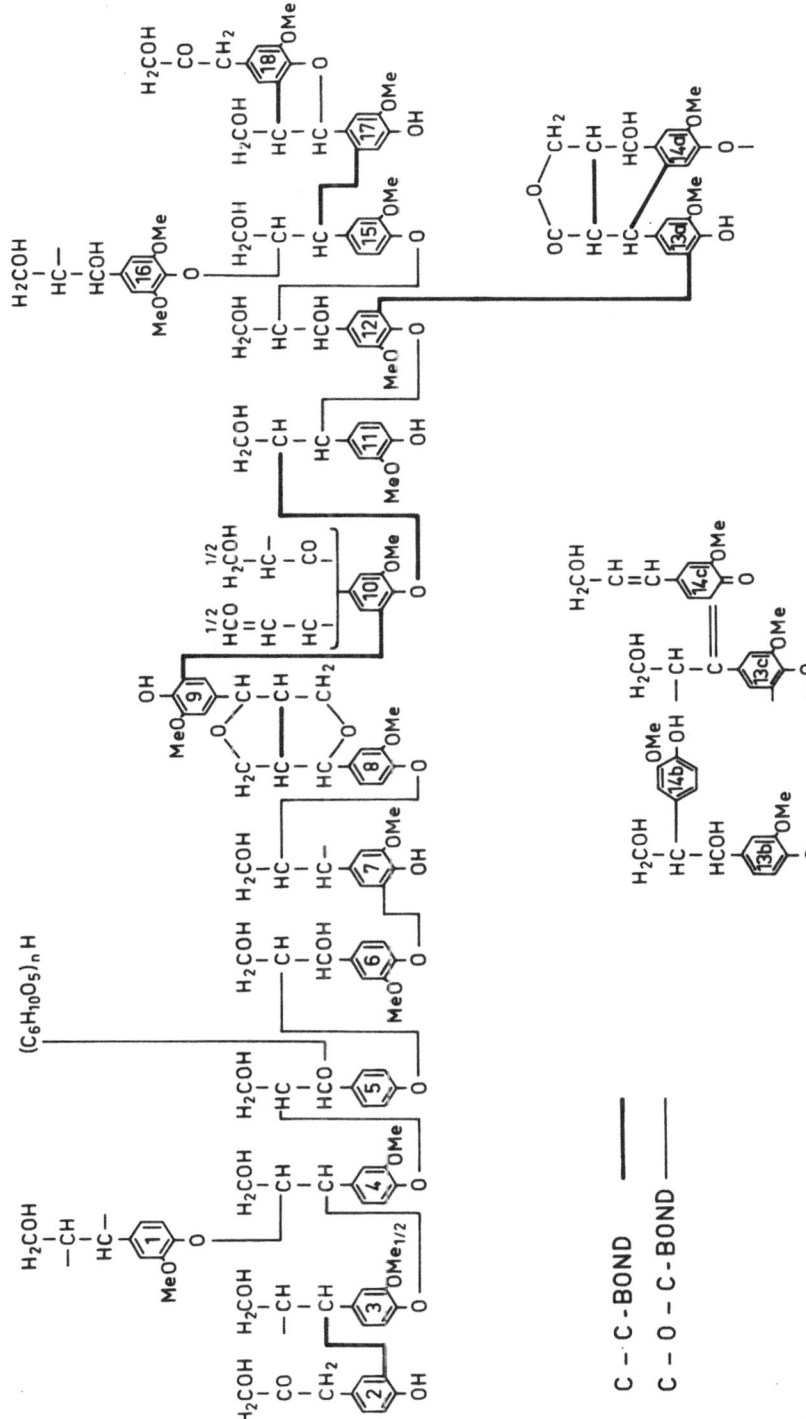

Figure 10. The structural formula of coniferous lignin according to Freudenberg

through enzyme-catalysed oxidative coupling of radicals, the bonds between the C_9 units are distributed at random, giving lignin a structure similar to that of bakelite which by definition can be given no strict quantitative formula.

An examination of the formula in Fig. 10 reveals structures which are in conflict with kown properties of lignin. The chain structure of the molecule would probably involve coiling, but this contrasts with the fact that the intrinsic viscosity of lignin in solution is only one forthieth of that of cellulose with the same weight-average molecular weight [15]. Lignin in solution has often been assumed to have a spherical, compact molecular shape.

According to Luner and Kempf [21], the lignin molecule is compact but not spherical; it is rather lamellar in shape, a concept that Goring [16] expressed. The abundance of C_9 units combined only by ether bonds should make the molecule easily hydrolysable. However, one of the most typical properties of lignin is that it cannot be hydrolysed to any great extent.

Heterogeneity of Lignin

It is known that the results of experimental lignin chemistry are often contra-dictory and difficult to incorporate into a general picture of this substance. The reasons for these difficulties are largely connected with the lignin preparations used.

It has not been possible to isolate lignin quantitatively in an unaltered form. This may even be theoretically impossible as linkages between lignin and carbohydrates may require the breaking of covalent bonds prior to the extraction of lignin. The preparations used thus represent only a fraction of the total lignin and their structure has been affected by the isolation process. Furthermore, the lignin preparations used have often been employed without preceding analysis of their purity or their carbohydrate contents. Before any meaningful scientific work can be initiated, not only the purity but also the homogeneity of the preparation to be studied must be ensured. In lignin chemistry the demand for homogeneity has surprisingly enough not been taken into consideration *a priori*, as a result of the theory that lignin is one single polymer formed by oxidative coupling of the precursors present in the plant concerned. It has, however, been shown that lignin is probably not one homogeneous polymer but a mixture of about 20 per cent low-molecular-weight compounds resembling lignin, called "hemilignins" and about 80 per cent carbohydrate bound high-molecular-weight lignin [4]. For the carbohydrate bound part of lignin, Sarkanen has suggested the name "glycolignin" [22]. The hemilignins, which, like the glycolignin, are composed of phenylpropane units, are probably identical with the "cell wall lignin" demonstrated by Whiting and Goring [29].

Glycolignin — a Regular Polymer

By heating wood with an acidic solution of bisulphite, both the hemilignins and the glycolignin macromolecules are sulphonated and the hemilignin sulphonic acids are dissolved. The glycolignin is hydrolytically fragmented into glycolignin sulphonic acids of varying molecular weights. The bonds between glycolignin and hemicelluloses are simultaneously hydrolysed and the water-soluble glycolignin sulphonic acids are dissolved.

To the knowledge of the present authors no-one has hitherto been able to produce a pure unaltered glycolignin hydrolysed from its bonds with the hemicelluloses. The preparations available which are closest to glycolignin are the high-molecular-weight glycolignin sulphonic acids. These preparations can be prepared in a very pure state.

Glycolignin sulphonic acids have been studied at the Finnish Pulp and Paper Research Institute for a number of years [4–10]. These studies have revealed very surprisingly that glycolignin can be described as a regular polymer composed of identical repeating units, each of which is composed of 18 phenylpropane units. According to these results, the repeating unit is a dense, crosslinked structure about 3 nm wide and 1.4 nm thick. Linked together, the repeating units form a lamella. Fig. 11 shows the structure of the repeating unit in two dimensions. A space filling model of the repeating unit is shown in Fig. 12.

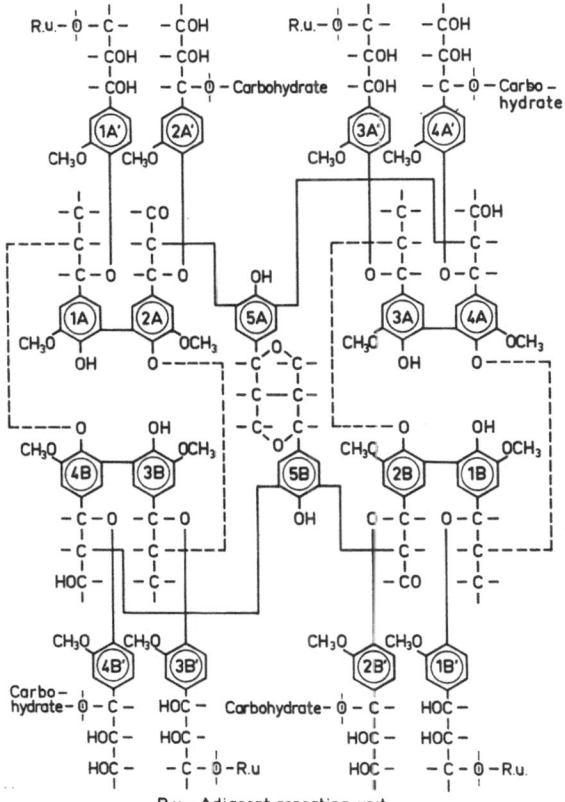

Figure 11. Repeating unit of spruce lignin (8)

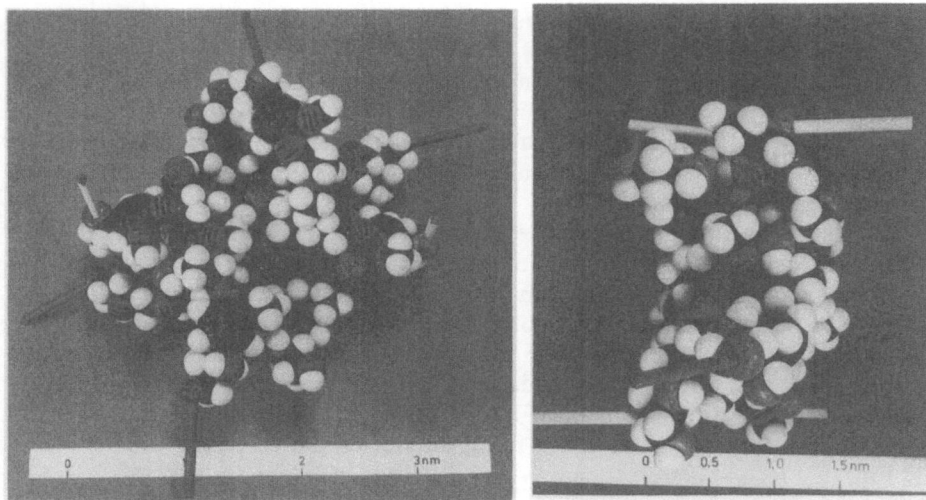

Figure 12. Three-dimensional model of the lignin repeating unit. The four dark rods show the bonds to adjacent repeating units, the four light rods (two on each side) show the bonds to carbohydrates

Steric Factors in the Enzyme Catalysed Reactions

If the results showing lignin as a regular polymer composed of structurally identical repeating units can be proved to be principally correct, the biosynthesis of lignin from its precursors will need to take place not only under the influence of one single phenol oxidase but also under the influence of hitherto unknown regulating mechanisms. This is obvious from the fact that an ordered molecular structure containing a great number of asymmetric carbon atoms in combination with lack of optical activity requires symmetry elements in the molecule.

That some sort of regulating mechanism is present in the enzymatic polymerization is indicated by the fact that coniferous wood contains a group of phenolic compounds called lignans made up of two phenylpropane units linked by the beta carbon atoms of the side chains. Their biosynthesis undoubtedly involves oxidative coupling of two phenylpropane units. One such lignan is pinoresinol the formation of which is shown in Figs. 6 and 7. However, whereas oxidative coupling *in vitro* leads to equal amounts of (+)- and (−)-pinoresinol, only (+)-pinoresinol is formed *in vivo*. The same applies to the other naturally occuring lignans, which are all optically active. The biosynthesis of the natural products must therefore involve asymmetric induction at some stage [26].

In a recent study Ulla Westermark [27] drew attention to steric considerations in the polymerization of lignin. According to its accepted mode of reaction the enzyme requires direct contact with its substrate. To form the lignin polymer the enzyme thus has to be localised within the carbohydrate skeleton of the cell wall already formed, and for every coupling it must dehydrogenate both the growing polymer and the monomer. The limited space within the carbohydrate skeleton casts doubt on whether it is possible for an enzyme to act within the cell wall. Westermark reported that an enzymatically-generated superoxide radical in aqueous solution can act as a single electron oxidant of coniferyl alcohol, giving the same products as phenol oxidases. This means that a superoxide-generating enzyme can be localized outside the cell wall and produce a radical which can diffuse into the cell wall and achieve lignification there.

Whereas the cell wall may constitute steric hindrance to polymerization at the beginning of the lignification, the bulky and dence lignin macromolecule may itself constitute similar hindrance at a late stage, as well as in the enzymatic degradation of lignin.

It has been known for years that phenol oxidase producing white-rot fungi decompose lignin in plant tissue (cf. 19) but in the theories concerning the formation of lignin the enzyme catalyzed degradation have not been taken into account. In 1970 Sundman and Selin [23] showed that the utilisation of ligno-sulphonates by microorganisms is accompanied by polymerization to dark components of high molecular weight, and in 1976 Hiroi, Eriksson and Stenlund [18] studied the influence of the white-rot fungus *Pleurotus ostreatus* on purified lignosulphonates with narrow molecular weight distributions. The results showed that polymerization and degradation occurred at the same time. When the experiments were performed in the absence of cellulose the polymerization was very extensive, although degradation also occurred. It was found that low-molecular-weight lignosulphonates were more rapidly degraded than the high-molecular-weight lignosulphonates. However, in the presence of cellulose the degradation rate was higher and the lignosulphonates were degraded at the same rate irrespective of their molecular weight. The increased degradation of lignin in the presence of cellulose is explained by the finding of Westermark and Eriksson [28] according to which cellobiose, which is formed by enzymatic hydrolysis of cellulose in the presence of cellobiose: quinone oxidoreductase, reduces the quinones formed, thereby preventing the polymerization.

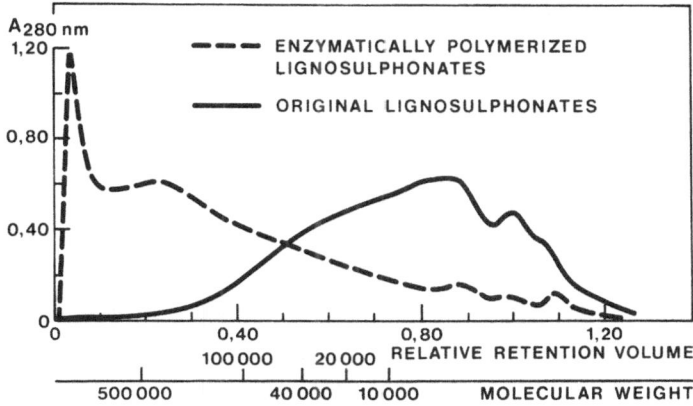

Figure 13. Enzymatic polymerization of lignosulphonates by *Trametes hirsutus*

The observation that the lignosulphonates were degraded at the same rate irrespective of their molecular weight may be of principal interest. It seems logical to assume that if lignin is a spherical molecule the degradation rate should vary with the molecular weight — a low-molecular-weight lignin should be degraded more rapidly than a high-molecular-weight lignin derivative. If lignin has a lamellar structure the degradation rate should be independent of the molecular weight. However, polymerization experiments with crude extracellular enzymes obtained by filtering off the mycelium of the white-rot fungus *Trametes hirsutus* [11] revealed onyl polymerization but no degradation as can be seen from Fig. 13. This experiment, which was performed with the total lignosulphonates from an acid bisulphite cook on spurce wood revealed that the amount of lignosulphonates with molecular weights exceeding 40000 increased during the polymerization due to the polymerization of low-molecular-weight molecular species.

In conclusion it may be said that the enzyme catalyzed reactions of lignin and its precursors may be far more complicated than hitherto considered.

References

1. Adler E (1961) Ueber den Stand der Ligninforschung. Das Papier 15:604–609
2. Cousin H, Hérissey H (1908) Oxydation de l'eugénol par le ferment oxydant des champignons et par le perchlorure de fer; obtention du dehydrodieugénol. Compt Rend 146:1413–1415. Oxydation de l'isoeugénol. Sur le dehydrodiisoéugenol. Ibid 147:247–249
3. Erdtman H (1933) Dehydrierungen in der Coniferylreihe II. Dehydroiisoéugenol. Ann 503:283–294
4. Forss K, Fremer K-E (1964) The dissolution of wood components under different conditions of sulfite pulping. Tappi 47:485–493
5. Forss K, Fremer K-E (1965) The repeating unit in spruce lignin. Paperi ja Puu — Papper och Trä 47:443–454
6. Forss K, Fremer K-E (1975) A structural model for coniferous lignin and its applicability to the description of the acid bisulfite cook. Symposium on Enzymatic Hydrolysis of Cellulose, Aulanko, Finland, 12–14 March 1975, pp 41–63
7. Forss K, Fremer K-E (1976) Application of a structural model of coniferous lignin in the description of the lignin reactions in acid bisulphite cooking. 1976 Canadian Wood Chemistry Symposium. Tech Sect, and the Chemical Institute of Canada, Mont Gabriel PQ, September 1–3, 1976, Extended Abstracts pp 87–92
8. Forss K, Fremer K-E (1981) Some properties of lignosulphonic acids — a stereochemical discussion. The Ekman-Days 1981, International Symposium on Wood and Pulping Chemistry, Stockholm, June 9–12, 1981. Vol 4:29–38
9. Forss K, Fremer K-E, Stenlund B (1966) Spruce lignin and its reactions in sulfite cooking 1. The structure of lignin. Paperi ja Puu — Papper och Trä 48:565–574
10. Forss K, Fremer K-E, Stenlund B (1966) Spruce lignin and its reactions in sulfite cooking II. The reactions in sulfite cooking. Paperi ja Puu — Papper och Trä 48:669–671, 763–676
11. Forss K, Jokinen K, Savolainen M (1982) Work to be published
12. Freudenberg K (1965) Lignin: its constitution and formation from p-hydroxycinnamyl alcohols. Science 148:595–600
13. Freudenberg K, Neish AC (1968) Constitution and Biosynthesis of Lignin. Springer-Verlag, Berlin Heidelberg New York
14. Freudenberg K, Sohns F (1933) Zur Kenntnis des Lignins. Ber 66:262–269
15. Goring DAI (1971) Polymer properties of lignin and lignin derivatives. In: Sarkanen KV, Ludwig CH (eds) Lignins. Occurrence, formation, structure and reactions. Wiley Interscience, New York, p 703

16. Goring DAI (1977) A speculative picture of the delignification process. In: Arthur JC (ed) Cellulose Chemistry and Technology, ACS Symp Ser 48, Washington DC, pp 273–277
17. Harkin JM (1967) Lignin — a natural polymeric product of phenol oxidation. In: Taylor WI, Battersby AR (eds) Oxidative Coupling of Phenols. Marcel Dekker Inc, New York, pp 243–321
18. Hiroi T, Eriksson K-E, Stenlund B (1976) Microbiological degradation of lignin, part 2. Influence of cellulose upon the degradation of calcium lignosulfonate of various molecular sizes by the white-rot fungus *Pleurotus ostreatus*. Svensk Papperstidn 79:162–166
19. Kirk TK (1971) Effects of microorganisms on lignin. Ann Rev Phytopathology 9:185–210
20. Klason P (1897) Om sulfitcellulosaframställningens teori samt om granvedens eteriska olja. Svensk Kem Tidskr 9:133–140.
21. Luner P, Kempf U (1970) Properties of lignin monolayers at the air-water interface. Tappi 53:2069–2076
22. Sarkanen KV (1980) EUCHEM, The Structure and Reactivity of Lignin, Helsinki, June 9–10. 1980, personal communication
23. Sundman V, Selin J-F (1970) Microbial utilization of lignosulfonates of various molecular sizes. Paperi ja Puu — Papper och Trä 52:473–479
24. Tiemann F (1875) Ueber Coniferylalkohol, das bei Einwirkung von Emulsin auf Coniferin neben Traubenzucker entstehende Spaltungsprodukt, sowie Aethyl- und Methylvanillin, Ber 8:1127–1136
25. Tiemann F, Mendelsohn B (1975) Bestandteile des Holzteerkreosots. Ber 8:1136–1144
26. Weinges K, Spänig R (1967) Lignans and cyclolignans. In: Taylor WI, Battersby AR (eds) Oxidative Coupling of Phenols. Marcel Dekker Inc, New York, p 353
27. Westermark U (1982) Involvement of superoxide radical and divalent cations in lignification of plant cell walls. The Ninth Cellulose Conference at the State University of New York, College of Environmental Science and Forestry, Syracuse, New York May 24–27, 1982
28. Westermark U, Eriksson K-E (1974) Cellobiose: quinone oxidoreductase, a new wood-degrading enzyme from white-rot fungi. Acta Chem Scand B 28:209–214
29. Whiting P, Goring DAI (1981) The morphological origin of milled wood lignin. Svensk Papperstidn 84:R120–R122

Application of Enzymes to Organic Synthesis

Otto Andresen and Poul B. Poulsen

Introduction

Historically, chemistry has been divided into two areas, organic and inorganic chemistry. Organic chemistry was concerned with the chemicals which could be found in living organisms. It was believed that these substances possessed a special vital element making it impossible to prepare organic chemicals from inorganic material.

This was disproved for the first time in 1828 when Wöhler succeeded in converting ammonium cyanate, an inorganic compound, to urea which is indisputably organic.

The following years saw a gradual rejection of the established opinions on what characterizes an organic chemical. More and more organic chemicals were prepared "synthetically" and, each time, it could be shown that the synthetic product had the same effect on living organisms as the corresponding product isolated from natural sources. As a result, organic chemistry was redefined to be the chemistry of carbon compounds (Fig. 1).

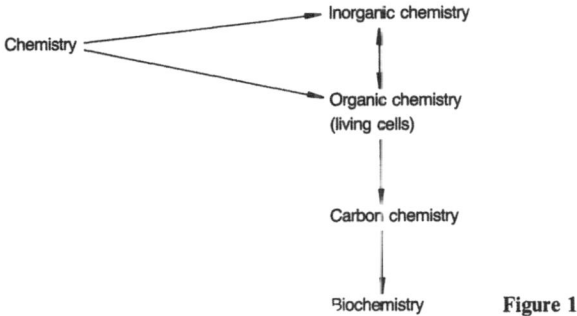

Figure 1.

With time, an extremely large number of organic compounds with no immediate connection to living organisms were described. It was natural to regard the part of organic chemistry which is directly connected with life as a separate science. This became known as biochemistry.

The processes used by the pioneers of organic chemistry were, in principle, the same as those used in inorganic chemistry. Harsh reaction conditions were frequently employed. These include high temperatures and pressures as well as aggressive chemicals. Any living organism would rapidly be killed by such conditions. In contrast, living organisms can produce the same or similar chemical products, but under much milder reaction conditions.

The key to this difference was found to be the highly specialized catalytic function of enzymes.

Enzymes for technical synthesis

In the following sections, some examples of technical synthesis of organic chemicals by enzymatic processes will be discussed. Some of these examples have already been put into practice, others are still at the development stage. The term "synthesis" is used in its broadest sense, namely conversion of one chemical to another. Production of organic chemicals by fermentation will not be considered, nor will processes that have academic interest only.

Enzymatic or non-enzymatic

When should enzymes be considered for use in organic synthesis and when avoided? This depends on the special advantages and disadvantages associated with enzymes. Enzymes have the advantage of being able to differentiate between chemicals of closely related structures, e.g. between stereoisomers, and of being effective at low temperatures and at neutral pH values. This gives fewer byproducts, a more simple purification process, and an improved quality of the final product. Furthermore, enzymes are non-toxic and readily degradable (non-polluting), and they can be made in unlimited quantities (Fig. 2).

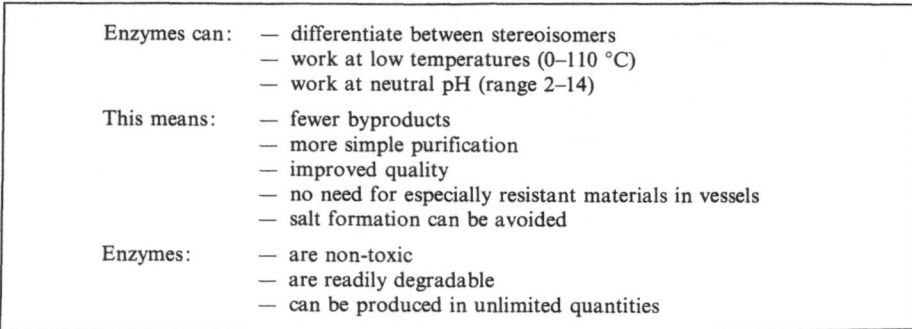

Figure 2. Advantages of enzymes in organic synthesis

Enzymes can often replace strong acids and bases. This means that use of especially resistant materials in reaction vessels can be avoided which saves money. It also means that the presence of large amounts of salts which otherwise would have to be removed after the reaction can be avoided. This can contribute both to cost savings and to a reduction of environmental pollution.

```
Enzymes are fragile and sensitive to:
— high temperatures
— extreme pH-values
— aggressive chemicals
— certain metal ions
```

Figure 3. The limitations of enzyme use in organic synthesis

Enzymes have their limitations in the relatively fragile nature of the amino acid building blocks, and the even more fragile tertiary structure of the protein molecule. These factors make enzymes sensitive to high temperature, extreme pH values, aggressive chemicals, and in some cases to organic solvents (Fig. 3).

It will not be possible to apply enzymes to reactions involving chemicals that denature proteins, i.e. destroy their tertiary structure. Also, the presence of chemicals that inhibit the catalytic properties of the enzymes must be avoided. There are a number of processes, particularly in the petrochemical industry, which are so highly developed that one cannot expect further improvement by use of enzymes. Finally, the cost of enzymes will exclude their application in process steps that function well with cheap chemicals.

Now let us look at some examples which can demonstrate the possibilities of using enzymes in organic synthesis.

Examples 1 and 2: Semisynthetic penicillins

Production of semisynthetic penicillins by means of enzymes instead of organic chemicals is perhaps the oldest and most wellknown example.

All penicillins consist of a 6-aminopenicillanic acid group (6-APA) combined with different side chains.

The structure and type of side chain exert a strong influence on many of the most important properties of the penicillin, such as antibiotic spectrum, resorption characteristics, binding to tissue and thus also serum concentration as function of dosage, and acid stability.

The only penicillins which can be produced in high yields by fermentation are Penicillin-V (phenoxyacetyl-6-APA) and Penicillin-G (phenylacetyl-6-APA). (Fig. 4). However, microorganisms resistant to these two penicillins have always existed and more and more resistant mutants have appeared. In order to expand the antibiotic effect of penicillin towards the resistant strains, the V and G side chains are removed and new side chains are coupled to the 6-APA. Several semi-synthetic penicillins, in particular ampicillin and amoxycillin, have proved to be effective towards the former resistant strains.

The V and G side chains can be removed both by an enzymatic process and by a chemical process (Fig. 5). At least one of the organic processes consists of protecting the carboxylic group as a silylester and then form an iminoether as an intermediate step in a butanol solvent at —40 °C. The yields are high but the process costs for the low temperature, the regeneration of the protecting groups, and the solvent are high too.

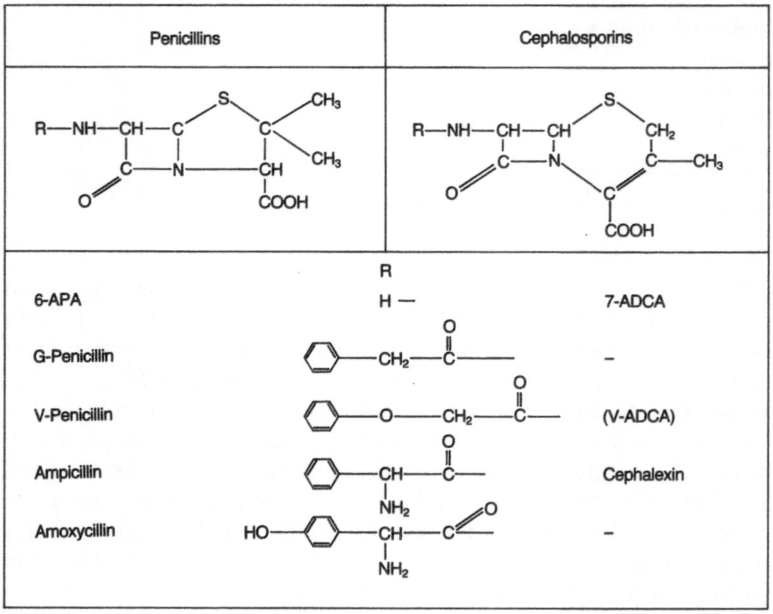

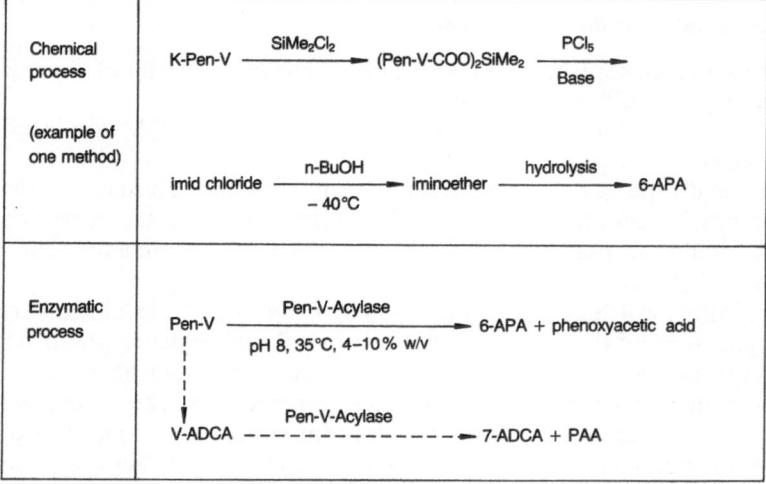

Figure 4. Beta-lactam antibiotics

Figure 5. Conversion of Pen-V to 6-APA

Since 1960, attempts have been made to convert especially Pen-G to 6-APA by an enzymatic route. From around 1966, the Penicillin-G-Acylase has been immobilized, which improved the process considerably, and the enzymatic route is now the one generally used in the industry. The annual production of 6-APA by means of enzymes has now exceeded 3000 tons.

The enzymatic process is used at a temperature of 35 °C and is carried out in water solutions. The typical productivity of the "Novo Semacylase" will be around 250–500 kg 6-APA produced per kg immobilized enzyme.

The enzyme is interesting for another reason; it is specific towards the side chain only. That means that a Pen-G-Acylase will only hydrolyze Pen-G and not Pen-V, and vice versa with the Pen-V-Acylase. This also means that, if the V-chain for example is coupled to another group than 6-APA, then the enzyme can be used also in this process. This is expected to have practical consequences in the near future for the hydrolysis of the cephalosporins V- or G-ADCA to 7-ADCA.

If we assume that we have produced the 6-APA and 7-ADCA, then new problems will arise. First, we need a new side chain, secondly, we have to couple the side chain to 6-APA or to 7-ADCA in order to obtain the potent anti-biotic.

If the object is to produce e.g. ampicillin or cephalexin, then we will have to use the side chain D-phenylglycine. It has been found that the D-form coupled to either 6-APA or 7-ADCA is much more active than the L-form.

So we have a stereospecificity problem because, when phenylglycine is synthesized, it will normally consist of a racemic mixture of 50 % D-phenylglycine and 50 % L-phenylglycine.

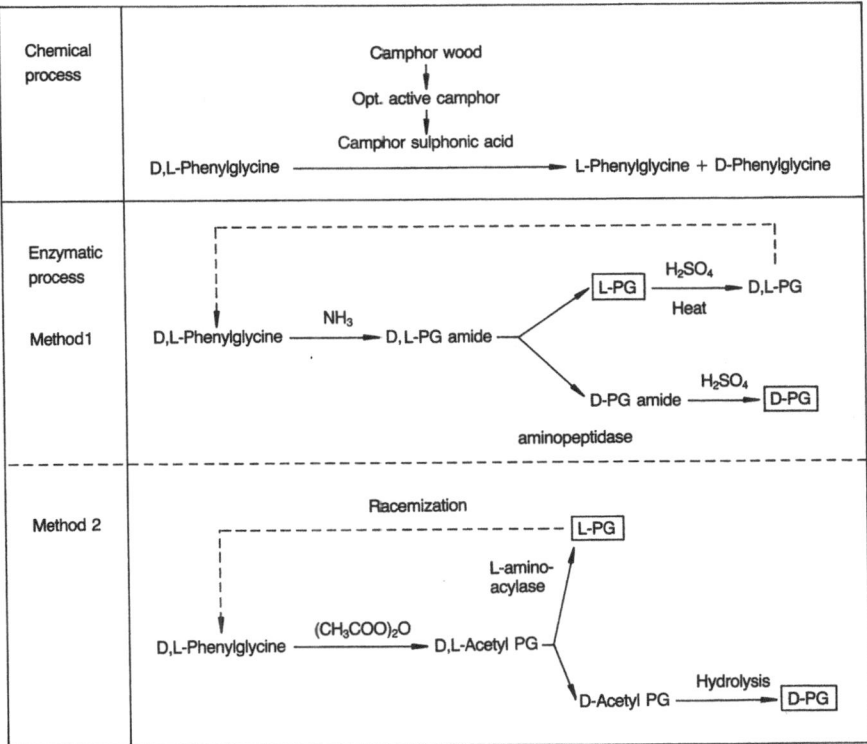

Figure 6. Conversion of D,L-phenylglycine to D-phenylglycine

Up to now, these isomers have only been produced industrially by means of a chemical process utilizing camphor sulphonic acid. This acid must be prepared from optically active camphor, which again is extracted from the wood of camphor trees, and the supply is thus both variable and uncertain.

Several enzymatic process possibilities seem to exist for the process engineer.

One process is the following, carried out by means of the enzyme aminopeptidase:

The D- and L-phenylglycine is converted to the amide forms. The D- and L-phenylglycinamides are then treated with the aminopeptidase (around 40 °C and in water), which stereospecifically only hydrolyzes the L-isomer of the phenylglycinamide to the free L-phenylglycine and ammonia. The D-phenylglycinamide can then be separated from the L-phenylglycine by traditional chemical means. Finally, the pure D-phenylglycinamide is hydrolyzed with sulphuric acid to the desired product, D-phenylglycine. The unwanted L-phenylglycine is racemized by heating and then recycled.

The aminopeptidase can also be used to produce isomers of other amino acids.

Another process is to utilize the enzyme amino acylase: The racemic mixture of the preferred amino acid is acylated. The enzyme amino acylase will then again stereospecifically split only the N-acetyl-L-amino acid which is racemized and recycled. The N-acetyl-D-amino acid is chemically deacetylated and the D-amino acid isolated.

Some proteases can also be used to resolve racemic acylated aminoesters.

The final step in the production of semisynthetic penicillins and cephalosporins is to couple the amino acid isomer with the nuclei, 6-APA or 7-ADCA. This process can in principle utilize the same splitting enzyme but under changed conditions.

It has been found that a Pen-G-Acylase will split Pen-G at pH 7.5, but that it will synthesize Pen-G at e.g. pH 5.0. However, none of these enzymatic routes is used industrially at the present time as we still do not have any acylases which are specific for the side chains in question (e.g. D-phenylglycine).

Example 3: Aspartase

Production of aminoacids is also done industrially using of enzymes. For example, an immobilized aspartase is used for the production of L-aspartic acid. In contrast to the processes mentioned above, this process involves a synthesis step in the traditional sense of the word, namely coupling of ammonia to fumaric acid (Fig. 7). This results in production of only the L-form of the amino acid. A different enzyme can then be used, if desired, to decarboxylate the aspartic acid, yielding the amino acid alanine.

Example No. 4: Nitrilase

Another interesting enzyme is nitrilase. The hydrolysis reactions of nitriles are illustrated in Fig. 8.

In the traditional acid or base catalyzed hydrolysis, it is normally not possible to stop the reaction at the amide stage. Before hydrolysis of the nitrile to the

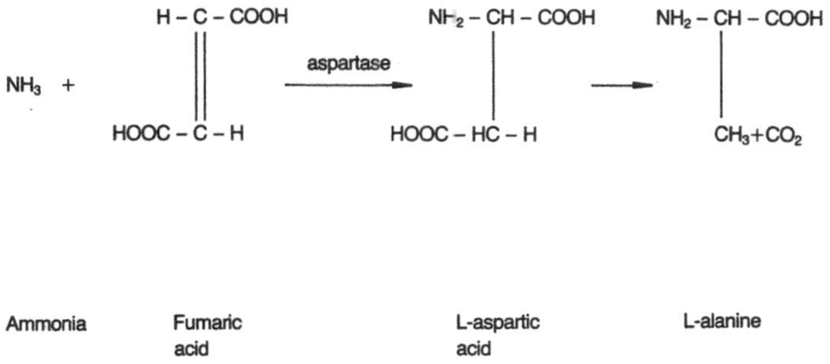

$$NH_3 \; + \quad \begin{matrix} H-C-COOH \\ \| \\ HOOC-C-H \end{matrix} \quad \xrightarrow{\text{aspartase}} \quad \begin{matrix} NH_2-CH-COOH \\ | \\ HOOC-HC-H \end{matrix} \quad \longrightarrow \quad \begin{matrix} NH_2-CH-COOH \\ | \\ CH_3+CO_2 \end{matrix}$$

Ammonia	Fumaric acid		L-aspartic acid	L-alanine

Figure 7. Enzymatic synthesis of L-aspartic acid and L-alanine

amide is complete, part of the amide will be hydrolyzed further to the carboxylic acid. Thus, in order to prepare an amide from a nitrile, it is often necessary to make an undesirable detour round the carboxylic acid and ester forms.

The crude preparations of nitrilases will also normally hydrolyze all the way to the carboxylic acid. However, a mutant of a nitrilase-producing organism which has lost the enzyme that catalyzes the last step of the nitrilase reaction has now been found.

The remaining enzyme, which more correctly should be called a nitrile hydratase because it catalyzes the uptake of one molecule of water, can catalyze the single reaction step from nitrile to amide. Thus, the four steps required by conventional chemical methods (Fig. 8) are avoided.

It should be mentioned that (expensive) inorganic catalysts have now been developed that can be used to produce acrylamide from acrylonitrile by a similar

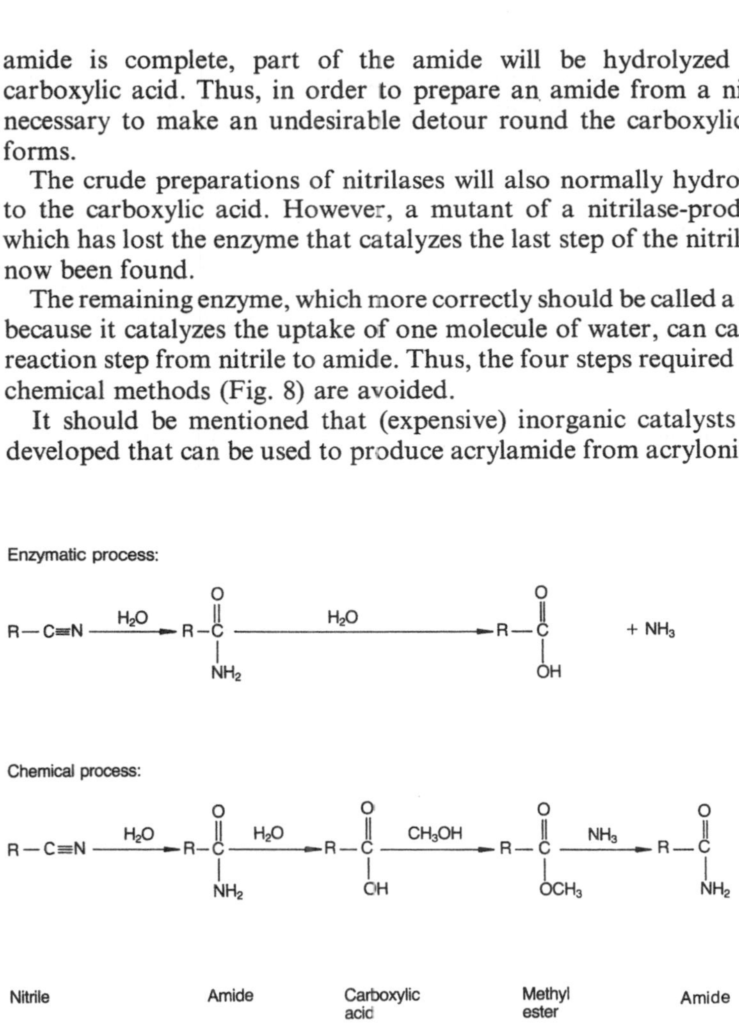

Enzymatic process:

$$R-C\equiv N \xrightarrow{H_2O} R-\overset{\overset{\displaystyle O}{\|}}{\underset{\underset{\displaystyle NH_2}{|}}{C}} \xrightarrow{H_2O} R-\overset{\overset{\displaystyle O}{\|}}{\underset{\underset{\displaystyle OH}{|}}{C}} \;\; + NH_3$$

Chemical process:

$$R-C\equiv N \xrightarrow{H_2O} R-\overset{\overset{\displaystyle O}{\|}}{\underset{\underset{\displaystyle NH_2}{|}}{C}} \xrightarrow{H_2O} R-\overset{\overset{\displaystyle O}{\|}}{\underset{\underset{\displaystyle OH}{|}}{C}} \xrightarrow{CH_3OH} R-\overset{\overset{\displaystyle O}{\|}}{\underset{\underset{\displaystyle OCH_3}{|}}{C}} \xrightarrow{NH_3} R-\overset{\overset{\displaystyle O}{\|}}{C}$$

Nitrile	Amide	Carboxylic acid	Methyl ester	Amide

Figure 8. Conversion of a nitrile to amide and carboxylic acid

one-step process. Nevertheless, the Japanese chemical corporation, Nitto Chemical Industry, will introduce an enzyme-catalyzed process for the production of this amide. Their process gives an important energy saving as the reaction temperature is reduced from 120 °C to below 10 °C. As a result of this low temperature and the use of an immobilized enzyme, this process gives, according to Nitto, a product that is so pure that it can be polymerized without subsequent purification.

Example No. 5: Urease

This enzyme, which hydrolyzes urea to ammonia, was developed for use in purification of waste water from urea factories. Actually, it does not belong to the group of synthesis enzymes, but urease may also find use within the organic chemical industry.

If a plant has a urea disposal problem, then a possible way to solve it is by a chemical treatment which can reduce the urea concentration to around 50 ppm; however, this creates another problem: formation of isocyanate as a side reaction.

By means of the enzyme urease, urea can be hydrolyzed down to a level around 5 ppm without giving rise to isocyanate problems. A typical enzyme product will contain 10,000 units (each unit decomposing 1 micromole urea/min. at 42 °C, pH 9.2, 0.1 M urea) per gram, which means that 1 g enzyme product can hydrolyze around 0.5 g urea per minute under the above mentioned conditions.

Example No. 6: Esterases

It is striking that the majority of the processes described so far involve hydrolysis of a carbon-nitrogen bond and that the enzymes involved are, in this respect, related to the proteinases. However, only a relatively small part of organic synthesis involves reactions of C—N bonds. It follows that the major part of organic chemistry still is unexplored territory as regards the technical application of enzymes.

One area which is currently receiving a great deal of interest, as revealed by the large number of new patents, is the enzymatic synthesis of esters. These compounds of alcohols and acids are widely distributed in nature and enzymes that can hydrolyze them are of course equally widely distributed.

Enzymes do not differ from other catalysts as their function is to accelerate the rate of attainment of a chemical equilibrium. In the case of the hydrolysis process in question, the equilibrium can be written as follows (Fig. 9):

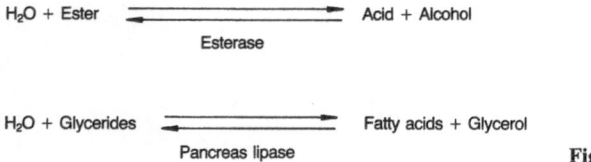

Figure 9. Esterases

The equilibrium depends on the concentration of the reacting chemicals. If water is not only a reactant but also the solvent, it will be present in very high concentrations, and this will tend to displace the equilibrium towards the right.

However, by reducing the concentration of water to the minimum necessary to keep the enzyme active and perhaps also by using one of the products as solvent, the equilibrium is displaced to the left.

In 1909, Pottevin illustrated this principle by showing that a low concentration of water encouraged the synthesis of glycerides (esters of fatty acids and glycerol) by pancreas lipase whereas the reaction is reversed when the water concentration is increased.

A number of companies have recently patented similar processes aimed at synthesizing fats that are "tailor made" for particular purposes. The patents have been concentrated on a replacement for cocoa butter. This particular fat is in heavy demand and is therefore expensive. It has a sharp melting point at a temperature just below that found in the mouth.

Again, enzymes offer the unique advantage of specificity, in this case with regard to the positions on the glycerol molecule where the fatty acid should be replaced.

Enzyme catalyzed synthesis is also indicated for esters whose component alcohols and acids are unstable in the presence of acids or at elevated temperatures and thus would be decomposed during the traditional ester synthesis.

The idea of reversing an enzyme catalyzed hydrolysis by lowering the concentration of water can also be applied to the proteinases. In this way, proteins and peptides can be synthesized from amino acids, or existing proteins can be modified. This principle is used in "Novo's" method for conversion of porcine insulin to human insulin and in the synthetic method developed by the Japanese company Toyo Soda for production of an artificial sweetener, the dipeptide Aspartame.

Other reactions

Other interesting areas, where a need may appear for the enzymes' specificity and their ability to facilitate reactions of low temperature, are oxidations and hydroxylations.

The future

What can be achieved with enzymes in organic synthesis? Apart from the limitations already mentioned, there appear to be fine prospects in the field.

The applications of enzymes are not limited to the compounds that are found in living organisms or take part in reactions associated with life, e.g. some plastics can be decomposed by certain kinds of bacteria, albeit slowly.

Large problems can, however, be foreseen if a synthetic step requires energy. In living organisms, cofactors are involved in the transfer of chemical energy from one substance to another; energy that originates as solar energy trapped by the photosynthetic processes of plants.

In spite of considerable efforts, cofactor mediated enzyme syntheses have not yet become a commercial reality. The reason is that cofactors are expensive to produce, and that they, being co-substrates, are consumed during the reaction. This makes regeneration of the cofactor necessary. Cofactor regeneration processes have so far been too expensive except in the case of the most costly speciality chemicals, such as steroids for use in contraceptives.

One method that is judged to have a good chance of becoming economically acceptable is application of immobilized whole microorganisms whose metabolism is sufficiently intact after the immobilization to ensure regeneration of cofactors when supplied with the appropriate nutrients.

There are still many practical difficulties to be overcome, in particular concerning transport of reactants and products in and out of the immobilized biocatalysts, the stability of the semi-living system and the risk of infection. It seems likely that a number of years will elapse before an enzyme-catalyzed synthesis of energy-rich compounds from energy-rich raw materials becomes a commercial reality.

— Increasing demand for enviromental protection
— Increasing demand for safety at work
— Increasing prices for energy and oil-based raw materials
— Increasing scarcity and prices of natural raw materials
— Less by-products means better yields

Figure 10. Factors creating a demand for enzymatic processes

The increasing demand for environmental protection and safety at work, increasing prices for energy and petrochemical based raw materials, increasing scarcity and increasing prices of natural raw materials are among the most important factors that will create a demand for new enzymatic processes (Fig. 10). Established chemical synthesis methods can quickly become unacceptable if, for example, it is shown that the product contains toxic or carcinogenic impurities, formed because of the process conditions. This could also create the need for the specific catalysis provided by enzymes.

It will be demands arising for these reasons combined with the continuous advances in biotechnology which eventually will dictate when and to what extent enzymes in large scale will be applied to organic synthesis.

When it is realized by the organic chemical industry that enzymes in many cases are able to solve the problems just mentioned in a profitable way, there is little doubt that this industry will join the starch industry and the detergent industry as a major consumer of technical enzymes.

Preparation of Isotopically-Labeled Amino Acids with L-Methionine γ-Lyase

Kenji Soda, Nobuyoshi Esaki, Hidehiko Tanaka and Seiji Sawada

Summary

L-Methionine γ-lyase catalyzes the exchange of α- and β-hydrogens of L-methionine and S-methyl-L-cysteine with deuterium or tritium as solvents. The rate of α-hydrogen exchange with deuterium was about forty times faster than that of the elimination reactions. The deuterium and tritium were exchanged also with the α- and β-hydrogens of the straight-chain amino acids which do not undergo the elimination: L-alanine, L-α-aminobutyrate, L-norvaline, and L-norleucine. These enzymatic hydrogen-exchange reactions facilitate specific labeling of the L-amino acids with deuterium and tritium.

Introduction

Specifically labeled amino acids are useful for the study of the mechanism of various chemical reactions and biological processes. The labeled amino acids, however, are not easily available. We have purified L-methionine γ-lyase (EC 4.4.1.11) to homogeneity from *Pseudomonas putida* to elucidate its physicochemical and enzymological properties (Tanaka et al., 1976; 1977). Here we describe the simple methods for specific labeling of various L-amino acids with ^{2}H and ^{3}H. These methods are based on the hydrogen exchange reactions catalyzed by L-methionine γ-lyase.

Materials and Methods

L-Methionine γ-lyase was purified to homogeneity from a cell-free extract of *Pseudomonas putida* as described previously (Tanaka et al., 1976). The enzymatic elimination reactions were followed by determination of α-keto acids formed with 3-methyl-2-benzothiazolone hydrazone HCl (Soda, 1968). The exchange of substrate proton with solvent deuteron was conducted in an NMR tube containing 100 µmol of the substrate amino acid, 100 µmol of potassium phosphate buffer (p^{2}H 8.6), and enzyme in 0.5 ml of ^{2}H$_2$O.

Results

Deuterium-labeling of methionine

We observed the ^{1}H-NMR spectral change of L-methionine during incubation with L-methionine γ-lyase in ^{2}H$_2$O. Peaks of the α and β protons were lost with time, and the triplet of γ protons was transformed into a singlet. This shows that the α proton and both the β protons were exchanged with deuterium of the solvent. The α-proton exchange occurred about 40 times faster than the α,γ-elimination. Thus, deuterium-labeled methionine can be prepared effectively by the present enzymatic method.

Deuterium-labeling of the other amino acids

When S-methyl-L-cysteine was used as a substrate, the α and β hydrogens were completely exchanged with the deuterium of the solvent. These hydrogen exchanges proceeded much faster than the α,β-elimination in a similar manner to that of L-methionine. A series of straight-chain amino acids that are not substrates for elimination and replacement reactions also underwent the α- and β-hydrogen exchange. None of the following amino acids were susceptible to the enzyme: the D-isomers of the above amino acids, and the L-isomers of α-methyl-methionine, N-acetylmethionine, acidic, basic and branched-chain amino acids. However, glycine, L-tryptophan, and L-phenylalanine slowly underwent the α-hydrogen exchange.

Tritium-labeling of amino acids

Attempts were made to label amino acids with tritium by the action of L-methionine γ-lyase. After incubation in ^3H_2O, amino acids were isolated and crystallized. Specific radioactivity of the amino acids isolated was determined as shown in Table 1. The labeled positions of substrates were analyzed with an amino acid racemase with low substrate specificity (Soda and Osumi, 1971) which catalyzes the exchange of the α-hydrogen of various amino acids with deuterium or tritium of the solvents, but not β-proton exchange (Soda et al., 1976).

Table 1. Specific Radioactivity of Tritiated Amino Acids Prepared

Amino Acids	Specific Radioactivity (dpm/μmol)
L-Methionine	8,450
D-Methionine	150
S-Methyl-L-cysteine	8,600
Glycine	3,900
L-α-Aminobutyrate	6,940
L-Norvaline	8,000
L-Norleucine	7,540

The amino acids tritiated with L-methionine γ-lyase were incubated with amino acid racemase in ^1H_2O. About one third of the radioactivity was lost during incubation except for glycine, in which all the radioactivity was lost. These results indicate that L-[α-^{3}H, β-$^3H_2$]amino acids prepared with L-methionine γ-lyase were converted into DL-[β-$^3H_2$]amino acids by incubation with amino acid racemase in ^1H_2O.

Reaction mechanism

In Fig. 1, a mechanism is proposed for L-methionine γ-lyase-catalyzed incorporation of deuterium (or tritium) of solvent into the α and β positions of various amino acids. After transaldimination from the enzyme-coenzyme Schiff base to the coenzyme-substrate Schiff base, the α hydrogen is abstracted by a basic side chain at the active site, and an α-carbanion is formed. The hydrogen attached to the base is probably exchanged with a solvent deuterium (or tritium) rapidly, and an amino acid species deuterated (or tritiated) in the α-position is labelated through the reversal of the process. The β-hydrogen exchange probably occurs through a reversible tautomerization of the ketimine to the enamine.

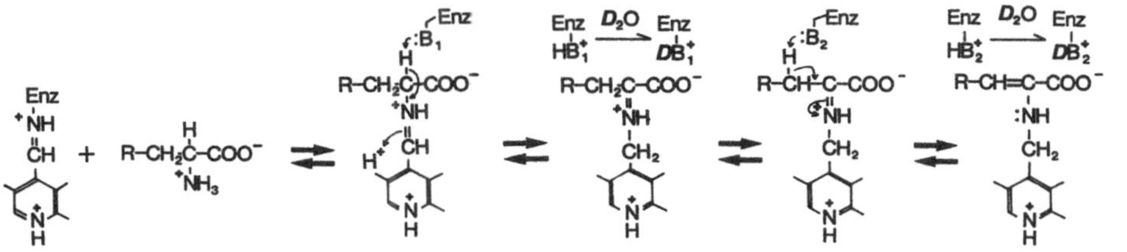

Figure 1. Proposed mechanism of L-methionine γ-lyase-catalyzed incorporation of denterium (or tritium) into the α and β positions of L-amino acids

References

Soda K (1968) Anal Biochem 25, 228–235
Soda K and Osumi T (1971) Methods in Enzymology, vol 17B (Tabor H and Tabor CW eds), Academic Press, New York, pp 629–636
Soda K, Ohmori Y, Yagi T and Sawada S (1976) The Abstract of the Annual Meeting of Agricultural Chemical Society of Japan (kyoto), p 404
Tanaka H, Esaki N, Yamamoto T and Soda K (1976) FEBS Lett 66:307–311
Tanaka H, Esaki N and Soda K (1977) Biochemistry 16:100–106

V. Immobilized Enzymes and Cells: Application

Use of the Enzyme Thermistor as a Flow Analyzer in Biotechnology

Bengt Danielsson

Summary

A continuous-flower analyzer, enzyme thermistor, based on the use of immobilized enzymes in a simple flow calorimeter has been applied in various types of analysis in biotechnology. The applications reported include: enzymatic determination of metabolites in bioreactors, continuous monitoring of metabolites, recording of thermograms of microorganisms and monitoring of enzymic activity, especially in chromatographic processes.

Introduction

Calorimetry has long been recognized as a valuable tool in biochemical analysis because of its general applicability — almost all processes produce or consume heat. Especially enzymes can be conveniently studied with this technique since most enzymatic reactions are accompanied by considerable heat evolution (in the range of 5–100 kJ/mol) (Spink and Wadsö, 1976). More extensive use of calorimetry has been hampered by the cost and complexity of the conventional apparatus and by time-consuming procedures. In recent years, however, cheaper and less sophisticated instruments have emerged, involving simpler and quicker procedures for routine analysis (Grime, 1980).

The combination of such simple devices with immobilized enzymes appears particularly attractive, as apart from the specificity and the obvious possibility of repeated use of the immobilized biocatalyst, the advantages gained by such an arrangement include higher sensitivity, quicker response, probable stanilization of the enzyme and the possibility of applying such analytical devices in continuous flow operations (Mosbach and Danielsson, 1981). The enzyme thermistor (ET) developed in our laboratory measures with a thermistor the temperature changes in a small column containing immobilized enzyme(s) (Danielsson et al. 1981a). Since most enzymic reactions are highly exothermic only one enzyme is usually required to give the ET-analysis sufficient sensitivity. This is in contrast with many other enzymatic techniques, which have involved enzyme sequences of up to four or sometimes even more enzymes in order to give a measurable change in e.g. color or pH. The choice of enzyme for use in ET analysis is, by consequence, simpler and can preclude such enzymes requiring co-factors or other expensive reagents. Moreover, the analysis can be performed on crude samples irrespective of color or turbidity, which is of particular importance in biotechnological applications. Furthermore, the ET-unit can be used for monitoring enzymatic or microbial activity in, for instance, fermentation broths.

Experimental

Apparatus. The equipment required for ET-analysis includes a pump, a sampling valve, the calorimeter unit, a Wheatstone bridge including temperature control for the calorimeter, and a recorder. A schematic drawing of the calorimeter is shown in Fig. 1. A detailed description of the system was recently given by Danielsson et al. (1981a). The sample/buffer is pumped through the ET-unit with a peristaltic pump at a flow rate of 0.5–3 ml/min. The fluid first enters the

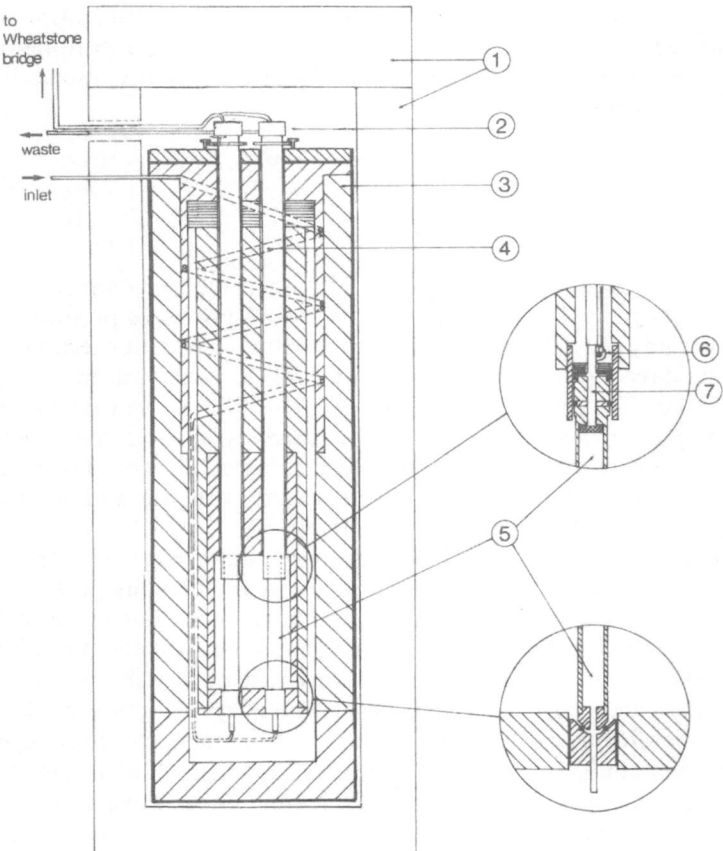

Figure 1. Enzyme thermistor with aluminium constant temperature jacket. (1) Polyurethane insulation. (2) Plexiglas tube with bayonet lock for column insertion. (3) Thermostated aluminium cylinder, height approximately 250 mm. (4) Heat exchanger. (5) Enzyme column.(6) Thermistor attached to a goldcapillary. (7)Column outlet

heat exchanger tube that is mounted between two concentrically arranged, thermostatically controlled (usually 30.0 °C) and well insulated aluminium cylinders, which provide a constant temperature environment for the enzyme column (0.2–1 ml). The temperature at the outlet of the column is continuously monitored by a thermistor connected to a sensitive Wheatstone bridge. At the highest amplification, the recorder output is 100 mV for a temperature change of 1 m°C (10^{-3} °C). For most measurements, a full- scale sensitivity of 10–20 m°C is adequate and permits determination of substrate concentrations around 0.1 mM within $\pm 1\%$ for reactions with high enthalpy (—80–100 kJ/mol). There are two parallel fluid lines which could be used either independently or with one of them as a reference.

Procedure. Sample volumes of 0.25–1 ml can be introduced with a chromatography injection valve. Thermal steady state will not be obtained for short sample pulses, but the enzymic reaction will generate a temperature peak, the height of which is normally taken as a measure of substrate concentration. The linear range of this relationship is often quite wide, typically 10^{-5}–10^{-1} M, when not limited by reactant concentrations. The area under the temperature peak or the initial slope of the peak are also useful measures of substrate concentration (Danielsson et al. 1981a).

If the pulse length is sufficiently increased (>1–5 min.) the temperature response will eventually reach a constant value, proportional to the concentration. Consequently, the ET can be used for continuous monitoring of the substrate level.

When a large number of samples is to be analyzed, we prefer to use an automatic sample changer controlled by a desk-top computer which can also be used for calculating concentrations based on recordings of ΔT, slope or area. The number of measured samples per hour depends on the flow rate and the sample volume, and is 10–15 for sample volumes of 0.5–1 ml. This figure can be increased to 60 if the sample volume is reduced to 0.1 ml and the flow rate increased to 2–3 mL/min.

Enzyme Column. The enzyme column is inserted into the apparatus with a plexiglas tube containing the outlet tubing and the temperature sensor. Change of columns is thereby a simple and rapid procedure. Plastic columns of different diameters (<8 mm i.d.) and bed heights (<30 mm) can be used. A commonly used column type is 4 mm i.d. with a bed height of 20 mm (0.25 ml). Nylon tubing wound around a special adaptor can also be used as enzyme support. Nylon tubing is especially useful with crude samples containing particulates, but it suffers from low enzyme loading capacity (Mattiasson et al., 1981). Thus, the normally employed carrier material is CPG (controlled pore glass) which offers high enzyme coupling capacity, good mechanical, chemical and microbial stability/resistance as well as relatively simple coupling procedures. Other materials, such as "Sepharose" and polyacrylamide gel, have also been used (Danielsson et al. 1981a). Normally a large excess of enzyme is applied — typically 100 units, if affordable. This ensures good operational stability, i.e. unchanged performance over long series of samples or long periods of continuous monitoring and an expected column life of several weeks or even months. Consequently, the enzyme preparations employed should have a high specific activity. High specificity of analysis requires highly purified enzymes, since virtually all enzymatic activities are detected by a calorimeter, even undesired activity. Enzymes obtained from e.g. Boehringer or Sigma for analytical purposes are usually quite adequate for the use of the ET.

Applications

This paper will mainly deal with the use of the ET for applications of biotechnological interest. Determinations of metabolites in discrete samples as well as in continuous analysis will be discussed. Some studies on process control based on the ET-signal will be related. The use of the ET-apparatus as a flow calorimeter for monitoring of microbial activity will be described, and finally the determination of enzymatic activities in flow streams will be exemplified by studies on chromatographic processes. Although initial work on the ET and similar thermal enzyme probes focussed on clinical applications, such as determinations of glucose and urea in serum samples, the special features of enzyme calorimetric analysis, specificity and simple procedures directly with crude samples appear even more attractive in biotechnology. In fermentation analysis, for example, there is a lack of techniques for continuous measurement of specific metabolites that could be used for controlling the process. Our studies so far indicate that the ET could doubtlessly help in filling this gap, as has been shown to be the case with carbohydrates. These compounds are of primary concern in most biotechnological processes and have been the subject of many ET studies (Mattiasson and Danielsson, 1982; Danielsson et al. 1981 b). Several ET procedures have been based on the determination of glucose after hydrolysis of, for instance, a disaccharide, but in some cases direct procedures with specific enzymes have been suggested. Glucose can be determined with high sensitivity using glucose oxidase. It is advantageous to coimmobilize this enzyme with catalase since catalase will reduce the oxygen demand, thereby extending the linear measuring range from 0.5 mM up to maybe 0.8 mM. Furthermore the removal of the hydrogen peroxide by catalase will protect the oxidase from detoriation and, finally, the high enthalpy of the catalase reaction (-100 kJ/mol) will approximately double the temperature response. The limit of detection will then be very low, in the μmolar range. The operational stability of the glucose oxidase/catalase system is very high and the same column should last for months if not clogged. This ET assay can be regarded as a fundamental determination and the base for many other coupled reactions, its only drawback being the limited linear concentration range. This drawback, however, is shared with any glucose oxidase based assay. The practical consequence is a demand for extensive dilution (maybe 1000-fold) for biotechnological samples.

The determination of lactose, described by Mattiasson and Danielsson (1982) and that of cellobiose, described by Danielsson et al. (1981 b), represent assays based on a glucose oxidase/catalase thermistor. In these studies lactose was first split into glucose and galactose by lactase and cellobiose into two glucose units by β-glucosidase. The enthalpy of the hydrolysis of a disaccharide is usually low and so the overall performance of these three-enzyme systems is favored by placing the immobilized disaccharidase in a precolumn before the ET,

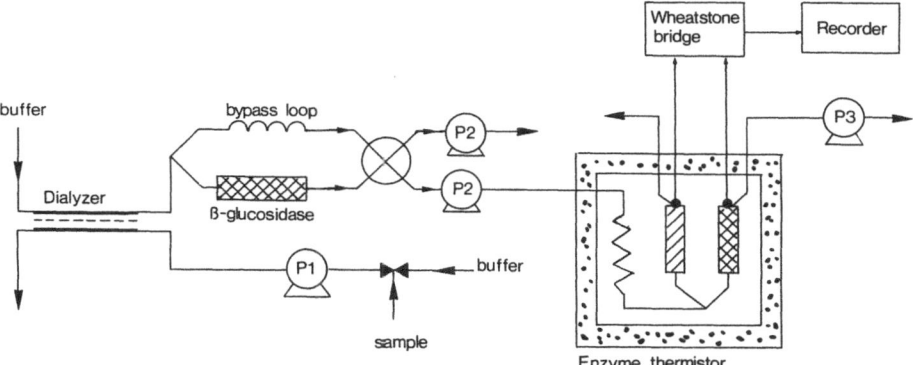

Figure 2. Experimental set-up for determination of cellobiose with a split-flow ET containing an inactive reference column. The peristaltic pump, P3, was adjusted to give the same flow in both columns. The set-up includes an optional Technicon 24-in. dialyzer

thereby not "diluting" the heat produced in the glucose oxidase/catalase column. This arrangement also enables simple determination of background glucose present in the sample by switching the disaccharidase column in and out of the flow system as described by Danielsson et al. (1981 b), see Fig. 2. For lactose a working range of 0.25–5 mM was obtained, while that of cellobiose was 0.1–10 mM. Cellobiose concentrations over 10 mM could be directly determined with a β-glucosidase thermistor.

It is interesting to note that a much larger enthalpy change is evolved in the hydrolysis of sucrose by invertase (Mattiasson and Danielsson, 1982). Therefore, a specific direct determination of sucrose in the range 1–200 mM is possible with invertase as only enzyme. Although moderate, the sensitivity of this assay is sufficient for most biotechnological applications. This is a nice example for the versatility of the ET, since this reaction cannot easily be measured by other techniques such as spectrophotometry or by electrodes. Furthermore, the enzyme is highly stable, which makes continuous sucrose monitoring very reliable (Mandenius et al. 1981). For the determination of oxalate, we have studied two enzymes: oxalate decarboxylase and oxalate oxidase (Danielsson et al. 1981a). The latter enzyme turned out to work better in the ET; it had better stability and a higher response. A study on the use of oxalate oxidase in the ET for analysis of urine as well as various fruit juices was recently finished in our laboratory (Winquist et al., 1983). Oxalate could be directly determined in 5–10-fold diluted urine with just enough sensitivity for normal specimens — the response was linear with concentration over the range 0.01–0.5 mM with a rather high sensitivity (10 m °C/mM). The method appeared quite adequate for measurements of fruit juices and vegetable extracts.

In a recent investigation, Guilbault et al. (1982) compared the enzyme electrode with ET assays for alcohols based on the alcohol oxidase from *Candida boidinii*. The stability of the alcohol oxidase in the ET was dramatically increased by coimmobilizing with catalase. Ethanol, propanol, butanol and (even better) methanol were found to give good responses with both techniques. In many cases, however, ethanol is the predominant component, and then both methods are quite useful alternatives in biotechnological analysis. The good long-term stability of the ET makes it particularly suitable for continuous monitoring. Typical calibration curves for different substrates are shown in Fig. 3.

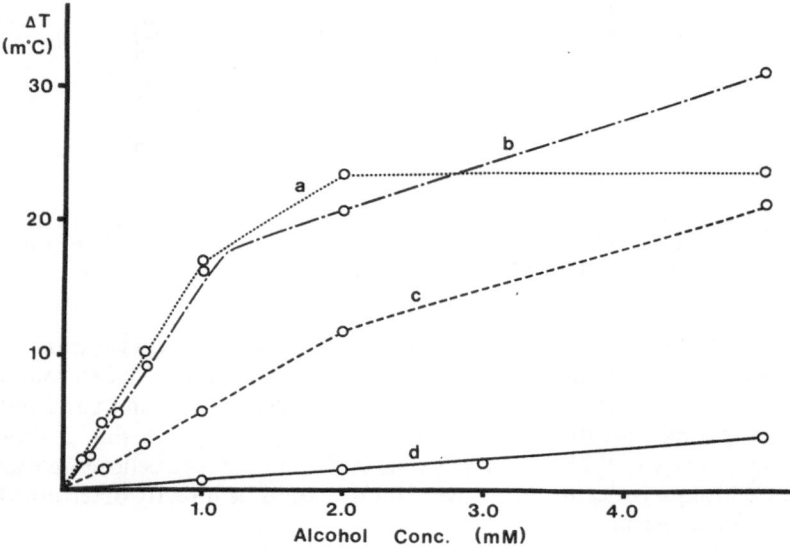

Figure 3. Calibration plots for (A) methanol, (B) ethanol, (C) butanol, and (D) propanol obtained with an alcohol oxidase/catalase column measuring the total heat change. Buffer: 0.2 M sodium-phosphate pH 7.5; Flow rate 0.70 ml/min; Sample volume: 0.5 ml (From Guilbault et al. 1982)

Excellent results have been obtained with ET analysis of penicillins and other antibiotics in fermentation solutions. The concentration range of these substances is suitable for this kind of analysis and the simple sample preparation is a further advantage. One company routinely runs ET assay of penicillin V; 250 samples per day (30 per hour) with only weekly changes of enzyme columns. The β-lactamases used in the ET can be obtained with different specificities, including those having a more pronounced specificity for cephalosporins (Danielsson et al. 1979). Various possibilities for sample treatment and enzyme immobilization for penicillin determination in fermentation broths were described by Mattiasson et al. (1981). Fig. 4 illustrates the different operating ranges of CPG and nylon tubing-bound penicillinase. While the nylon tubing better withstands fouling, its enzyme binding capacity is low and the CPG-column gives higher sensitivity and a wider measuring range.

Several other metabolites have been successfully measured by the ET and similar thermal probes including determination of lactate by lactate oxidase and of triglycerides by lipoprotein lipase. Table 1 is a compilation of such determinations of biotechnological interest.

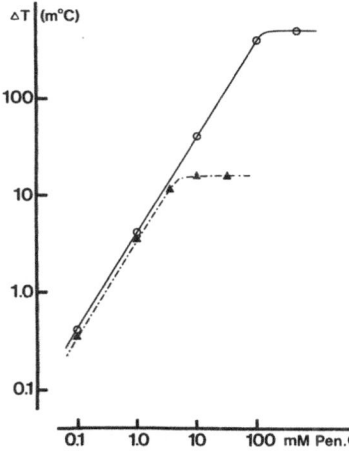

ΔT (m°C)

0.1   1.0   10   100 mM Pen.G

Figure 4. Calibration curves (loglog-scale) for penicillin G. The linear ranges obtained for CPG-bound (O) and for nylon tubing-coupled penicillinase (▲) are compared. (From Mattiasson et al. 1981)

Table 1. Substances analyzed with enzyme thermistors and related thermal probes

Substance	Enzyme	Conc. range (mM)	Reference
Ascorbic acid	Ascorbic acid oxidase	0.05 –0.6	Mattiasson & Danielsson, 1982
Cellobiose	β-glucosidase + glucose oxidase + catalase	0.05 –5	Danielsson et al., 1981 b
Cephalosporin	Cephalosporinase	0.005–10	Danielsson et al., 1979 a
Ethanol	Alcohol oxidase	0.01 –1	Guilbault et al., 1982
Galactose	Galactose oxidase	0.01 –1	Mattiasson & Danielsson, 1982
Glucose	Glucose oxidase + catalase	0.002–0.8	Danielsson et al., 1977 Schmidt et al., 1976
Glucose	Hexokinase	0.5 –25	Bowers & Carr, 1976
Lactate	Lactate oxidase	0.01 –1	Danielsson et al., 1981 a
Lactose	Lactase + glucose oxidase + catalase	0.05 –10	Mattiasson & Danielsson, 1982
Oxalic acid	Oxalate oxidase	0.005–0.5	Winquist et al., submitted
Penicillin G	Penicillinase (β-lactamase)	0.01 –500	Mattiasson et al. 1981
Sucrose	Invertase	0.05 –100	Mattiasson & Danielsson, 1982
Triglycerides	Lipoprotein lipase	0.1 –5	Satoh et al., 1981
Urea	Urease	0.01 –500	Bowers et al., 1976; Danielsson et al., 1976

Continuous Monitoring. As already pointed out, temperature measurements with a thermal probe such as a thermistor are naturally made in a continuous manner and the ET signal is generally related to the substrate level over a very wide range. By applying a large excess of enzyme on columns with good mechanical durabilities, the system allows for good operational stability and gives constant responses for a given substrate concentration over long periods of time. Only when very small concentration changes are to be detected at sensitive measuring ranges (e.g. 10 m °C), it is necessary to recheck the baseline every 4–6 hours. However, this drawback is easily eliminated in an automated system in which a valve can be programmed to introduce a blank solution for baseline control at suitable intervals. If a second valve is included for standard samples, the system can also be automatically calibrated whenever required. In our own system, the computer used for process control can be used for supervising the calibration procedures as well.

In our first study on the use of the ET for process control in a model system, the flow rate through a plug-flow reactor containing β-galactosidase immobilized on Sepharose was regulated to give a constant glucose level in the effluent as measured by a glucose oxidase/catalase ET (Danielsson et al. 1979b). The enzyme thermistor signal was converted by a PID-controller to a control signal for the pump feeding the enzyme reactor. It was shown that variations in the concentration of the incoming lactose solution (whey) as well as changes in the catalytic capacity due to fouling could be compensated for continuously.

Our current studies on process control are focussed on pilot plant reactors for ethanol fermentation. The principal design of one of our model systems is shown in Fig. 5. The ultimate goal of our research is to obtain simultaneous control of the substrate flow to the fermenter and of the ethanol recovery in the effluent in order to maintain the substrate concentration at an optimal level. Another point of interest is the final utilization of the remaining substrate in the effluent by a plug flow reactor containing immobilized yeast of a special type, resistent to higher ethanol concentrations. The main fermenter is a continuous stirred tank reactor loaded with baker's yeast immobilized in alginate beads. The total

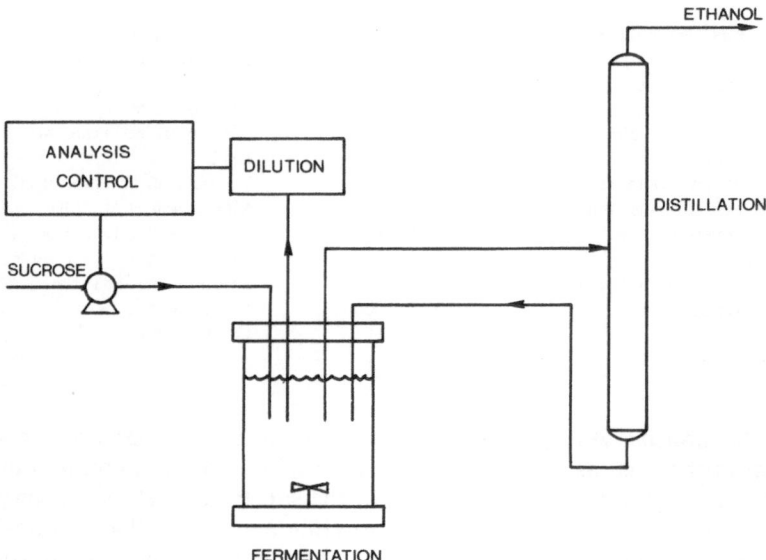

Figure 5. Model system for studying the enzyme thermistor as a monitor and control instrument for fermentation processes. See the text for details

system is operated with a PDP-11 computer (Digital Equipment), while sub-systems have been studied using simpler computers, type hobby computer or even along controllers. Thus, Mandenius et al. (1980) showed that the sucrose feed to a fermenter could be adequately controlled by an invertase thermistor in connection with a PI-controller. The control task was facilitated by the fact that no nutrients were added during the fermentation, resulting in a rather low metabolic activity. In a subsequent study on a more efficient fermenter, Mandenius et al. (1981) found that the simple analog controller sometimes ended up in oscillation. However, decreasing the proportional band to 1 % resulted in an on-off control that reduced the oscillation to an acceptable value, $\pm 10 \%$ of the set level (Fig. 6).

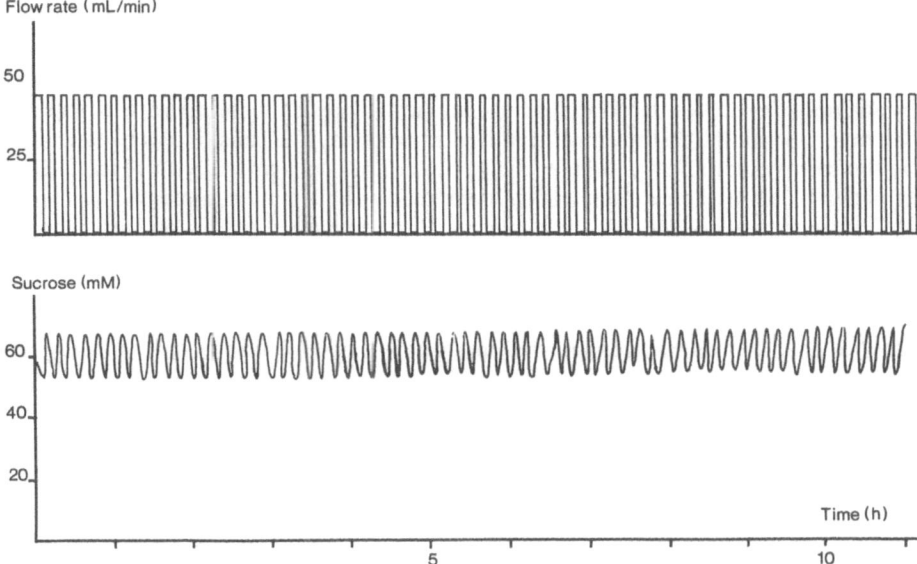

Figure 6. On-off control. The upper curve shows the sucrose pump flow. The lower curve shows the sucrose concentration as measured by the invertase-ET. (From Mandenius et al. 1981)

More sophisticated control, however, requires a computer, especially when control of the recovery of ethanol is taken into account and when control of the final utilization of substrate in the additional column reactor is included.

Microbial Thermogram: Aside from the possibility of continuously monitoring product (e.g. penicillin formed in a fermentation process), substrates (e.g. lactose consumption) or enzyme activity, the ET unit is also able to simultaneously register the overall thermal behavior of such a microbial process. This information can be recorded as a thermogram (timepower curve) which provides additional valuable information. In order to monitor microbial activity, the microorganisms can either be immobilized and used in a column, resulting in a "microbe thermistor" (recently used by Mandenius et al. (1981) in the study of the effect of ethanol on yeast metabolism) or the solution of microorganisms can be continuously drawn from the fermenter to a calorimetric unit (in the same way as with enzyme activity assays described below) in order to provide information on the metabolic conditions in the fermenter (Danielsson et al. 1981 a).

Enzyme Activity Measurements. Various techniques for measuring enzyme activities in solution by the ET-unit were discussed by Danielsson et al. (1981 a). The apparatus can be directly used for this purpose simply by replacing the enzyme column with an inert column or a reaction coil. Both the substrate solution and the enzyme solution (the sample) are mixed and after temperature equilibration (to eliminate mixing heats) the mixture is pumped through the reaction coil. The temperature increase during this passage is a measure of the enzymic activity (Danielsson and Mosbach, 1979). The sensitivity was about 0.1 units ml^{-1}. The same sensitivity (at least) can be obtained with an alternative technique involving biospecific absorption. If an enzyme solution is pumped through an enzyme thermistor equipped e.g. with a concanavalin A-Sepharose-column the enzyme will bind to the column and can subsequently be quantified by adding an excess of substrate to the enzyme. The resulting temperature response will be proportional to the amount of bound enzyme. The sensitivity is determined by the sample volume applied. Finally, the column is washed with glycine-HCL in order to elute the enzyme and is then ready for a new sample.

Chromatographic Effluent Monitoring. The enzyme thermistor has recently been successfully applied as an instrument for the specific monitoring of some common chromatographic procedures: gel filtration, ion exchange chromatography and affinity chromatography (Danielsson et al. 1981 c). Since a calorimeter can be used for direct, continuous monitoring of enzymic activities on crude samples, it allows for the possibility of localizing a specific component in a complex chromatogram as, for instance, in the initial steps of an enzyme purification scheme (Fig. 7). Furthermore, elution in affinity chromatography

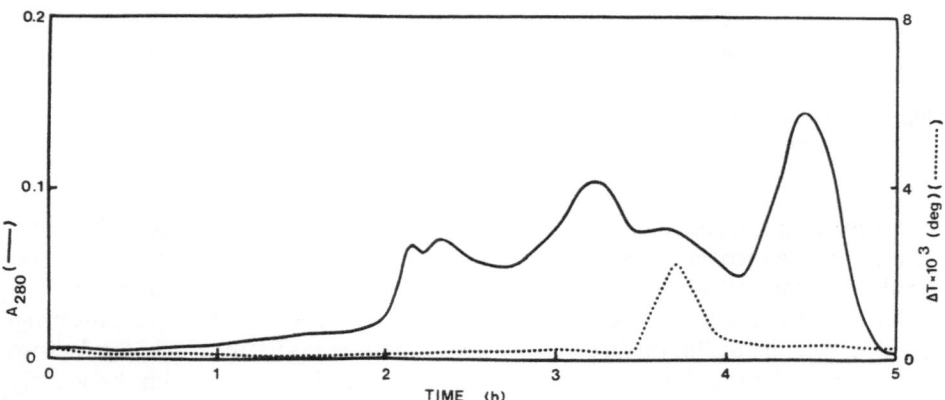

Figure 7. Gel filtration of 1 ml crude yeast extract on an Ultro-gel AcA 44 column (53 × 2.5 cm) eluted with 0.2 M Tris-HCL, 0.0133 M MgCl$_2$, pH 7.8, at a flow rate of 0.75 ml min^{-1}. For thermal analysis, the effluent was mixed with a substrate solution containing 0.54 M glucose, 0.011 M ATP, flow rate 0.2 ml^{-1}. The UV absorbance at 280 nm was registered with a Unicord S. Reprinted from Danielsson et al. 1981

is frequently accomplished with coenzymes having a strong UV-absorbance, thereby often precluding direct on-line assay of a particular enzyme (either by UV-monitoring or by spectrophotometric monitoring of changes in NAD(P)H concentration). Calorimetric detection of the eluted enzyme activity should thus have definite advantages in this area as well (Fig. 8).

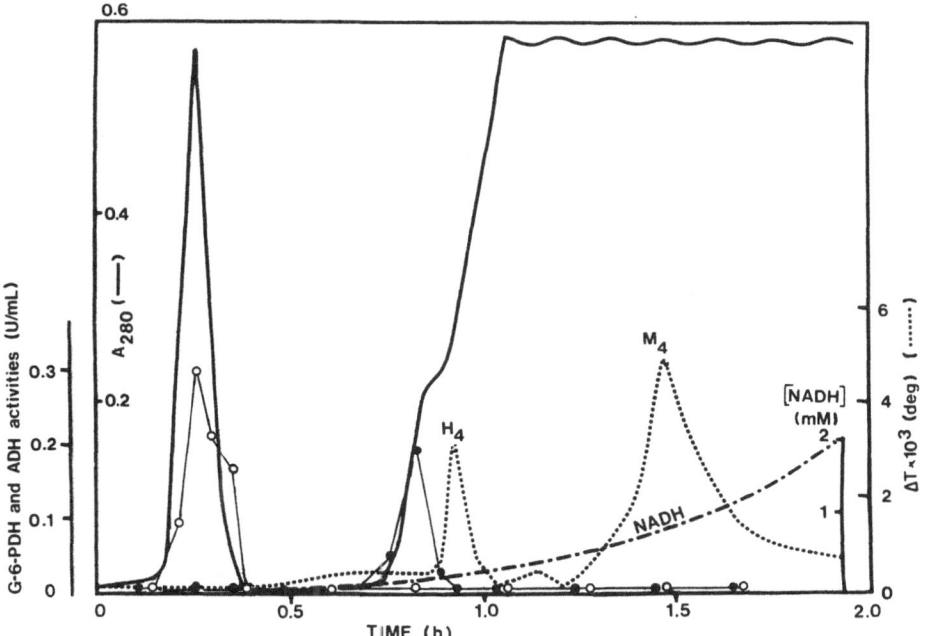

Figure 8. Affinity chromatography of LDH isoenzymes on a column (4.5 × 1.2 cm) of 5'-AMP-Sepharose.
Sample: 5.0 mg BSA, 3,8 mg ADH, 0.06 mg G-6-PDH, 0.1 mg LDH-M_4, and 0.1 mg LDH-H_4 applied in 0.5 mL of the buffer 0.1 M sodium phosphate, pH 7.5. Elution was by a concave NADH gradient as indicated. Flow rate: 0.58 ml/min. The substrate for the ET analysis was 12 mM pyruvate, 4 mM NADH introduced at 0.11 ml/min. ADH (●) and G-6-PDH (○) were assayed spectrophotometrically at 340 nm. (From Danielsson et al. 1981 c)

Conclusion

The ET is a routine bioanalytical instrument that can be used to specifically measure metabolites in biological samples of more than approximately 0.1 mM in concentration. The ET can be used for measuring discrete samples with a speed of up to 60 samples per hour as well as for continuous monitoring. The favorable performance of the ET in process control has been demonstrated. Furthermore, the ET-unit can be used for monitoring total microbial activity in fermentation solutions. In a similar way, it can measure enzyme activity in solution, for instance, in enzyme purification processes.

Acknowledgements

Thanks are due to M-B. Larsson, C-F. Mandenius, B. Mattiasson, K. Mosbach, L. Persson and F. Winquist for their valuable contributions to the development of the enzyme thermistor techniques.

The generous financial support by the National Swedish Board for Technical Development is gratefully acknowledged.

References

Bowers LD and Carr PW (1976). An immobilized-enzyme flow-enthalpimetric analyzer: application to glucose determination by direct phosporylation catalyzed by hexokinase. *Clin Chem*, 22:1427–1433

Bowers LD, Canning LM Jr, Sayers CN and Carr PW (1976). Rapid-flow enthalpimetric determination of urea in serum, use of an immobilized urease reactor. *Clin Chem*, 22:1314–1318

Danielsson B and Mosbach K (1979). Determination of enzyme activities with the enzyme thermistor unit. *FEBS Lett*, 101:47–50

Danielsson B, Gadd K, Mattiasson B and Mosbach K (1977). Enzyme thermistor determination of glucose in serum using immobilized glucose oxidase. *Clin Chim Acta*, 81:163–175

Danielsson B, Mattiasson B and Mosbach K (1979a). Enzyme thermistor analysis in clinical chemistry and biotechnology. *Pure & Appl Chem*, 51:1443–1457

Danielsson B, Mattiasson B, Karlsson R and Winquist F (1979b). Use of enzyme thermistor in continuous measurements and enzyme reactor control. *Biotechnol and Bioeng*, 21:1749–1766

Danielsson B, Mattiasson B and Mosbach K (1981a). Enzyme thermistor devices and their analytical applications. *Applied Biochem and Bioeng* Vol 3:97–143

Danielsson B, Rieke E, Mattiasson B, Winquist F and Mosbach K (1981b). Determination by the enzyme thermistor of cellobiose formed on degradation of cellulose. *Appl Biochem and Biotechnol*, 6:207–222

Danielsson B, Buelow L, Lowe CR, Satoh I and Mosbach K (1981c). Evaluation of the enzyme thermistor as a specific detector for chromatographic procedures. *Anal Biochem*, 117:84–93

Grime JK (1980). Biochemical and clinical analysis by enthalpimetric measurements — a realistic alternative approach? *Anal Chim Acta*, 118:191–225

Guilbault GG, Danielsson B and Mosbach K. A comparison of enzyme electrode and thermistor probes for assay of alcohols using alcohol oxidase. Submitted

Mandenius CF, Danielsson B and Mattiasson B (1980). Enzyme thermistor control of the sucrose concentration at a fermentation with immobilized yeast. *Acta Chem Scand*, 34B:463–465

Mandenius CF, Danielsson B and Mattiasson B (1981). Process control of an ethanol fermentation with an enzyme thermistor as a sucrose sensor. *Biotechnol Lett*, 3:629–634

Mattiasson B and Danielsson B (1982). Calorimetric analysis of sugars and sugar derivatives with aid of an enzyme thermistor. *Carbohydr Res*, 102:273–282

Mattiasson B, Danielsson B, Winquist F, Nilsson H and Mosbach K (1981). Enzyme thermistor analysis of penicillin in standard and in fermentation broth. *Appl Environ Microbiol*, 41:903–908

Mosbach K and Danielsson B (1981). Thermal bioanalyzers in flow streams — enzyme thermistor devices. *Anal Chem*, 53:83A–94A

Satoh I, Danielsson B and Mosbach K (1981). Triglyceride determination with use of an enzyme thermistor. *Anal Chim Acta*, 131:255–262

Schmidt H-L, Krisan G and Grenner G (1976). Microcalorimetric methods for substrate determinations in flow systems with immobilized enzymes. *Biocim Biophys Acta*, 429:283–290

Spink C and Wadsö I (1976). *Methods of Biochemical Analysis*, 23:1–159

Winquist F, Danielsson B, Malpote J-Y, Larsson M-B and Persson L (1982). Oxalate determination . . . Submitted

Preparation, Properties and Possible Application of Coimmobilized Biocatalysts

W. Hartmeier

Summary

Recently, coimmobilizates have been developed which combine the biocatalytic properties of whole cells or parts of the cells and additional enzymes. The new method presented leads to very small immobilized particles with extraordinarily high specific activities and neglegible diffusion barriers. Enzymatic properties and results of laboratory scale application tests are given for the system of *Aspergillus niger* mycelium with internal glucose oxidase and catalase and additionally bound glucoamylase. Concepts for oxygen removal from beer by means of this system are developed. The system of living *Saccharomyces cerevisiae* cells and additional pepsin envelop is shown to be highly advantageous for wine must fermentation. The data are discussed with regard to possible economical and technical feasibility.

Introduction

Immobilization of biocatalysts has gained considerable scientific and technical importance in the last decade. Binding of single enzymes and immobilization of whole cells are the two main streams within this field. Recent investigations brought new methods to coimmobilize whole microorganisms or other cells and additional enzymes (Hägerdal and Mosbach 1980; Hartmeier 1981). These new techniques promise to combine the biocatalytic properties of whole organisms or parts of the organisms and of enzymes from a possibly different source.

The present paper deals with a newly developed coimmobilization method and some of the catalytic properties of the coimmobilizates made. Some possible applications in the food industries should be examined in laboratory scale trials.

Methods of Coimmobilization

Figure 1 schematically presents the coimmobilization procedure worked out by Hägerdal and Mosbach (1980). The enzyme is first covalently bound to soluble sodium alginate by the carbodiimide method. Then the microbial cells (for example yeast cells) are admixed and the mixture of cells and enzyme bearing alginate is dropped into a calciumchloride solution. There, a ionotropic gel formation takes place, so that water insoluble beads result which contain the whole cells and the additionally bound enzyme as well.

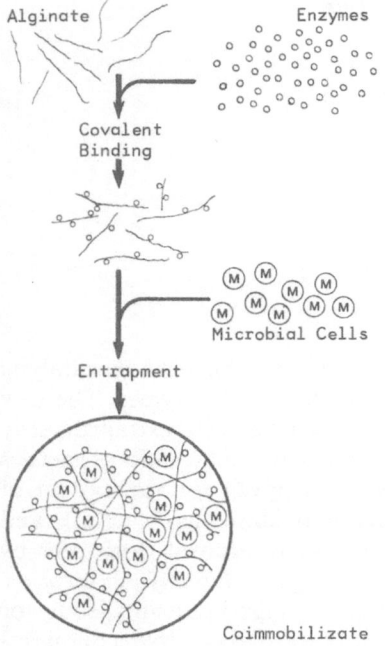

Figure 1. Coimmobilization in alginate beads

The method of alginate entrapping (see Fig. 1) is used with some slight modifications. So, the procedure of a Danish group (Svenson and Ottesen 1981, Godtfredsen et al. 1981) starts with binding enzymes to dextran particles. The dextran-enzyme conjugates are entrapped together with the microbial cells in alginate beads.

Our own method, shown in Fig. 2, is quite different from the alginate entrapping methods mentioned above. Our method starts with partly dewatered cells which are suspended in an aqueous solution of the enzyme to be bound. The cells rehydrate and the enzyme solution is sucked up by the cells. The high molecular enzyme naturally cannot enter the cells. The enzyme molecules are

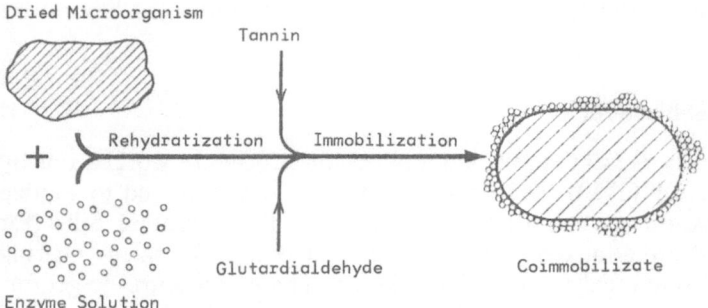

Figure 2. Coimmobilization by envelopping cells with enzymes

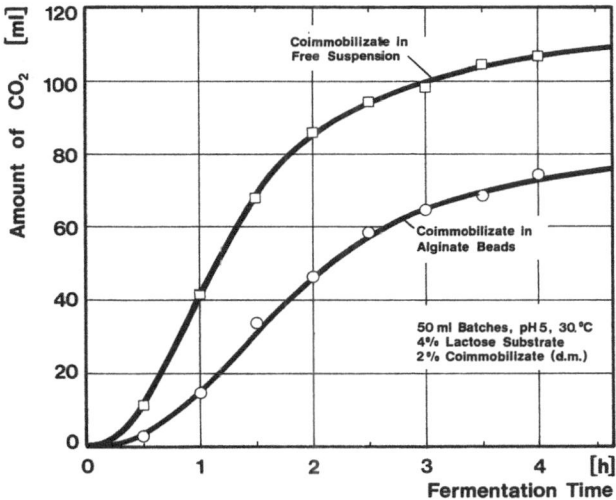

Figure 3. Fermentation with free and alginate-entrapped Coimmobilizate

so deposited on the cell walls, where they are bound and crosslinked by addition glutardialdehyde and tannin.

The coimmobilizate, resulting from the procedure shown in Fig. 2, consists in the majority of single cells with enzyme envelop. By making the enzyme molecules with fluorescein, it can be shown in the fluorescent microscope that a homogenious envelop is built up by this procedure.

Characteristics of the Coimmobilizates

Comparing the coimmobilizates made by the two methods described, the most striking difference between them is their extremely different particle size. Alginate entrapping generally leads to relatively big particles of 2 or 3 mm. In contrary to that, the enzyme envelopping method creates coimmobilizates in the magnitude of single cells or mycelia. So, in the case of enzyme envelopped yeast the typical particle size is in the range of 5 to 15 µm.

Together with the different particle size, a major difference in the diffusion hindrance of the alginate coimmobilizate and the enzyme enveloped cells can be observed. Fig. 3 gives a comparison of lactose fermentation with coimmobilizates of yeast and β-galactosidase (lactase). In both cases shown, we used the same amount of living yeast cells and of β-galactosidase bound. It can be seen that the alginate entrapped system works slower due to the rate limitation by the increased diffusion barriers.

Figure 4 gives the dependency of the specific fermentation rate from the lactose concentration. It also demonstrates that the reaction velocity of the alginate entrapped system is considerably reduced over a wide range of lactose concentrations compared with the freely suspended coimmobilizate.

For economical considerations it is most interesting to know how much of the activity set in is lost by the immobilization procedure. Unfortunately these data vary to a considerable extend with the kind of the biocatalyst treated.

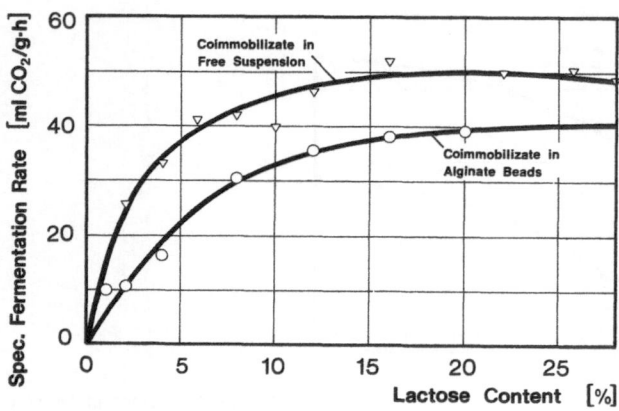

Figure 4. Fermentation rates of free and alginate-entrapped coimmobilizate as a function of lactose concentration

In case of β-galactosidase bound to yeast cells for example 70% of the enzyme activity added and about 70% of the fermentation activity of the native yeast remain active after binding. In case of the yeast-glucoamylase system shown in Fig. 5, only 13% of the glucoamylase (=amyloglucosidase) added are active on dextrin substrate after coimmobilization, but about 90% of the native yeast's fermenting power remains active.

Figure 5 shows some interesting phenomena. It first makes clear that on glucose substrate the fermentation rate is slightly diminished by the coimmobilization procedure. But it also shows that on maltodextrin, where native yeast cannot ferment, fermentation rates in the same magnitude like on glucose substrate are reached. At low substrate concentrations the fermentation velocity of the coimmobilizate on dextrin is even higher than the fermentation rate of the native yeast on glucose. This observation can possibly be explained by the fact that

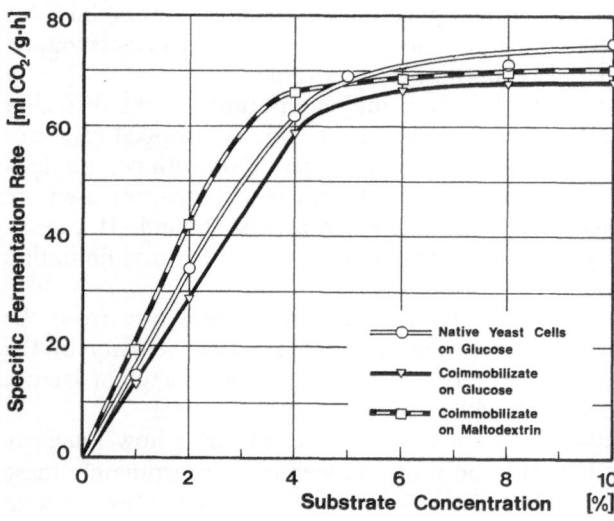

Figure 5. Fermentation rates of native and coimmobilized cells as a function of substrate concentration

the glucoamylase envelop sets free glucose from maltodextrin in the nearest proximity of the cells so that a higher glucose concentration occurs there than in the bulk of the solution.

Coimmobilizate of Mycelium and Glucoamylase

Systems of non-living cells and mycelia can be regarded as relatively easy. Therefor such a system of dead *Aspergillus niger* mycelium with additional glucoamylase as shown in Fig. 6 has been taken for model studies.

Mycelium with Intracellular
Glucose Oxidase

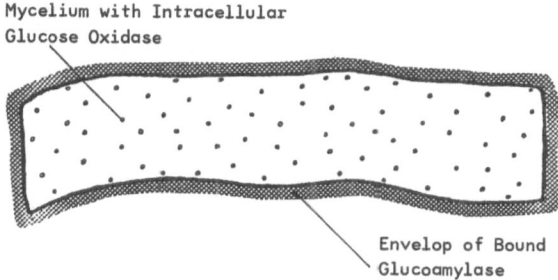

Envelop of Bound
Glucoamylase

Figure 6. Coimmobilizate of Aspergillus mycelium and glucoamylase

The intracellular glucose oxidase/catalase system of the mycelium oxidizes glucose to gluconic acid. The first step of the reaction is, as shown in Fig. 7, a dehydration of glucose to glucono-δ-lactone, which later hydrolyses spontaneously or is catalysed by lactonase to gluconic acid. The hydrogen, split off from the glucose is first deposited on FAD. The reoxidation of the formed $FAD \cdot H_2$ is carried out by transferring the hydrogen to molecular oxygen. The hydrogen peroxide formed in this way is split by catalase. Finally an overall reaction of the whole enzyme combination results, where 1 mol glucose and 0.5 mol molecular oxygen is combined to 1 mol of gluconic acid. The enzyme system is used

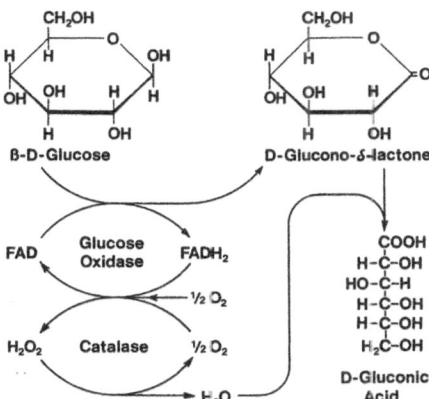

Figure 7. Reaction mechanism of the glucose oxidase/catalase system

industrially either for oxygen removal or for glucose removal from food products.

The application field we have in mind when coimmobilizing the system with glucoamylase is oxygen removal from beer. Oxygen is an undesired component when occurring in beer, because it causes turbidities and leads to undesired taste. The soluble glucose oxidase is able to eliminate oxygen from beer, but in most cases it is too expensive. So, the immobilized system could be advantageous because it possibly can be used repeatedly or continuously.

Some of the enzymatic characteristics of the coimmobilizate seem to predestine it for use in the brewery. The pH-optimum of the enzyme is very close to the pH of beer. The activation energy of the glucose oxidase $E_a = 32{,}6$ kJ/mol is very low so that the reaction proceeds with considerable velocity already at low temperatures being typical for the fermentation, storage and filtration cellars of breweries.

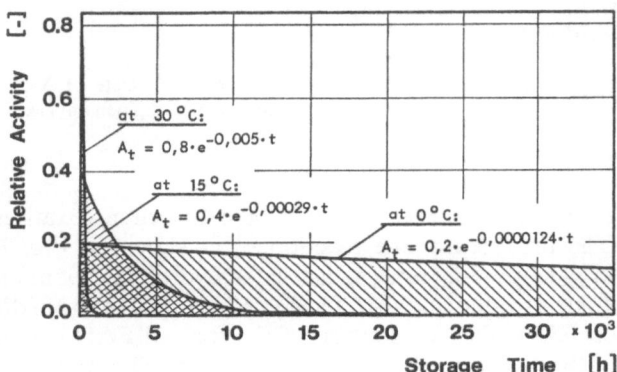

Figure 8. Glucose oxidase activity as a function of storage time in beer

Figure 8 shows the activity of glucose oxidase at different temperatures as a function of the time the enzyme is kept under this temperature. With higher temperature of for example 30 °C the reaction rate at the beginning of the storage is with 0.8 relatively high, but it rapidly decreases with the time the enzyme remains under 30 °C. At 0 °C the reaction rate at the beginning is only 0.2, but the enzyme remains stable under such a low temperature, so that the long-term stability and efficiency of the enzyme at 0 °C is assumed to be better than at 30 °C. From this consideration it can be derived that the optimal point of an activity versus temperature curve is always too high for long term application. For the purpose of brewing an application temperature of 0–5 °C is ideal since it is the normal temperature range used in breweries.

Application trials with our coimmobilizate were carried out in a packed bed reactor. Fig. 9 shows such a packed bed reactor like it is commercially available under the name frame reactor. It consists of a lot of chambers (frames), where the immobilized particles are packed in between filter sheets. For the investigation reported here a laboratory model of this reactor type has been used.

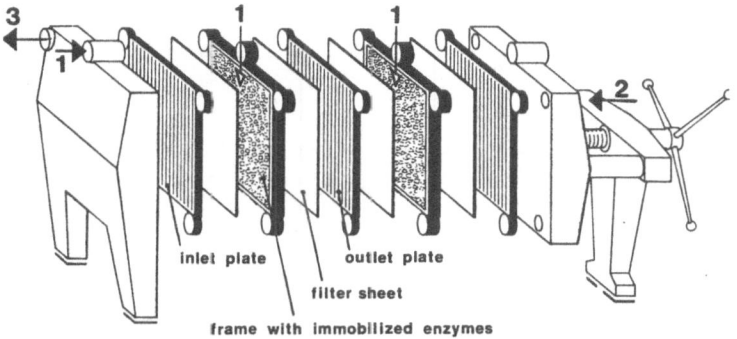

1 filling with immobilized enzymes   **2** substrate inlet   **3** product outlet

Figure 9. 40 × 40 cm frame reactor used for pilot plant trials

In order to examine the influence of different amounts of added amylo-glucosidase (= glucoamylase) a series of runs with subsequent treatment of beer first with the amyloglucosidase and second with the glucose oxidase were done (see Fig. 10).

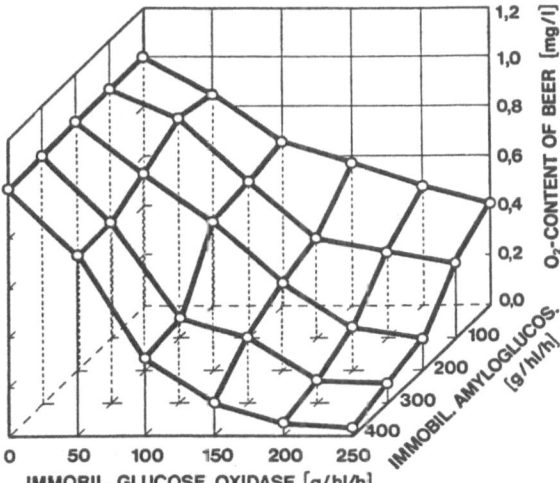

Figure 10. Oxygen content of a German beer treated with amyloglucosidase and glucose oxidase/catalase

Figure 10 makes clear that without any amyloglucosidase the oxygen content of the beer treated is only slightly decreased, in the given example to a maximum of 0.4 mg/l. With addition of amyloglucosidase very much better oxygen reduction is reached. Since beer normally contains only traces of glucose, the amyloglucosidase action increases the reaction with oxygen by setting free its contrapart glucose from the dextrins of the beer.

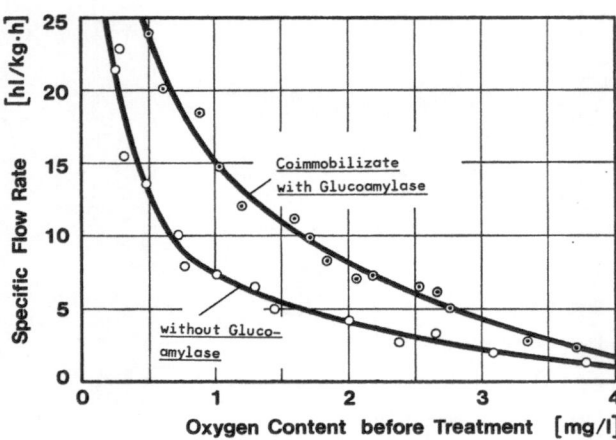

Figure 11. Flow rate to reduce oxygen in beer to 0.1 mg/l as a function of oxygen content before treatment

Figure 11 gives a comparison of the coimmobilizate with and without glucoamylase bound to its surface. It demonstrates that under the conditions given here, the specific flow rate through the packed bed can be doubled when using the glucoamylase containing coimmobilizate instead of the glucoamylase free system.

The results given in Fig. 11 always refer to a final oxygen concentration of 0.1 mg/l in the beer after treatment with the immobilized enzymes. The specific flow rate to get this low value of 0.1 mg/l is a function of the oxygen content of the beer before treatment. The more oxygen the beer contains, the smaller the flow rate must be in order to reduce it to the limit of 0.1 mg/l which is generally regarded as harmless.

Table 1. Relative Raw Material Costs per Activity for Different Forms of Glucose Oxidase

	Mycelium	Soluble	Immobilized	Coimmobilized
Raw Material	—	Mycelium	Soluble	Mycelium
Yield	—	60%	30%	50%
Costs/unit	100	167	557	200

Table 1 gives an impression of the cost situation in a very simplified manner. Only the costs of the glucose oxidase raw material is regarded, because glucose oxidase is by far the most expensive enzyme of the system. The normal purified soluble enzyme is made by extracting the *Aspergillus* mycelium with a yield of about 60% so that the raw material costs, which have been assumed to be 100 per unit in the mycelium, become 167 for the purified soluble quality. Conventionally immobilized glucose oxidase starts from purified soluble enzyme which is bound to carriers with yields of about 30%. That leads to raw material costs for the immobilized enzyme of 557 per unit. The striking advantage of the coimmobilizate is that it does not start from the expensive purified soluble enzyme but from the mycelium with a yield of 50 %. In this way the raw material costs of the coimmobilizate are only in the range of 200 per unit; that is less than half the raw material costs of the conventionally immobilized enzyme.

The cost consideration is the main reason which makes us believe that the coimmobilizate has very much better chances to be industrially used than the conventionally immobilized glucose oxidase.

Coimmobilizate of Living Yeast and Pepsin

Another coimmobilizate which could get industrial interest is pepsin bound to living wine yeast, like schematically shown in Fig. 12. Table 2 gives some of the characteristics of the pepsin component in this system.

Yeast Cell with Intracellular
Fermentation Enzymes

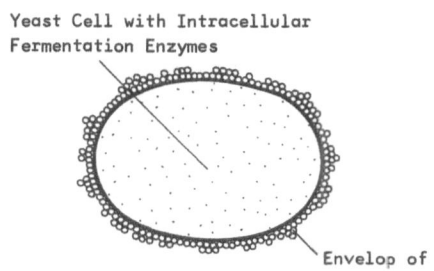

Envelop of
Bound Pepsin

Figure 12. Coimmobilizate of yeast and additional pepsin

Table 2. Characteristics of the coimmobilized pepsin

Proteolytic Activity	Anson-u/g	0.5
pH-Optimum	–	3.0
Activation Energy	kJ/mol	57
Inactivation Energy	kJ/mol	218
Half Life at 40 °C	d	21

Most of the characteristics of the coimmobilized pepsin are similar to those of the conventionally immobilized enzyme given in a further publication (Hartmeier 1979). The pH-optimum is slightly shifted to the alcaline, so that it nearly coincides with the normal pH of wine, being near 3.5.

The stability of the coimmobilized pepsin of 21 days at 40 °C makes sure that the enzyme will survive the must fermentation which is normally carried out at very much lower temperatures between 15 and 30 °C.

The course of a batchwise wine must fermentation is presented in Fig. 13. It gives the alcohol formation and height of foam cover when fermenting a German Silvaner must of 58 °Oe. The tests were done in 5 l batches with 10 g coimmobilizate dry mass respectively 10 g normal dry wine yeast per hl must.

Since wine musts by nature differ very much the results especially with regard to foam formation vary from must to must. The best results were obtained with protein rich musts. There, a major advantage of using the coimmobilizate instead

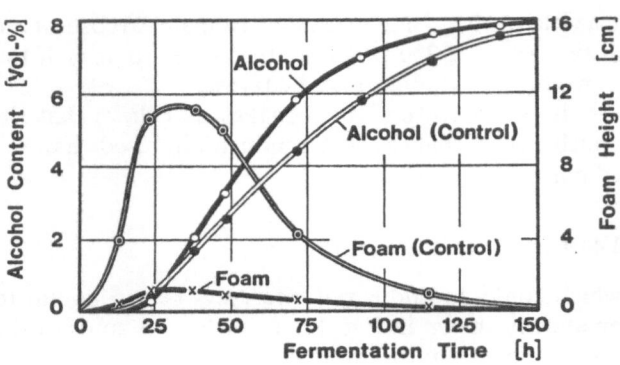

Figure 13. Alcohol formation and height of foam cover during must fermentation

of normal yeast was that wines with high stability against protein turbidity resulted. A further treatment of wines that were made with coimmobilizate, with betonite was not necessary. The main advantages of the coimmobilizate application for must fermentation can be summed up as follows:
— increased fermentation velocity,
— reduced foam formation during the fermentation,
— better taste and coulor characteristics of the wine,
— increased protein stability of the wine.

The amino acid content of the wines made with coimmobilized yeast-pepsin system was not markedly increased. It is assumed that most of the amino acids liberated are directly used by the yeast. The increased fermentation rate is possibly a result of the better nutrition of the yeast by the amino acids set free from the proteins.

Wine makers more and more tend to use special dry wine yeasts because the yeast population normally occuring on the grapes is sometimes a risk for the fermentation. The coimmobilizates of yeast and pepsin could possibly bring further advantages like increased protein stability without use of other finings. Further investigation in larger scale is necessary to prove the economical feasibility of the coimmobilizate for wine making.

Other Coimmobilizates

The two examples above treated in detail should give an impression what coimmobilization means and how coimmobilizates could possibly be used. There are a lot of further possibilities to combine whole cells or parts of whole cells together with single enzymes or groups of enzymes. Some of the most promising examples of yeast/enzyme coimmobilization are given in Table 3.

A realistic estimation whether or not coimmobilizates are economically and technically feasible for industrial application is not yet possible. But this new group of biocatalysts should be submitted to further investigation since it opens some new possibilities of combined action which could not be entered by immobilized cells and immobilized enzymes alone.

Table 3. Coimmobilizates of yeasts with enzymes for food technology

Yeast Species	Enzyme added	Application Field
Sacch. cerevisiae	β-galactosidase	alcohol from whey
Sacch. cerevisiae	pectinase	wine must fermentation
Sacch. cerevisiae	β-glucosidase	cellobiose fermentation
Sacch. uvarum	glucoamylase	dietetic beer making
Sacch. uvarum	β-glucanase	low viscosity beer
Sacch. uvarum	protease	chill proofed beer

Acknowledgements

I am grateful to my coworkers, Mrs. Etienne D. Jankovic and Mr. I. Mücke, for contributing the data of figures 3, 4 and 5.

References

Godtfredsen SE, Ottesen M, Svensson B (1981) Application of immobilized yeast and yeast coimmobilized with amyloglucosidase in the brewing process. Proceedings EBC-Congress. Kopenhagen. pp 505–511

Hägerdal B, Mosbach K (1980) The production of ethanol from cellobiose using baker's yeast coimmobilized with β-glucosidase. In: Linko P, Larinkari J (eds) Food process engineering vol 2. Applied Science Publishers. London. pp 129–132

Hartmeier W (1979) Immobilized pepsin: properties and use to prevent haze formation in beer and wine. Biotechnol Letters 1:225–230

Hartmeier W (1981) Basic trials on the conversion of cellulosic material to ethanol using yeast coimmobilized with cellulolytic enzymes. In: Moo-Young M (ed) Advances in biotechnology vol 3. Pergamon. Toronto. pp 377–382

Svensson B, Ottesen M (1981) Entrapment of chemical derivatives of glycoamylase in calcium alginate gels. Carlsberg Res Commun 46:13–24

New Developments in the Field of Cell Immobilization — Formation of Biocatalysts by Ionotropic Gelation

Klaus-Dieter Vorlop and Joachim Klein

Summary

Different ionotropic gelation processes, applicable to obtain biocatalytic systems by entrapment of whole cells are described. The formation of spherical alginate, chitosan and chitosan-alginate biocatalysts and its physical properties like swelling behaviour in ionic media and the porosity of the networks are discussed. The applicability of the various immobilization procedures for resting, viable and multiplying cells is demonstrated. Finally it is shown that the method of ionotropic gelation can also be used as an intermediate step toward the preparation of other covalent stabilized biocatalysts.

Introduction

In the last years the immobilization of whole cells has proven to be an established method. A number of reviews have appeared in the literature (Durand and Navarro 1978; Klein and Wagner 1978; Abbott 1978; Klein and Wagner 1979;

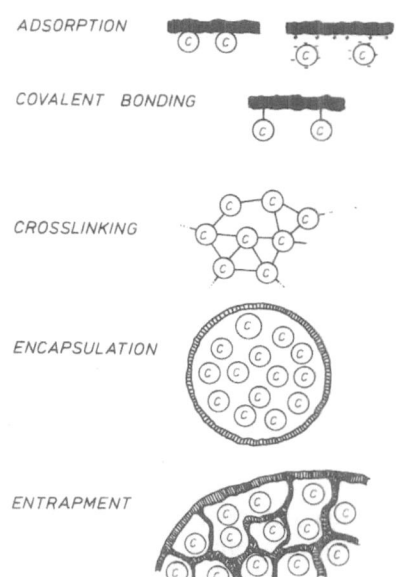

Figure 1. Methods for whole cell immobilization

Chibata and Tosa 1981; Mosbach 1982). The immobilization of whole cells is advantageous compared to immobilized enzymes where the reaction requires a multi-enzyme system or where the enzyme isolation is too expensive. Advantages of immobilized cells are the simple product isolation, that they can be used repeatedly and in continuous processes. One of the most important point is the observation that immobilized cells are much more stable than free cells suspended in a liquid.

A variety of techniques has been developed from simple flocculation to specific covalent attachment. One possible categorization is given in Fig. 1. Beads, fibers, membranes or granules with different size can be obtained.

Entrapment within a network is by far the most widely applied immobilization method, first described by Mosbach et al. (1966) who developed the poly-acrylamide immobilization method. At present, different carriers were tested like polyacrylamide, polymethacrylamide, agar, collagen, polyurethane, poly-epoxid, eudragit®, polystyrene, cellulose, cellulosetriacetate etc. A preparation which provides extremely mild immobilization conditions is the entrapment in ionotropic gels. This network type has first been applied for whole cell entrapment by Hackel et al. (1975), using Al-alginate biocatalysts.

Alginate Biocatalysts

Alginates are naturally occuring substances which are extracted from brown seaweed, but also produced by some species of bacteria as extracellular poly-saccharides. Alginates are polymers which consist of d-mannuronic and 1-gul-uronic acid joined by 1,4 linkages. They are block copolymers composed of M blocks, G blocks and MG blocks (Fig. 2). They are commercial products (~ 20 DM/kg), mainly used in the food industry.

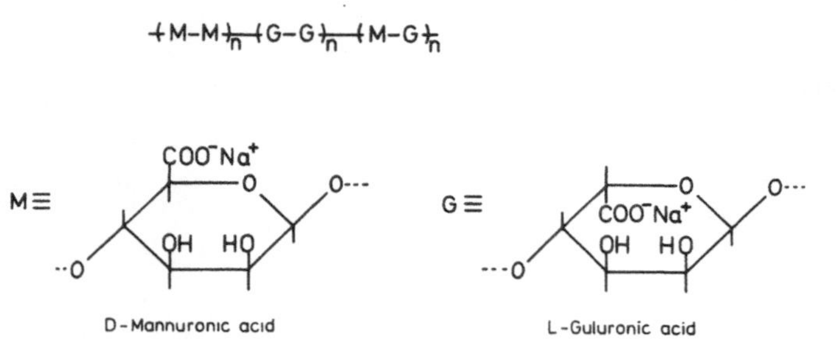

Figure 2. Chemical structure of alginates

By dropwise injection of a Na-alginate sol into a solution containing multi-valent counterions like Ca^{2+}, Co^{2+}, Zn^{2+}, Mn^{2+}, Ba^{2+}, Fe^{2+}, Al^{3+}, Fe^{3+} etc. (not Mg^{2+}) beads were formed due to the mechanism of ionotropic gelation. The ionic network formation procedure was originally developed by Thiele (1954) and our laboratory was the first to adopt and modify this technique to be applicable for whole immobilization.

The process for the formation of spherical cell biocatalysts can be divided into 6 steps:

1. Dissolving of alginate in water to obtain a highly viscose solution but capable for droplet formation.
2. Autoclaving of the alginate solution. Only a short time autoclaving is possible (max. 20 min., 121 °C) otherwise the alginate polymer will be reduced.
3. Addition of cell mass and dispersion (up to 0.3 g wet cells/ml).
4. Dropping this suspension through a capillary tube into a cross-linking solution. In the normal case a 500–100 mM $CaCl_2$-solution is used, but a bead formation is possible down to 5 mM.
5. Hardening of beads in this solution (30–60 min).
6. Separation of biocatalysts and washing.

Biocatalysts with good mechanical properties and cell loading up to 0.3 g wet cells/ ml catalyst can be obtained. Recently, the network properties could be drastically improved by introducing an intermediate step of gel drying (Klein and Wagner 1979). During this course of preparation the particle radius decreases without reswelling to its original dimension when contacted with the cross-linking solution again. The consequence is an increase of cell concentration per unit volume up to 0.8 g wet cells/ml catalyst which leads also to higher activity per unit volume. The mechanical properties are much more better then those of none-dried beads.

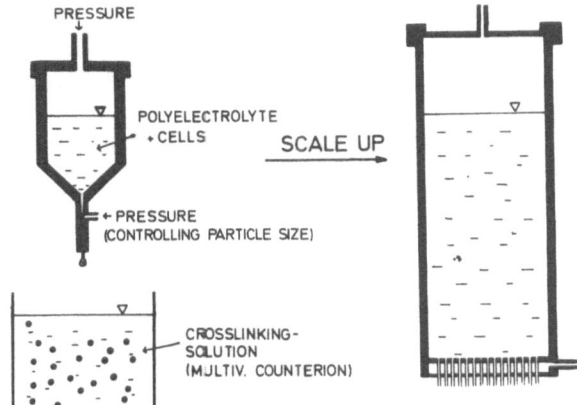

Figure 3. Technology of bead formation

Biocatalysts with a diameter from 0.1–4 mm can be produced easily (Fig. 3). For that the alginate cell suspension is filled into a cylindrical reservoir and quickly dropped by compressed air (0.5–1.5 bar) through a nozzle into the cross-linking solution. The small particles can be obtained by a controlled blow off stream of pressurised air concentric to the extrusion nozzle. To produce greater amounts of biocatalysts we developed a system with 42 outlets. With this immobilization apparatus 3–5 kg biocatalysts per hour can be obtained (Vorlop et al. 1980). Using this methods biocatalysts with a small size distribution were obtained (Fig. 4).

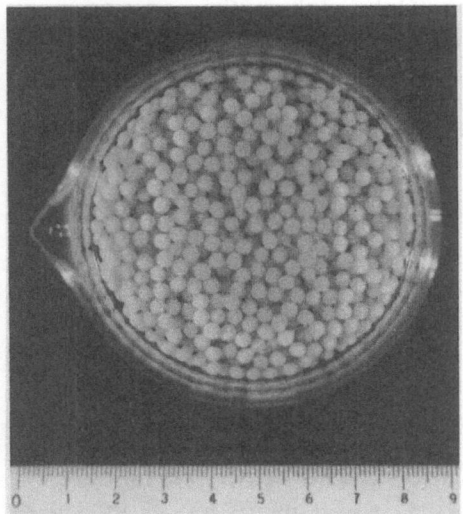

Figure 4. Ca-alginate biocatalysts

A disadvantage of this type of network is its reversibility. The ionotropic gel is not stable in the presence of high concentrated electrolyte solutions (K^+, Na^+ ...). The electrolyte tolerance of the gel depends on the selectivity coefficients of an alginate for a metal in comparison with sodium. This selectivity coefficient depends on the type of alginate, especially on its M/G ratio (Haug and Smidsrod 1965).

In swelling experiments we could show that alginates containing more guluronic acid fragments are much more stable in electrolyte solutions than those containing more mannuronic acid groups (Fig. 5). Using G-type alginates, as a rule

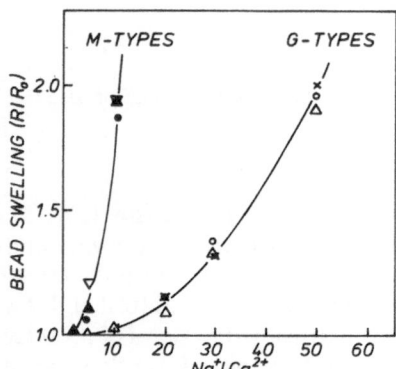

Figure 5. Swelling behaviour of dried Ca-alginate beads (2 beads in 50 ml solution (Na^+ + Ca^{2+} = 0.1 M), 48 h)
● Manucol·LF; ▲ Manucol DH; ▽ Manucol DM; ○ Manugel DLB; △ Protanol LF 20/60; + Manugel DMB

of thumb one can say that the ratio electrolytes (molare sum of Na^+, K^+, NH_4^+, Mg^{2+} ...) to Ca^{2+} is not allowed to be more than 20:1.

A disadvantage of these gels is also the instability in phosphate-buffer solutions, where the counterions (Ca^{2+} ...) will be precipitated and the matrix will redissolve. This effect may be a little lowered by using strontium or barium as cross-linking ions (Paul and Vignais 1980) but will be unacceptable where the product of the immobilized cells is used in food or pharmaceutical products.

There are buffer solutions which did not precipitate with the cross-linking ions (Fig. 6). Zwitterionic buffers show significant advantages over the other buffers, no inhibitory properties were observed in biological systems (Brimbla 1981). In the presence of multivalent cations the ionotropic gels are stable in the summerized buffer solutions.

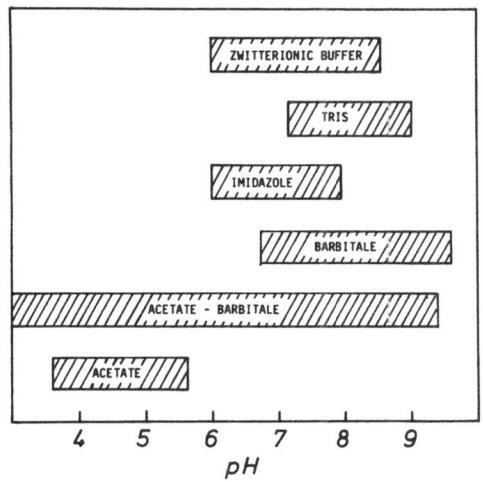

Figure 6. Buffers for alginate biocatalysts

After testing different types of alginates, the following seem to be the most suitable (with regard to autoclaving, cell concentration, mechanical stability, swelling in electrolyte solutions):
1. Protanal-LF 20/60 (2.5–3.5%), Protanal-LF 60 (2.5–3.5%) from Hellmut Carroux, Neuer Wall 37, 2000 Hamburg 11.
2. Manugel DMB (3–4%), Manugel DLB (3–4%) from Kelco/AIL International GmbH, Rödingsmarkt 52, 2000 Hamburg 11.

The simple and cheap immobilization in alginates is now one of the most popular methods and has been used for different applications: degradation of phenol (Klein et al. 1979), light production with luminous bacteria (Makiguchi et al. 1980), oxidation of cholesterol (Duarte and Lilly 1980), sucrose inversion (Linko et al. 1980), conversion of MeOH to CH_4 (Scherer et al. 1981), production of gluconic acid (Vorlop et al. 1980), etc.

A very important point of immobilized cells is the amount of cells which are still enzymatically active after the immobilization procedure. A loss of activity which can be observed in most the cases is caused by the immobilization procedure itself (enzyme desactivation) or by diffusion limitation. For the formation of gluconic acid from glucose and oxygen the influence of diffusion limitation mainly caused by oxygen is illustrated in Fig. 7. The activity of immobilized *Acetobacter sp.* cells (high glucose dehydrogenase activity) increases with decreasing particle radius and cell concentration.

In the case of alginate biocatalysts only nontoxic materials are used and entrapment of very sensitive cells like plant cells (Brodelius et al. 1979), protoplasts (Scheurich et al. 1980), animal cells (Nilson and Mosbach 1980) and red blood cells (Pilwat et al. 1980) has been successful. A reincubation is also possible and was confirmed by several authors (Klein et al. 1976, Kierstan et al. 1977, Ohlson et al. 1979).

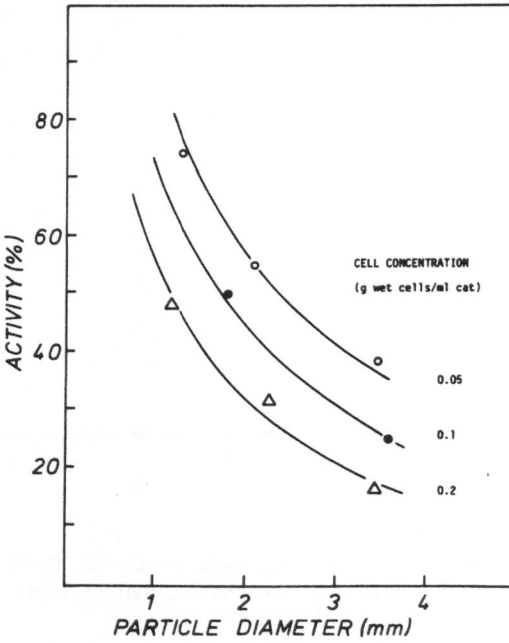

Figure 7. The influence of particle diameter and cell concentration on the observable gluconic acid productivity of Ca-alginate immobilized Acetobacter sp. cells.
$100\% = 35$ U/g wet cells (free cells), $(T = 30\,°C$, pH 6, $[O_2] \sim 7.4$ mg/l, 5% Glucose, 1% CaCl$_2$)

Very interesting systems are the continious cultures with growing and producing immobilized cells. World-wide extensively studied is the production of ethanol with immobilized yeast and *Zymomonas mobilis* cells. In our laboratory immobilized yeast cells worked more than 72 days without loss of activity (Becke 1982). Such immobilized yeast cells which are in a growth state are shown in Fig. 8.

Biocatalysts with other Natural and Synthetic Polyanions

The ionotropic gelation of other natural and synthetic polyanions containing carboxylic (Hackel 1976) or phosphonic groups (Vorlop and Klein 1982) with

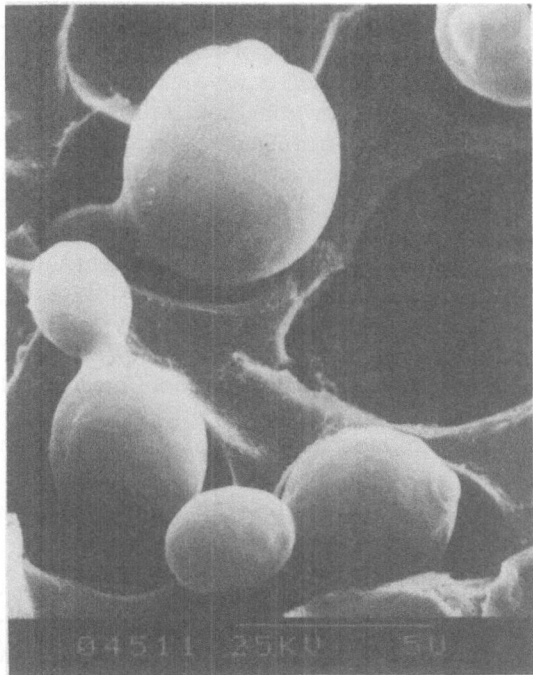

Figure 8. Ca-alginate immobilized Saccharomyces cerevisiae cells

POLYANIONS		COUNTERIONS
$-COO^-$	ALGINATE	Ca^{2+}, Al^{3+}, Zn^{2+}, Co^{2+}, Ba^{2+}, Fe^{2+}, Fe^{3+} ....
	PECTIN	Ca^{2+}, Al^{3+}, Zn^{2+}, Co^{2+},... and Mg^{2+}
	CARBOXYMETHYL-CELLULOSE	Ca^{2+}, Al^{3+} ......
	CARBOXY-GUAR-GUM	Ca^{2+}, Al^{3+} ......
	COPOLY-STYRENE-MALEIC ACID	Al^{3+} ......
$-PO_3^{2-}$	PHOSPHO-GUAR-GUM	Ca^{2+}, Al^{3+} ......
$-SO_3^{2-}$	CARRAGEENAN	K^+, Ca^{2+} ......
	FURCELLARAN	K^+, Ca^{2+} ......

Figure 9. Polymer/counterion systems for the ionotropic gelation of polyanions

multivalent cations is also possible. Very intersting is a new development by which biocatalysts can be obtained by the cross-linking of a special pectin with Mg^{2+}-ions. This may be an interesting carrier for cells with glucose isomerase activity (Vorlop and Klein 1982). Another well-known method is the formation of carrageenan biocatalysts (Takata et al. 1977). Carrageenans are natural sulfate groups containing polysaccharides, which are extracted from red seaweed. Especially kappa carrageenan gels react most readily and strongly in the presence of K^+-ions. In the presence of K^+ these gels are stable in phosphate-buffer and other electrolyte solutions, but sensitive against high Na^+-concentration. This method has now been used for the production of 2,3 butanediol, 1-isoleucine, 1-malic acid etc. (Chibata and Tosa 1981). A reason which till now prevents a common application of this method is the immobilization temperature of 37–55 °C, which is necessary to obtain biocatalysts with good mechanical properties.

The various possibilities of the formation of biocatalysts by ionotropic gelation is summarized in Fig. 9.

Chitosan Biocatalysts

Not only polyanions like alginate or carrageenan, but also polycations can be used for the formation of ionotropic gels (Thiele 1954). By application of this fact, we developed an immobilization process for getting spherical chitosan biocatalysts (Vorlop and Klein 1981; Klein et al. 1981).

Chitosan is a partially de-acetylated chitin formed by the reaction of chitin with concentrated alkali. It is a high molcular weight linear polymer, which consists of glucosamine in 1,4 linkage (Fig. 10). Chitosan is a commercial product for industrial purposes (~ 15 US\$/kg). In 1977 a quantity of more than 300,000 t was produced.

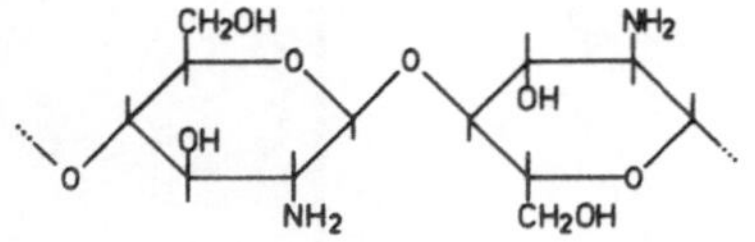

Figure 10. Chemical structure of chitosan

Chitosan is soluble in several diluted organic acids like acetic, formic, adipic acid, etc., but its solubility in mineral acids is limited (soluble in HCl). Using chitosan solutions (pH < 6), a crosslinking will occur with multivalent anionic counterions (Fig. 11). Counterions which will be preferred only depend on the compatibility with the cells.

A typical example of cell immobilization is described as follows:
A $\sim 1.4\%$ (w/v) viscose chitosan-acetate solution was obtained by mixing 1.4 g high viscosity chitosan (Chugai Boyeki Europe Office, Klosterstr. 30, 4000 Düsseldorf 1) with 98 ml H_2O and 0.8 ml acetic acid while stirring. 6 g wet E. coli cells were suspended in 6 ml 0.05 M acetate-buffer solution (pH ~ 5) and mixed with the chitosan-acetate solution. This suspension was drop-

POLYCATION	COUNTERIONS
CHITOSAN $-NH_3^+$	TRIPOLYPHOSPHATE PYROPHOSPHATE TETRAPOLYPHOSPHATE OCTAPOLYPHOSPHATE HEXAMETAPHOSPHATE $\left[Fe(CN)_6\right]^{4-}$, $\left[Fe(CN)_6\right]^{3-}$ POLY-ALDEHYDO-CARBONIC ACID POLY-1-HYDROXY-1-SULFONATE-PROPENE-2

Figure 11. Possible counterions for the ionotropic gelation of chitosan

ped (Figure 3) into a gently stirred 1.5% (w/v) Na-tripolyphosphate solution (pH 8.1). After a treatment of 2 h (for shrinking and hardening) the beads were collected on a nylon net. We obtained 36 g wet biocatalysts with a cell loading of 0.16 g wet cells/ml wet catalyst.

It should be added that immobilization under sterile conditions caused no problems. In this case chitosan is suspended in H_2O and autoclaved under normal conditions. By later addition of sterile acetic acid the sterile chitosan solution can be obtained.

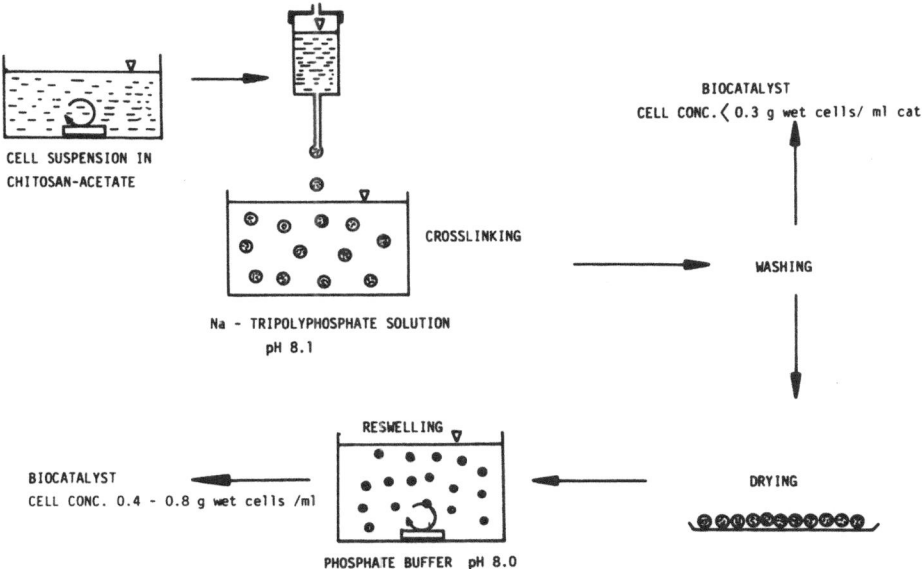

Figure 12. Schematic diagram for the preparation of chitosan biocatalysts

The cross-linking with tripolyphosphate down to 0.2% (w/v) is possible and the presence of phosphate does not spoil the effect. Possible is also a short time cross-linking with tripolyphosphate (15–30 min). Afterwards the beads must be stirred for ~2–4 h in a phosphate solution for shrinking and hardening.

The general scheme for the formation of chitosan biocatalysts is presented in Fig. 12.

Biocatalysts with good mechanical properties and cell loading up to 0.3 g wet cells/ml biocatalyst can be obtained. Recently the network properties could be drastically improved by introducing an intermediate step of gel drying. The consequence is also an increase of cell concentration.

The stability of the beads in buffer solutions especially phosphate buffer is a feature for biochemical purposes (Table 1). By addition of tripolyphosphate to gluconic or lactic acid solutions chitosan gels are also stable.

Table 1. Swelling behavior of chitosan beads
(2 beads in 50 ml, 48 h)

SOLUTION	pH	REMARKS
0.1 M K-PHOSPHATE-BUFFER	6	NO CHANGE
	7	NO CHANGE
	8	NO CHANGE
0.1 M CITRATE-BUFFER	3	NO CHANGE
	4	NO CHANGE
	5	NO CHANGE
0.1 M ACETATE-BUFFER	4	DISSOLVED
	5	DISSOLVED
	6	DISSOLVED
GLUCONIC ACID (15 %)	6	SWOLLEN
+ 0.01 M TRIPOLYPHOSPHATE	6	NO CHANGE
LACTIC ACID (15 %)	6	SWOLLEN
+ 0.01 M TRIPOLYPHOSPHATE	6	NO CHANGE

Figure 13 is a SEM of the chitosan matrix, showing the microporous bead surface and the larger cavities within the matrix. This is an ideal structure for biocatalysts because a washout of cells is not possible, but there are good diffusional properties for substrates and products. The diffusivity of indole and 1-tryptophan within the pure none dried gel is 58% respectively 54% of its value in water (Vorlop and Klein 1982). Substrate diffusion becomes more restricted in chitosan biocatalysts (Table 2). Observable is a decrease of the diffusivity by

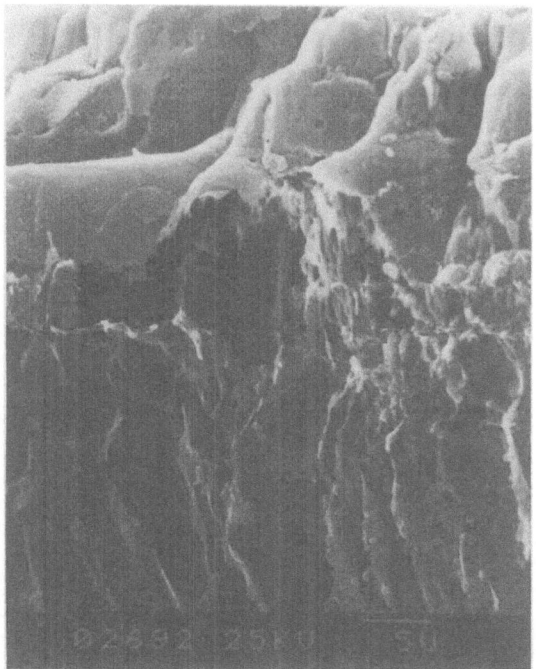

Figure 13. SEM of a critical point dried chitosan bead

increasing cell concentration. That may be caused by the additional barrier of the cells.

E. coli cells with tryptophan synthetase activity were immobilized by the chitosan method (Vorlop and Klein 1981). The obtained biocatalysts exhibit good activities (Table 3). A major part of the activity loss is caused by transport limitation. For the production of l-tryptophan from indole and serine in a continous-flow stirred tank reactor the biocatalyst showed a half life of 30 days.

In order to demonstrate the living state of the immobilized *E. coli* cells, the cell loading of the beads was chosen low (0.0015 g wet cells/ml cat) allowing subsequent growth. The cells were entrapped under sterile conditions directly from the culture broth. The incubation in growth medium led to increase of cell con-

Table 2. Diffusion coefficients (37 °C) in chitosan beads, from effusion of l-tryptophan

	CELL CONCENTRATION (g wet cells/ ml cat)	DIFFUSION COEFFICIENT (cm^2/s)$\cdot 10^6$	D_{eff} (%) of value in water
H_2O	–	8.9	100
CHITOSAN	0	4.8	54
CHITOSAN	0.17	3.7	42
CHITOSAN	0.3	2.8	31

Table 3. Activity of free and immobilized E. coli B 10 cells with tryptophan synthetase (T = 37 °C, 0.1 M phosphate buffer pH 8, 0.2%
indole, 0.2% l-serine, 1 mg pyridoxal-5-phosphate/100 ml)

TYPE	CELL CONCENTRATION (g wet cells/ ml cat)	ACTIVITY (mg trp/ h g cat)	(%)
FREE CELLS	-	176	100
FREE CELLS (DRIED, 5h)	-	159	88
CHITOSAN BEADS ∅ 2 mm	0.11	11	57
DRIED CHITOSAN BEADS ∅ 1.3 mm	0.43	24	32
∅ 0.7 mm	0.52	35	41

centration in the matrix (Fig. 14a, b) and increase of activity (Fig. 15)
(Wagner et al. 1981, Vorlop et al. 1981).

Very interesting is the observation that inside the beads near the surface the
cell concentration is much higher (Fig. 16) than in the middle of the beads
(Fig. 14b). A reason may be the better supply of the growing cells with
oxygen and other nutrients. The chitosan method has now been used for the
formation of 6-APA from Pen-G (Vorlop et al. 1980), from Pen-V (Kluge et al.
1982) and for the production of 1-serine from glycine and MeOH (Behrendt
1981).

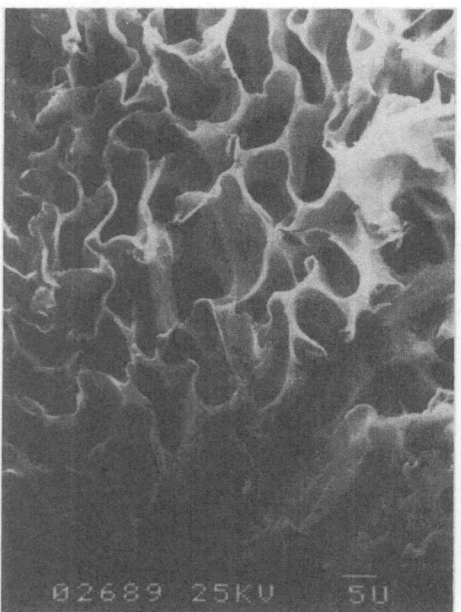 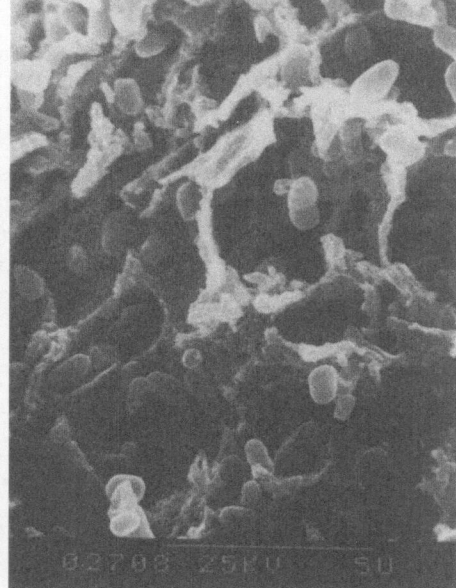

Figure 14a. Chitosan bead before incubation
Figure 14b. Chitosan bead with E. coli cells after incubation

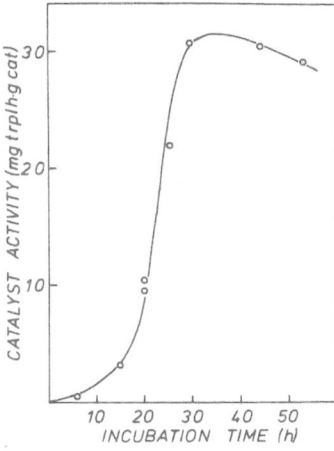

Figure 15. Growth of E. coli B 10 cells in chitosan beads, expressed as tryptophan productivity

Figure 16. E. coli B 10 cells within a chitosan bead, near the bead surface

Biocatalysis by Polyelectrolyte Complex (PEC) Formation

Beside the ionotropic gelation of polyelectrolytes with low molecular weight counterions, a polyelectrolyte complex (PEC) formation is possible. PECs are ionically cross-linked hydrogels formed by the co-reaction of two highly but oppositely charged polyelectrolytes. When concentrated solutions of these poly-

electrolytes are contacted, a thin interfacial film (<1000 Å) of PEC is rapidly formed which completely blocks further interpolymer reactions. These membranes posses unique properties that have led to a variety of applications:
1. high permeability to water and substrates
2. excellent chemical stability
3. excellent biological compatability
4. controllable ion-exchange capacity
5. only limited swelling or shrinking in electrolyte solutions, PEC will not dissolve.

The major commercial applications are ultrafiltration and dialysis membranes, battery separators, and biomedical prostetic material (blood conduits). The properties of PEC were extensively examined by Michaels (1965).

Only thin films can be formed by the known technique, so that a direct entrapment of cells within PEC was not possible. Lim and Sun (1980) used this technique for the *encapsulation* of pancreas cells. By a polycation treatment of preformed calcium alginate beads a surface-coated material was obtained. In this case the cell-suspension was encased in a thin semipermeable membrane consisting of alginate and polylysine. A comparable method was used for the immobilization of yeast cells (Veliky and Williams 1981).

To obtain PEC biocatalysts with good mechanical properties we developed a method for the direct *entrapment* of whole cells by PEC beadformation (Vorlop and Klein 1982). To prevent an immediate flocculation by the direct mixing of oppositely charged polyelectrolytes we suspended a fine grained, water insoluble polyelectrolyte (Ca-alginate) in a solution of an oppositely charged polyelectrolyte (chitosan). By the addition of Na-phosphate, Ca-phosphate is precipitated and the water-soluble Na-alginate is formed so that the ionotropic gelation of the polyelectrolytes can take place.

A typical example of cell immobilization is described as follows:

4 ml of a 10% (w/v) Ca-alginate (Kelco/AIL) suspension in 1% (w/v) CaCl$_2$-solution was mixed with 26 ml of a 2.5% (w/v) low viscosity chitosan (Chugai Boyeki ...) solution. 1 g wet cells in 1 ml 0.05 m acetate-buffer solution (pH $\sim$ 5) were added while stirring. This suspension was dropped into a gently stirred 1.5% (w/v) Na-tripolyphosphate solution (pH 8). After a treatment of 2 h the beads were collected and washed with 0.9% NaCl solution.

Table 4. Swelling behaviour of alginate/chitosan beads in a 0.1 M Na-acetate solution (2 beads in 50 ml, 48 h)

pH	PARTICLE DIAMETER (mm)
4	4.2
5	4.0
6	3.6

These beads were stable in all the solutions summerized in Table 1. A very important observation is the stability in acetate buffer solutions (Table 4) where the normal chitosan beads are fully dissolved. The principle of this immobilization method which is in an early stage of development, can surely be used for the cross-linking of other natural and synthetic polyelectrolytes.

Covalent Stabilization of Ionotropic Gel Biocatalysis

There is a variety of combined techniques leading to covalent cross-linked biocatalysts which uses the ionotropic gelation as a bead forming step.

One of these methods is the entrapment of cells in epoxy beads. The covalent polymer network is formed by polycondensation of an epoxy resin with a poly-amin at room temperature. To prepare spherical epoxy biocatalysts with sufficient porosity the polycondensation process was combined with the ionotropic gelation technique (Klein and Eng 1979; Klein et al. 1980).

By using a precondensation step of epoxy resin and curing agent to lower the amount of toxic low molecular weight compounds, prior to cell addition it was possible to entrap whole yeast cells in a living state (Fig. 17). The epoxy beads showed high mechanical strength and excellent chemical stability. A residual EtOH productivity of 21% was observed and a repeated and continuous re-incubation of the biocatalyst was possible (Klein and Kressdorf 1982).

To avoid the toxicity of low molecular weight chemicals we developed a method which uses polymers only (Vorlop and Klein 1982). In this case an Eudragit dispersion (Röhm 2837 A or E 30 D from Röhm Pharma GmbH, Post-fach 4168, 6100 Darmstadt) mixed with alginate or chitosan is used for the

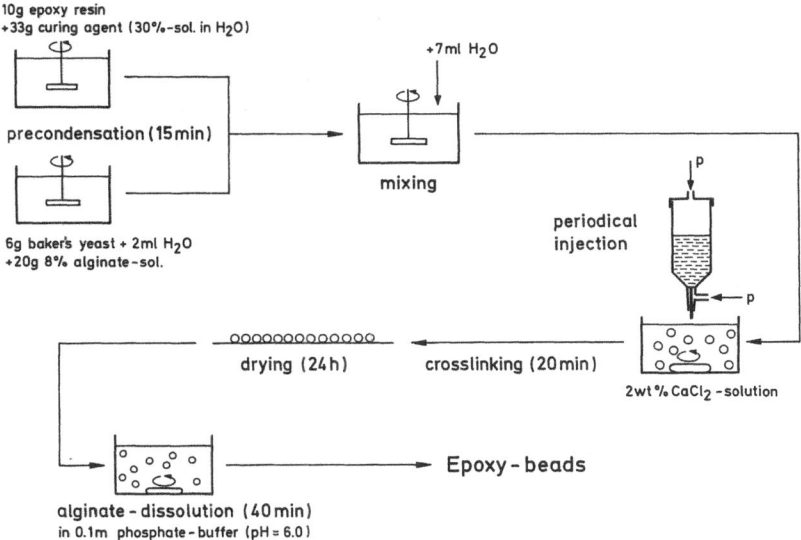

Figure 17. Schematic diagram for the preparation of spherical epoxy biocatalysts

entrapment of cells. A typical example of cell immobilization is described as follows:

First   2 g yeast + 2 g H_2O
          2 g Na-alginate solution (3 % Protanal LF 20/60)
          5.5 g Eudragit® dispersion (2837 A)*

were mixed while stirring. This mixture was dropped into a gently stirred 1 % $CaCl_2$ (w/v) solution. The beads were isolated after 40 min and dried using mild ventilation. After the drying process the biocatalysts were washed with 0.1 M phosphate buffer (pH 7) to dissolve the alginate.

The biocatalysts showed good mechanical properties. A residual EtOH productivity of 64 % by a cell loading of 0.33 g yeast/ml biocatalyst was observed.

Covalent stabilization methods of alginate gels were also developed by Birnbaum et al. (1981). Alginate beads containing entrapped cells were treated with polyethyleneimine followed by glutaraldehyde. Also used were alginate active esters or aldehyde alginate, mixed with cells and extruded into calcium chloride solution. The beads were cross-linked with polyethyleneimine. These covalent cross-linked alginate biocatalysts remained stable in phosphate buffer solutions.

References

Abbott BJ (1976) Preparation of pharmaceutical compounds by immobilized enzymes and cells. Adv Appl Microbiol 20:203–257

Becke JW (1982) Ethanolproduktion mit immobilisierten Zellen. Diplom Thesis, TU Braunschweig, F R G

Behrendt U (1981) Versuche zur Optimierung der L-Serin-Bildung mit dem methylotrophen Bakterium Pseudomonas 3ab. Dissertation, TU Braunschweig, F R G

Birnbaum S, Pendleton R, Larsson PO, Mosbach K (1981) Covalent stabilization of alginate gel for the entrapment of living whole cells. Biotechnol Lett 3(8):393–400

Brimble TW (1981) The developing role of zwitterionic buffers. Kontakte (Darmstadt) 1:37–43

Brodelius P, Deus B, Mosbach K, Zenk MH (1979) Immobilized plant cells for production and transformation of natural products. FEBS Lett 103(1):93–97

Chibata I, Tosa T (1981) Use of immobilized cells. Annu Rev Biophys Bioeng 10:197–216

Duarte JMC, Lilly MD (1980) The use of free and immobilized cells in the presence of organic solvents: the oxidation of cholesterol by Nocardia rhodochrous. Enzyme Eng 5:363–367

Durand G, Navarro JM (1978) Immobilized microbial cells. Process Biochemistry 13(9):14–23

Hackel U, Klein J, Megnet R, Wagner F (1975) Immobilization of microbial cells in polymer matrices. European J Appl Microbiol 1:291–293

Hackel U (1976) Polymereinschluß von Mikroorganismen. Dissertation, TU Braunschweig, F R G

Kierstan M, Bucke C (1977) The immobilization of microbial cells, subcellular organelles, and enzymes in calcium alginate gels. Biotechnol Bioeng 19:387–397

Klein J, Hackel U, Schara P, Washausen P, Wagner F (1976) Polymer entrapment of microbial cells: phenol degradation by Candida tropicalis. Abstracts- 5th Intern Ferment Symp Berlin: page 295

Klein J, Wagner F (1978) Immobilized whole cells. In Dechema Monogr 82 — Biotechnology —. Verlag Chemie, Weinheim New-York: 142–164

Klein J, Eng H (1979) Immobilization of microbial cells in epoxy carrier systems. Biotechnol Lett 1:171–176

---

* By using E30D the cell and alginate amount must be lower!

Klein J, Wagner F (1979) Immobilized whole cells. In: Bucholz K (ed) Dechema Monogr 84 — Characterization of immobilized biocatalysts —. Verlag Chemie, Weinheim New-York: 265–335

Klein J, Wagner F, Eng H, Vorlop KD (1980) Ger. Offen Nr 2835874

Klein J, Wagner F, Vorlop KD (1981) Ger Offen Nr 3005632 and 3005633

Klein J, Kressdorf B (1982) Immobilization of living whole cells in an epoxy matrix. Biotechnol Lett 4:375–380

Kluge M, Klein J, Wagner F (1982) Production of 6-aminopenicillanic acid by immobilized Pleurotus ostreatus. Biotechnol Lett 4(5):293–296

Lim F, Sun AM (1980) Microencapsulated islets as bioartificial endocrine pancreas. Science 210:908–910

Linko YY, Weckstrom L, Linko P (1980) Sucrose inversion by immobilized Saccharomyces cerevisiae yeast cells. In: Linko P, Larinkari J (eds) Food process Engineering Vol 2, Applied Science Publishers, Barking: 81–91

Makiguchi N, Arita M, Asai Y (1980) Preparation of luminous materials containing immobilized luminous bacterial cells. J Ferment Technol 58:167–169

Michaels AS (1965) Polyelectrolyte complexes. Industrial and Engineering Chemistry 57(10):32–40

Mosbach K, Mosbach R (1966) Entrapment of enzymes and microorganisms in synthetic cross-linked polymers and their application in column techniques. Acta Chem Scand 20:2807–2810

Mosbach K (1982) Use of immobilized cells with special emphasis on the formation of products form3d by multistep enzyme systems and coenzymmes. J Chem Tech Biotechnol 32:179–188

Nilsson K, Mosbach K (1980) Preparation of immobilized animal cells. FEBS Lett 118(1):145–150

Ohlson S, Larsson PO, Mosbach K (1979) Steroid transformation by living cells immobilized in calcium alginate. European J Appl Microbiol Biotechnol 7(2):103–110

Paul F, Vignais PM (1980) Photophosphorylation in bacterial chromatophores entrapped in alginate gel: improvement of the physical and biochemical properties of gel beads with barium as gel-inducing agent. Enzyme Microb Technol 2:281–287

Pilwat G, Washausen P, Klein J, Zimmermann U (1980) Immobilization of human red blood cells. Z Naturforsch 35c:352–356

Scherer P, Kluge M, Klein J, Sahm H (1981) Immobilization of the methanogenic bacterium Methanosarcina barkeri. Biotechnol Bioeng 23:1057–1065

Scheurich P, Schnabl H, Zimmermann U, Klein J (1980) Immobilization of individual protoplasts. Biochim Biophys Acta 598:645–651

Smidsrød, Haug A (1965) The effect on divalent metals on the properties of alginate solutions. Acta Chem Scand 19:329–340 and 341–351

Thiele H (1954) Ordered coagulation and gel formation. Discuss Faraday Soc 18:294 u 301

Veliky IA, Williams RE (1981) The production of ethanol by Saccharomyces cerevisiae immobilized in polycation-stabilized calcium alginate gels. Biotechnol Lett 3(6):275–280

Vorlop KD, Klein J, Wagner F (1980) Immobilization of microbial cells by polymer precipitation. Abstracts- 6th Intern Ferment Symp London (Canada): page 122

Vorlop KD, Klein J (1981) Formation of spherical chitosan biocatalysts by ionotropic gelation. Biotechnol Lett 3(1):9–14

Vorlop KD, Lang S, Klein J, Wagner F (1981) Reincubation behaviour of immobilized tryptophan producing Escherichia coli cells. Abstracts- 2nd Eur Congr Biotechnol, Eastbourne (England): page 49

Vorlop KD, Klein J (1982) unpublished results

Wagner F, Lang S, Bang WG, Vorlop KD, Klein J (1982) Production of l-tryptophan with immobilized cells. Enzyme Eng 6 (in press)

Immobilized Enzyme and Cell Technology to Produce Peptide Antibiotics

Erick J. Vandamme

I. Introduction

Already in the late 1800's, industrial application of enzymes had become a reality; ever since people have tried to reuse these enzymes. The concept of using microbial enzymes, attached to solid surfaces, as stabilised reusable catalysts — now commonly named "immobilised enzymes" — was indeed proposed early in the 1900's: however, attempts to adsorb enzymes on active coal as a means of enzyme stabilisation were initially rather unsuccessful.

Relatively stable enzyme systems became available in the 1950's, but industry did not show interest as to their scale-up. Only in the 1960's — stimulated by Katchalski's group at Rehovot in Israël — did industry get closer involved with immobilised biocatalysts (Mosbach, 1971; Mosbach, 1976; Brodelius, 1978; Barker, 1980).

Indeed, commercial realisation of this concept occurred in 1969 in Japan: Chibata and collaborators immobilised *Aspergillus oryzae* amino-acylase which is used for continuous industrial production of L-amino acids from acyl-DL-amino acids at Tanabe Seiyaku Co Ltd., Osaka, Japan (Chibata et al., 1972; Chibata and Tosa, 1977).

At that time, two other immobilised enzyme systems reached industrial pilot plant scale-level: in England, immobilised penicillin acylase, which is now commercially used to prepare 6-aminopenicillanic acid from penicillin G or V and in the USA, immobilised glucose isomerase, now worldwide used to convert glucose into fructose.

These industrial applications have provided the impetus for the enormous research activity during the last decade in a science area which is now designated as enzyme engineering or enzyme biotechnology. These realisation also confirmed the old speculation that large scale industrial use of microbial enzymes could have large economic benefits (Skinner, 1975).

Several prognoses generally agreed on the vast scope and positive value of this development, but have also forecasted that still 20 to 30 years more were needed to fullfill its potential, especially for those processes, involving complex multienzyme catalytic reactions (Wiseman, 1980).

Indeed, technology is now available to accomplish the design and operation of sophisticated enzyme reactors, but generally running costs are yet too high for industrial application, except in a few specific cases, using single step enzyme reactions. These costs are mainly caused by enzyme loss during recovery, immobilisation and operation stages. Further development and improvement of enzyme extraction, purification and storage methods, carrier materials, immobilisation techniques and reactor designs, cofactor immobilisation and regeneration are needed to lead to a wider application of such enzymatic processes in industry with economic profit.

An exciting alternative approach to enzyme immobilisation is immobilisation of whole (living) microbial cells as a source of the desired enzyme(s) (Vandamme, 1976; Abbott, 1976; Abbott, 1977; Jack and Zajic, 1977; Venkatasubramanian and Vieth, 1979; Chibata and Tosa, 1981). Immobilisation of whole cells is not a novel concept but rather a duplication and refinement of phenomena observed in nature (microbial activity in soil, leaching of mineral ores) and in some industrial microbiological processes, where microorganisms or cells are bound to solid surfaces of form films (trickling filters, vinegar process, tissue culture). This concept has lately exploded into a massive research effort and in an immediate industrial application of immobilised microbial cell systems, also called "controlled catalytic biomass" (Venkatasabramanian and Vieth, 1979).

The first generation of catalytic bioprocesses with immobilised microbial cells has recently gained commercial status. Again, preferentially those immobilised cell-reactions which involve a single enzymatic step are preformed industrially (amino-acylase, penicillin acylase, glucose isomerase, aspartase, fumarase, lactase, . . .) though here multi-step reactions hold a great promise. However, it is gradually becoming apparent that several fundamental aspects and the real potential of immobilised cell systems could only be fully understood by the detailed study of these multi-complex reactions, involving entire metabolic pathways, multi-enzyme systems, cofactor and ATP requiring and generating systems, immobilised cell organelles or spores, artificially composed pathways, or co-immobilisation of different cell types (Hough and Lyons, 1972; Demain et al., 1976; Martin and Perlman, 1976; Buttler, 1977; Lowe, 1981).

II. Peptide Antibiotics and Immobilised Enzyme or Cell Technology

So far immobilised bio-catalysts (cells or enzyme) have been applied mainly in simple bioconversion reactions, but recently there is increasing interest to use such systems to produce complex fermentation products and to apply them in organic synthesis. This chapter concentrates on the use of immobilised enzymes and cells for transforming or synthesising peptide antiobiotics and evaluates the applicability of this concept in the field of peptide antibiotic bioconversions and fermentations (Katz and Demain, 1977; Vandamme, 1980, 1981).

Antibiotics are the most important complex compounds made by microbiological synthesis (Perlmann, 1977; Vandamme, 1983). Since the discovery of early antibiotics and of penicillin, more than 5500 natural microbial compounds have been described, all of which display antibiotic activity. About 150 of these are produced on an industrial scale by fermentation and find use in medical and agricultural practice (Berdy, 1974; Perlman, 1977; Vandamme, 1983). At present, all antibiotics are synthesised by microorganisms in classical fermentation processes. In several cases the natural microbial product can be chemically or enzymically converted into a semi-synthetic antibiotic with superior therapeutical properties (Vandamme, 1977; Sebek, 1980; Vandamme, 1980).

Microbial bioconversions of antibiotics can vary from extensive degradations to well-defined reactions. Most such bioconversion reactions result in an inactivation of the antibiotic compound, thereby excluding a practical application for the obtained bioconversion products (Abraham and Chain, 1940; Sebek and Perlman, 1971; Benveniste and Davies, 1973; Sebek, 1974, 1975, 1980; Vandamme, 1980). However, in a few cases useful compounds are formed which can subsequently be used for the industrial production of semi-synthetic antibiotics or even new antibiotic analogues (Sebek, 1980; Vandamme, 1980). The use of immobilised enzymes and cells will be described to convert peptide antibiotics (penicillins, cephalosporins and newer β-lactams) into useful derivatives (bioconversion).

The total organic or enzymic synthesis of antibiotics presents, even today, a laborious and uneconomical task, although conventional fermentation also presents problems: except for the relatively simple chloramphenicol and cellocidin molecules, indeed none of the 150 commercial antibiotics is made by chemical synthesis, despite the fact that chemical routes are known for many important antibiotics. The economics of multi-reaction chemical synthesis of such complicated biological molecules are just too unfavourable. Of particular significance in this regard is the fact that development of the new semi-synthetic penicillins did not start at the time the chemical synthesis route to the penicillin nucleus (6-APA) was discovered, but only later was it found that under certain conditions, the fungus *P. chrysogenum* would excrete 6-APA, which could be converted into new penicillins. As organic synthesis is practically impossible, antibiotic synthesis by traditional fermentation also has its drawbacks! Apart from the use of sophisticated production facilities, the impressive increase in antibiotic productivity in industrial fermentations is largely a result of forcing the microorganisms to overproduce a useful metabolite by mutation or mutasysnthesis, by directly influencing cell metabolism and cell environment, e.g. nutritional control, precursor addition or, recently, by genetic engineering. Most antibiotics are produced as secondary metabolites, i.e. their synthesis is delayed until the growth of the cells declines or stops. Consequently, such classical fermentation processes always have a "non-productive" phase. Once all the necessary enzymes are formed in the cell theoretically, a "linear" antibiotic production phase could be maintained over a long period if it were not for the fact that the involved enzymes are generally rapidly inactivated. Conversion of sugar substrates into antibiotic is rather inefficient, due to usage for growth, maintenance and the many side-reactions occuring in intact cells. Furthermore, strain degeneration, i.e. the selection of poorly producing strains which grow faster, causes a major problem in fermentation.

As a solution to these disadvantages, attempts have been made to replace fermentation or cellular synthesis by acellular processes, i.e. total enzymic synthesis *in vitro* (Hamilton et al., 1973; Demain and Wang, 1976). In such a process it is the ultimate aim to use isolated, stabilised, and immobilised enzymes, which in sequential reactions perform the total synthesis of an antibiotic upon addition of its precursors, ATP as an energy source and cofactors (Demain and Wang,

1976). This concept can be seen as an extension of the already well-known and industrially applied simple enzymic bioconversion of antibiotics (Sebek, 1980; Vandamme, 1980). This development, the acellular or cell-free total enzymic synthesis of antibiotics using immobilised enzymes, is yet to be exploited. Total enzymic synthesis of antibiotics will be discussed with the oligopeptide antibiotic gramicidin S as an example. Alternatively, immobilised living cells can be used to produce peptide antibiotics as opposed to conventional fermentation. Indeed, immobilised enzyme or cell technology could provide for an improved production (bioconversion or fermentation) of antibiotics.

III. Production of 6-Aminopenicillanic Acid (6-APA) and Reacylation of 6-APA into Penicillins with Immobilised Catalysts

a) Penicillin acylases as useful enzymes

Bioconversion of penicillin into 6-aminopenicillanic acid (6-APA), the penicillin nucleus, and its side chain acid (Figure 1) is an important bioconversion reaction; 6-APA is the starting compound for the industrial production of the semi-synthetic penicillins with superior therapeutical effectiveness (Vandamme and Voets, 1974; Vandamme, 1977; Rolinson, 1979; Vandamme, 1980). The microbial enzymes that hydrolyse penicillins into 6-APA or acylate 6-APA have been named penicillin acylases (EC 3.5.1.11.).

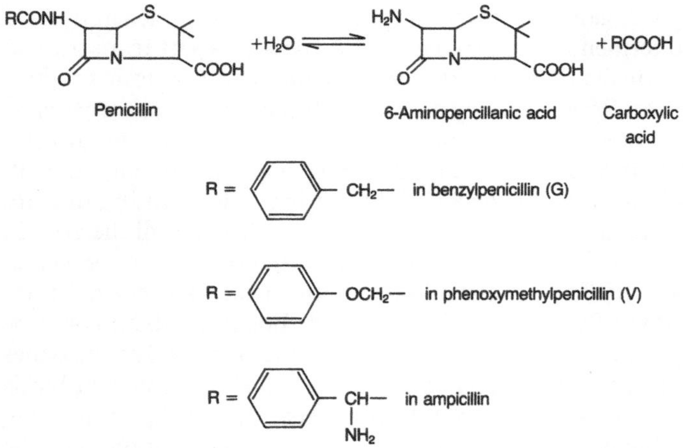

Figure 1. Formation and Reacylation of 6-Aminopenicillanic Acid (6-APA) by Penicillin Acylases: penicillin G, penicillin V and ampicillin acylase

   The worldwide demand for these semi-synthetic penicillins has brought 6-APA into a central position as a major pharmaceutical product. As many as 16 different semi-synthetic penicillins, all derived from 6-APA, are in widespread clinical use today (Rolinson, 1979). Bulk 6-APA production can now be achieved either by chemical (Weisenburger and Vanderhoeven, 1970; Fosker et al. 1971) or enzymic hydrolysis (Vandamme, 1980).

The advantage of either process has depended so far upon technological advances in individual companies and both systems have their own merits. However, with the introduction of methods of immobilisation of enzymes or cells, which promote stability and prolonged high activity that can result in re-use and continuous bioconversion, and with the ever increasing petrochemical stock and energy prices in mind, the enzymic splitting currently appears to display better economics (Vandamme, 1980; Savidge, personal communcation).

Penicillin acylase activity has now been demonstrated to occur in a wide range of bacteria, actinomycetes, yeasts and moulds (Vandamme, 1980; Vandamme and Voets, 1974; Vandamme, 1977). Penicillin V, penicillin G and ampicillin acylase are the three types of acylase now clearly recognised (Fig. 1) (Vandamme, 1980). At the moment highly productive (genetically engineered) penicillin acylase strains are used to produce 6-APA on an industrial scale, starting from either the biosynthetic penicillin G (benzylpenicillin) or V (phenoxymethylpenicillin). In addition, some penicillin acylases catalyse the reverse reaction too, and can acylate 6-APA to form penicillin compounds. Furthermore, certain cephalosporins, structural analogues of the penicillins, can also be deacylated or acylated into useful compounds.

Fermentation processes for penicillin acylase production by bacteria and fungi have been reviewed recently (Vandamme, 1980; Savidge and Cole, 1975). Characterisation and properties of these enzymes have recently been summarised by Vandamme and Voets (1974) and Vandamme (1977, 1980).

Some relevant characteristics are given below:

• Studies with purified enzymes have disclosed that penicillin V-acylases are highly specific enzymes with practically no activity towards penicillin G or other compounds, whether found in moulds (*Fusarium semitectum, Bovista plumbea (Pleurotus ostreatus), P. chrysogenum*) (Vandamme, 1980) or in bacteria such as *Erwinia aroideae* (Vandamme, 1980), *Pseudomonas acidovorans* (Lowe et al., 1981) or *Bacillus sphaericus* (Carlsen and Emborg, 1981).

Penicillin V-acylases produced by yeasts and actinomycetes have not yet been purified, but they all display preferential hydrolysis of the penicillin V molecule. Penicillin V-acylases are mainly intracellular enzymes, with optimal hydrolytic activity at pH 7–8; with a few exceptions known. None of the known purified fungal or bacterial penicillin-V-acylase display synthetic activity.

• In contrast to the penicillin V-acylase, the substrate specificity of penicillin G-acylase is rather broad and these enzymes can be considered as aspecific deacylating enzymes (Vandamme, 1980). The *Escherichia coli* ATCC 9637 and ATCC 11105 acylases, which are most studied, are specific for phenylacetylated compounds. Penicillin G-acylase from actinomycetes and fungi have not yet been fully characterised, but in addition to penicillin G they all hydrolyse a range of N-phenylacetyl-L-α-amino acids. Penicillin G-acylases are mainly intracellular enzymes with optimal hydrolytic action at pH 7–9. A few strains of *Streptomyces* and *Bacillus megaterium* are known to produce the enzyme extracellularly. Many bacterial penicillin G-acylases also catalyse at acid pH values (4.5–5.5) the synthesis of penicillins from 6-APA and phenylacetic acid or its derivatives. Profiting from this synthetic activity of penicillin G-acylases, enzymatic synthesis of many semi-synthetic penicillins (and cephalosporins) has been achieved (Svedas et al., 1980a, b). However, the rather low yield and the reversibility of these processes have so far precluded the industrial enzymic synthesis of semi-synthetic β-lactam antibiotics, although the use of immobilised acylases or cells for such synthesis looks promising.

● Ampicillin acylases have so far only been found in *Pseudomonas melanogenum* and *P. ovalis* strains; they do not hydrolyse nor synthesise penicillin G, V or related acylated compounds (Nara et al., 1972).

The industrial enzymic route for 6-APA production was developed around 1960. Initially, cell suspensions of active strains could only be used once and the productivity was very low, about 0.5 to 1 kg of 6-APA per kg *E. coli* suspension. Novel methods to immobilise *E. coli* cells increased productivity up to 50 kg of 6-APA per kg of immobilised catalyst. Current productivities are within the range 100–250 kg 6-APA/kg; however, no data have been published by the companies involved. It can be assumed though that about 60% of all 6-APA produced today is made by the immobilised catalyst route: this means that about 3000 tons of 6-APA are produced by 15 to 30 tons of immobilised penicillin acylase catalyst (Poulsen, 1981). Industrial 6-APA producers include

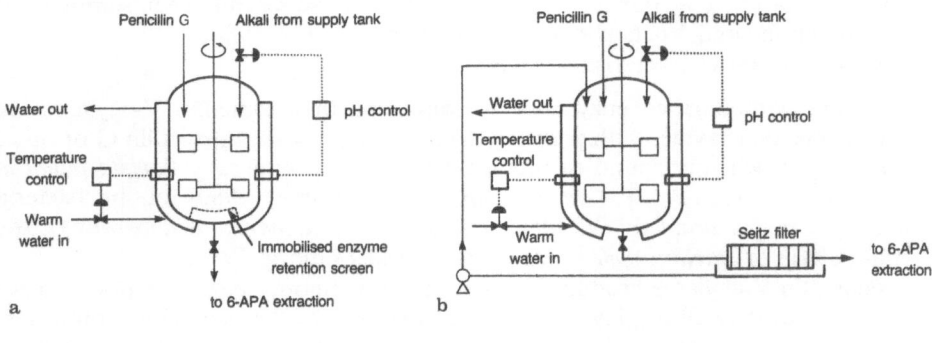

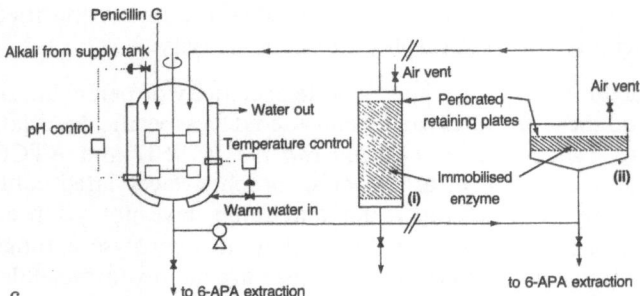

Figure 2. Immobilised Biocatalyst Reactors for 6-APA Production. (Reproduced with kind permission of T. A. Savidge)

Bayer, Gist-Brocades, Beecham, Glaxo, Astra, Biochemie, Antibiotics in Europe; Pfizer, Bristol-Meyers, Squibb, Wyeth in the USA, and Kyowa Hakko (Japan) and Yung Jin Pharmaceuticals (S. Korea) in the Far East.

Industrial application of immobilised penicillin acylases is still confronted with problems: upon penicillin hydrolysis into 6-APA, the side-chain acid is liberated which causes a decline in the pH. This pH change results in a slower reaction rate. A higher starting pH is not wanted due to β-lactam ring hydrolysis and inactivation, though it could increase the reaction rate. A strict pH control is thus necessary during this bioconversion process which is difficult to run in a continuous packed-bed reactor. Therefore, a batch reactor is generally used, but continuous multi-stage tubular or stirred tank reactors could allow for a continuous process (Carleysmith and Lilly, 1979; Park et al., 1982); such a process is then faced with another aspect of the catalytic properties of the enzyme: end-product inhibition. So in this particular case, it is in fact the reaction type which indicates the type of reactor and its control to be used. Immobilised enzyme or cell reactor types for enzymatic deacylation of penicillins or cephalosporins are presented in Fig. 2.

b) Immobilised enzymes as catalyst

6-APA-formation

Several attempts have been made to isolate and purify penicillin acylases and methods to stabilise and immobilise these enzyme preparations have been described (Abbott, 1976; Self et al., 1969; Szweczuk et al., 1979). Adsorption, cross-linking, covalent attachment and physical attachment have been used; both soluble and insoluble carriers have been tested out.

So far, few penicillin V-acylases have been immobilised. A crude *Rhodotorula glutinis* var. *glutinis* acylase preparation was entrapped in polyacrylamide gel and tested as to its capacity to act as a continuous immobilised enzyme reactor (Vandamme and Voets, 1973).

The Basidiomycete, *Bovista plumbea* (*P. ostrateus*), has been reported to be a highly suitable source of penicillin V-acylase since no substrate or endproduct inhibition occurs (Schneider and Rohr, 1976; Brandl and Knauseder, 1975). The enzyme appears to have excellent kinetic properties and its stability to purification, immobilisation and working conditions are reflected in a yield of 90% from 6% penicillin V by whole washed mycelium and 97% from 8% penicillin V by immobilised enzyme entrapped in cellulose triacetate in a cycled column reactor at 32 °C (Kluge et al., 1982).

No published reports are yet available on ampicillin acylase immobilisation. On the contrary, several penicillin G-acylases have been purified and immobilised. Self et al. (1969) extracted the penicillin acylase from *E. coli* ATCC 8637 strain, which was grown on phenylacetic acid and glutamate. They purified the enzyme by fractionation with streptomycin sulphate, ammonium sulphate, and polyethyleneglycol, followed by chromatography on DEAE-cellulose. The purification factor was about 100–200, and the overall yield was about 35%. The enzyme was chemically attached to triazine derivatives of cellulose and the kinetics of these immobilised penicillin acylase preparations were studied in a penicillin acylase reactor. High specific activities were obtained at an optimum pH of 7.5 and at a temperature of 37 °C. Adsorption of the substrate by the cellulose complex might have been responsible for the lower K_m values found as compared with the soluble enzyme. A DE 52-cellulose reactor and a cellulose sheet reactor were active during 11 weeks of operation without any loss of activity. The free enzyme, under the same conditions, lost 42% of its activity in 1 day. The *E. coli* NCIB 8,743A enzyme has also been immobilised by covalent binding to DEAE-cellulose, using 2-amino-4,6-dichloro-s-triazine. The preparation retained 45–81% of its activity observed before attachment. There was no evidence of diffusional limitation of the reaction rate. Using the immobilised enzyme, a K_m value of 6.3×10^{-4} M was calculated at 37 °C and pH 8.0. The optimum pH for penicillin G hydrolysis was 7.65, while the free enzyme displayed maximal activity at pH 8.2 (Warburton et al., 1972; Lilly et al., 1972). The use and stability of this immobilised enzyme preparation has been described in batch and continuous-flow stirred tank reactors (Warburton et al., 1973). Several other immobilisation methods involving covalent bond formation have been described, especially for *E. coli* penicillin acylase. Ekstrom et al. (1974) and Lagerlof et al. (1976) tested several coupling procedures involving a range of polysaccharides (amylose, α-glucan, dextran, carboxymethylcellulose, cellulose, Sepharose 4B, sulphoethyl-Sephadex, carboxymethyl-Sephadex, DEAE-Sephadex, Sephadex G-200) or polyacrylamide supports for *E. coli* penicillin acylase. Cyanogen bromide was used to bind the acylase to the polysaccharides. The Sephadex-acylase complex was used by the Swedish company Astra Läkemiddel for large-scale 6-APA production in stirred tank batch reactors. Up to 60 batch reactions were catalysed before significant losses of enzyme activity occurred, most probably due to proteolytic enzyme activity and microbial contamination. The enzyme was recovered by filtration. Other immobilisation methods include coupling of *E. coli* acylase to resins with glutaraldehyde (Savidge et al., 1974) or to copolymers containing acrylamide and ethylene maleic anhydride (Huper, 1973). Up to 98% of the added enzyme became covalently bound under optimised immobilisation conditions. Ryu et al. (1972a, b) compared the kinetic properties of the soluble and the bentonite- or diatomaceous-earth-immobilised form of extracellular enzyme of *B. megaterium* ATCC 14945 and found that they exhibited significantly different inhibition constants. A continuous enzyme reactor was designed, consisting of a continuous-flow stirred tank and an ultrafiltration unit; it enables recirculation of the enzyme and continuous removal of the end products. A kinetic model was derived that was based on the inhibition effects of the end products on the enzymic action. This model was then used to simulate the performance of a continuous enzyme reactor system and to optimise the productivity of the reactor in terms of process variations (Ryu et al., 1972a, b). An immobilisation process involving a combination of physical entrapment and covalent bond formation has also been described (Beecham Group, 1974). An *E. coli* acylase was here coupled to a water-soluble sucrose-epichlorohydrin copolymer with a molecular weight of 400000. This attachment procedure facilitated enzyme recovery through ultrafiltration. As well batch reactions as two-stage continuous reactions were applied and resulted in 6-APA yields in excess of 90%. Bead-form macroporous carriers based on glycidyl-methacrylate with ethylene dimethacrylate copolymers containing the reactive oxirane group were used for immobilisation of *E. coli* penicillin acylase by Drobnik et al. (1979). Direct binding through oxirane groups results in activity loss, while immobilisation on a glutaraldehyde-activated amino-carrier was most efficient, irrespective of whether the amino groups were formed by ammonia or by 1,6-diaminohexane treatment of the original oxirane carrier. Hydrazine or acide treatment gave lower immobilisation yields. Most enzyme activity was preserved by coupling the carbodiimide-activated enzyme to the carrier with alkyl or acylamino groups. The immobilised activity obtained is very stable in solution and upon freeze-drying with sucrose. These data might indicate how a combination of the unique functions of proteins with excellent physical properties of synthetic and semi-synthetic carriers might lead to new applications in enzyme biotechnology. An *E. coli* acylase preparation in a spherical granule form was obtained by Szewczuk et al. (1979) by copolymerisation of penicillin acylase, previously modified with maleic anhydride and acrylamide via diallyl malonate diamide as cross-linking agent. A semi-purified preparation of the penicillin acylase from *E. coli* ATCC 9637 cells was entrapped at Snam Progetti (Italy) in cellulose triacetate fibres which were packed in a column in order to obtain an immobilised enzyme reactor (Marconi et al., 1973). An enzyme solution was added dropwise to a solution of cellulose triacetate in methylenechloride to form an emulsion which was then extruded into a coagulating toluene bath. The acylase fibres were found to be quite stable but exhibited lower activity than the soluble enzyme. No loss of activity by casual contamination was found, mainly because the entrapped enzyme was protected against microbial proteolytic attack. Hindered substrate diffusion through the fibre matrix reduced the activity. A remarkable reduction of the hydrolysis time was observed and a conversion yield higher than 90 % was reached. It was found that the acylase fibres were a better catalyst at low temperature than the free enzyme. The fibre column reactor yielded 85 % hydrolysis of a 12 % solution of penicillin G in 190 min by recycling and using a high flow rate. During a 4 month period, only 20 % of the initial acticity was lost. Other polymers such as poly-γ-methylglutamate and ethylcellulose could also be used for fibre immobilisation, giving high activity, improved thermal stability and protection against bacterial contamination and proteolysis (Marconi et al., 1975; Dinelli et al., 1972).

The allergenicity of penicillins is due in part to the formation of penicilloylated proteins which can cause anaphylactic shock in man. Even crystalline 6-APA was found to contain protein impurities responsible for allergic reactions. This antigenic impurity found in 6-APA appeared to be derived from the free penicillin acylase preparations used for side chain removal from penicillin G and V. These immunological manifestations could be reduced by treatment of 6-APA with a bromo-acetyl cellulose-immobilised protease mixture from *Streptomyces griseus* (Shaltiel et al., 1970; Shaltiel et al., 1971). A comparison of the antigenicity of 6-APA, produced by immobilised acylase, and that produced by free enzyme or whole-cell suspension revealed that the 6-APA from the bound enzyme reactor was significantly less allergenic. Formation of these allergens in immobilised acylase reactors was prevented because the enzyme complex was stable and did not release protein into solution.

6-APA acylation

Penicillin G-acylases are also able to acylate 6-APA to produce penicillins. Though immobilised acylases have been used successfully in this respect, commercial application is postponed by low yields and the availability of simple acylation procedures. Self et al. (1969) tested the ability of the cellulose-immobilised *E. coli* ATCC 9637 acylase to carry out the reverse reaction (reacylation) by adjusting effluent from a penicillin hydrolysis experiment to pH 5.0 and passing it through the reactor again. The presence of benzyl-penicillin in the effluent confirmed that the reverse reaction was indeed occuring. The *Bacillus circulans* succinolated enzyme, bound to DEAE-Sephadex, was able to synthesise ampicillin from 6-APA and D-phenyl-methylglycinester with a 57% yield (Fig. 3) (Abbott, 1977). The fibre-entrapped *E. coli* acylase also catalysed the synthesis of ampicillin and amoxycillin from 6-APA and esters of D-phenylglycine or p-hydroxyphenylglycine, respectively, at pH 7.0 and 25 °C (Marconi et al., 1975). The acylase displays a preference for the D-configuration of the acyl donor.

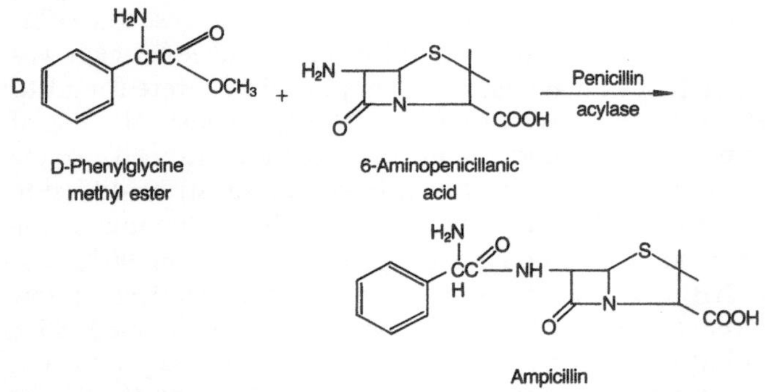

Figure 3. Synthesis of ampicillin from 6-APA and D-phenylglycine methylester by penicillin acylase

A two-step process can be envisaged to synthesise ampicillin from penicillin G, based on the reversibility of the acylase action and on choosing the proper acylase-type enzymes. After conventional *E. coli* acylase hydrolysis of penicillin G into 6-APA and phenylacetic acid, the free side-chain acid normally must be removed to prevent the reacylation reaction. This separation is not needed when, in a subsequent step, *P. melanogenum* or *P. ovalis* acylase is used, which is specific for phenylglycine acyl donors (Nara et al., 1972; 1973); by adding this acylase and a phenylglycine donor to the 6-APA phenylacetic acid mixture, only ampicillin can be formed. No reports on immobilisation or co-immobilisation of these two types of acylases have yet been published.

c) *Immobilised cells as catalyst*

Recently, the immobilisation of whole microbial cells and their applications have been the subject of increased interest (Chibata and Tosa, 1977, 1981;

Vandamme, 1976; Abbott, 1976; Jack and Zajic, 1977; Vandamme, 198). Such systems obviate the need for enzyme isolation prior to immobilisation and can catalyse a series of sequential reactions without the need of supplying cofactors. However, the intact cell walls or membranes may act as permeability and diffusion barriers, and unwanted side reactions might occur. Several attempts have already been made to apply this cell immobilisation technology for deacylation of penicillin G or V and reacylation of 6-APA. The first attempt to immobilise intact cells containing penicillin acylase was made by Dinelli (1972), who entrapped whole *E. coli* ATCC 9637 cells in wet-spun cellulose triacetate fibres. The fibres, containing 15 mg of wet cells g^{-1} of polymer, exhibited 80% of the activity of a similar amount of cells in free suspension. This novel system displayed a high and lasting activity over a period of 4 months. Chibata and his group (Sato et al., 1976) immobilised the same strain by entrapment of living cells in a polyacrylamide gel lattice and used an immobilised cell column for continuous production of 6-APA. The half-life of the enzyme was 17 days at 40 °C and 42 days at 30 °C. From the effluent of the column, 6-APA was obtained with a yield of 78%. They concluded that this procedure is much more advantageous than continuous methods using immobilised enzyme. In an attempt to develop a very efficient immobilised cell system for 6-APA production, Mayer et al. (1979) first increased the specific activity of the cell system itself and then selected the appropriate immobilisation technique. As to the selection of strains with improved acylase activity, they reported that an increase of specific enzyme activity in *E. coli* ATCC 11105 could be achieved by a temperature shift from 27 to 24 °C during fermentation at low dissolved oxygen tension. Mayer et al. (1979) were also the first to apply genetic engineering techniques by subcloning the *E. coli* ATCC 11105 penicillin G-acylase gene (using "cosmid packaging" for initial gene isolation) on multicopy plasmids such as pOP203-3 and pBR322. A cosmid hybrid *E. coli* 5K (pH M:12) strain was obtained with ten-fold higher activity: the hybrid strain does not need any more phenyl-acetate as an inducer, nor was the acylase sensitive to repression by carbohydrate carbon sources (Vandamme, 1980). These authors showed that this increase in free cell activity can be transferred to immobilised cell systems if preparations of sufficiently small particle size and high porosity can be obtained (Klein and Eng, 1979). Comparing different polymer entrapment immobilisation techniques (alginate, polymethylacrylate, polyurethane, epoxide entrapment) for whole *E. coli* ATCC 11105 cells, the epoxy bead process proved to be superior (40 days half-life time compared with 1 day for free suspended cells) due to its high cell loading capacity, its excellent mechanical properties, and appreciable level of efficiency. While part of the loss in activity, always observed with entrapment procedures, is due to deactivation of the enzymic function, a considerable influence of diffusional limitation may exist. A quantitative evaluation of this effect is based on the well-known Thiele modulus ($\emptyset$) which has also been shown to be applicable in entrapped immobilised whole cell catalysis (Klein et al., 1979). The immobilised cells remained viable and an increase in catalytic activity was observed upon reincubation. Similar observations were reported with polyacrylamide gel-immobilised *Kluyvera citrophila* cells (Morikawa et al., 1980).

Silicagel, alumina (acidic, neutral and alkaline), cellulose, charcoal and carb-oxymethylcellulose were compared as entrapment materials for washed *Strepto-myces* sp. mycelium by Amin et al. 1980).

Cellulose-immobilised cell columns could be used for 6-APA production at 38 °C at pH 7–8 for at least 30 days without substantial loss of activity. The penicillin acylase of this strain appears to be related to the ampicillin acylase type, which preferentially hydrolyses ampicillin. Glutaraldehyde-treated cells of *Proteus rettgeri*, bound to glycidylmetacrylate polymers could also convert penicillin G into 6-APA (Nelson, 1976). Penicillin V and G-acylase activity of intact fungal and *Streotpmyces* spores has been studied by Vezina et al. (1968), Singh et al., (1969), Vandamme et al. (1971), Vandamme (1972) and Amin et al. (1980). Immobilised spores of a *Fusarium* sp. were used to catalyse the bioconversion of penicillin V into 6-APA in flow reactors (Charles et al., 1980). Several of the tested catalysts not only gave higher yields (>85%) and productivities than did the corresponding free spores, but also had longer active half lifes (>2 weeks). A penicillin V-producing gram-negative bacterium, immo-bilised by Diers and Emborg (1980) with polyethylene imine could be reused 20 times at pH 7,5 at 40 °C to deacylate 3% penicillin V solutions.

Relatively few data have been published on the enzymic acylation of 6-APA using immobilised whole cells as catalysts. Commercial realisation of this process has been until now postponed by the availability of efficient chemical acylation methods and by the reversibility of the reaction which is furthermore subject to product or substrate inhibition (Vandamme, 1980). Intact cells of *Achromobacter* spp. and *B. megaterium*, immobilised by adsorption on DEAE-cellulose, have been used to synthesise ampicillin from 6-APA and D-phenylglycine methylester (Fig. 3) (Fujii et al. 1973, 1976). These cells could also acylate the cephalosporin nucleus 7-ACA and 7-ADCA, as does the fibre-entrapped *E. coli* acylase (see Figure 7). The introduction of methods of immobilisation of enzymes and cells which give stability and prolonged high activity that can result in a continuous bioconversion process, has clearly improved the economics of enzymic or microbial processes for 6-APA pro-duction.

However, the development of a commercially available penicillin acylase process for enzymatic synthesis of semi-synthetic penicillin (ampicillin, amoxy-cillin) from 6-APA is still doubtful due to low conversion ratios, difficult product separation and cost of side chain. In this respect, the ampicillin acylase and the α-amino acid ester hydrolase (see below) seem to offer more potential.

IV. Bioconversion of Cephalosporins with Immobilised Catalysts

a) Acylases and cephalosporin bioconversions

General aspects

Cephalosporins are penicillin analogues and can be synthesised by direct fermentation with *Cephalosporium acremonium* or by chemical ring expansion of penicillin (Abraham 1974). The major fermentation product, cephalosporin C, contains the 7-ACA nucleus and the side chain, α-aminoadipic acid (Fig. 4).

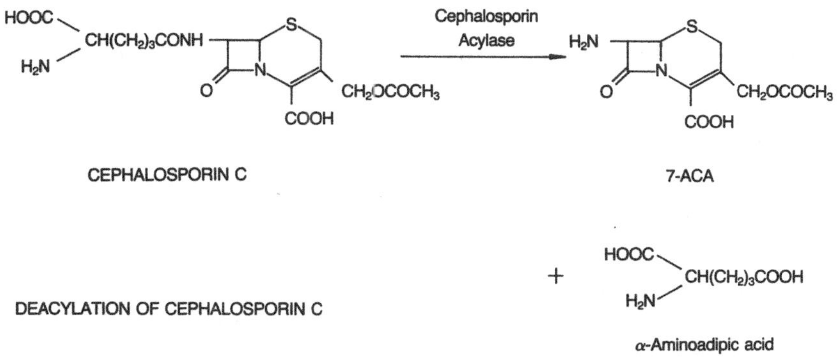

HOOC
CH(CH$_2$)$_3$CONH
H$_2$N

Cephalosporin
Acylase

H$_2$N

CH$_2$OCOCH$_3$

COOH

CH$_2$OCOCH$_3$

COOH

CEPHALOSPORIN C

7-ACA

DEACYLATION OF CEPHALOSPORIN C

+

HOOC
CH(CH$_2$)$_3$COOH
H$_2$N

α-Aminoadipic acid

Figure 4. Conversion of Cephalosporin C into 7-aminocephalosporanic acid (7-ACA) and α-amino-adipic acid

7-ACA derivatives with other side chains cannot be obtained by fermentation. The synthesis of semi-synthetic cephalosporins containing the 7-ACA nucleus thus depends upon removal of the α-amino adipic acid side chain: this can be accomplished with reagents such as PCl$_5$ or nitrosylchloride. Enzymic hydrolysis to yield 7-ACA has also recently been claimed (Figure 4). Chemical as well as enzymic acylation procedures can be used to couple other side chains to 7-ACA. Acylation of 7-ACA with 2-rhiophene acetic acid derivatives yields the therapeutically important cephalothin. For a long time, there has existed doubt about the occurrence of microbial acylases capable of hydrolysing cephalosporin C specifially into 7-aminocephalosporanic acid (7-ACA) and D-α-amino-adipic acid (Demain et al. 1963, Singh et al. 1980, Walton 1964a, b). However, recent Japanese patents claim direct enzymatic hydrolysis of cephalosporin C into 7-ACA and α-amino adipic acid with *Pseudomonas putida* (Niwa et al., 1977; Goi et al. 1978). An interesting two-step enzymic process for 7-ACA production was recently proposed by Fujii et al. (1979): cephalosporin C is first trans-formed by *Trigonopsis variabilis* CBS 4095 into glutaryl-7-ACA, which is then hydrolysed by *Comamonas* or *Pseudomonas* sp. into 7-ACA (Fig. 5).

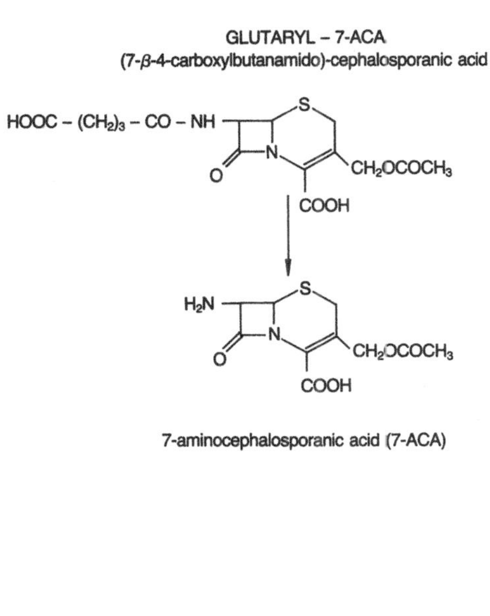

GLUTARYL – 7-ACA
(7-β-4-carboxylbutanamido)-cephalosporanic acid

HOOC – (CH$_2$)$_3$ – CO – NH

CH$_2$OCOCH$_3$

COOH

H$_2$N

CH$_2$OCOCH$_3$

COOH

7-aminocephalosporanic acid (7-ACA)

Figure 5. Bioconversion of 7-β-(4-carboxybutan-amido)cephalosporanic acid, (glutaryl-7-ACA) into 7-ACA

These researchers from Toyo Jozo Cy in Japan reported also in detail on the isolation and properties of *Pseudomonas* (*Pseudomonas putida*, *Pseudomonas* SY-77-1) strains, able to deacylate 7-β-)4-carboxy-butane-amido) cephalosporanic acid (glutaryl 7-ACA) into 7-ACA; glutaryl-7-ADCA was similarly hydrolysed into 7-ADCA. These strains specially hydrolysed cephalosporin compounds having aliphatic dicarboxylic acid in the acyl side chain (Shibuya et al. 1981); cephalosporin C was not hydrolysed. Glutaryl 7-ACA can be obtained by oxidative deamination from cephalosporin C by D-amino acid oxidase from fungi and pig kidney (Mazzeo and Romeo, 1972).

Production of the intracellular glutaryl-1-7-ACA-acylase is stimulated by glutaric acid, similar to the effect phenylacetic acid has on penicillin acylase induction. β-Lactamase deficient and acylase constitutive mutants were derived from the *Pseudomonas* SY-77-1-strain with improved activity (Ichikawa et al. 1981 a). Ichikawa et al. (1981 b) cystallized the glutaryl-7-ACA-acylase from cell-free extracts from *Pseudomonas*: the molecular weight of this periplasmic enzyme was estimated at 130.000 by Sephadex G-100 gelfiltration maximal activity was found in the broad pH range from 6,5 to 10.0 at 37 °C. Glutaryl-7-

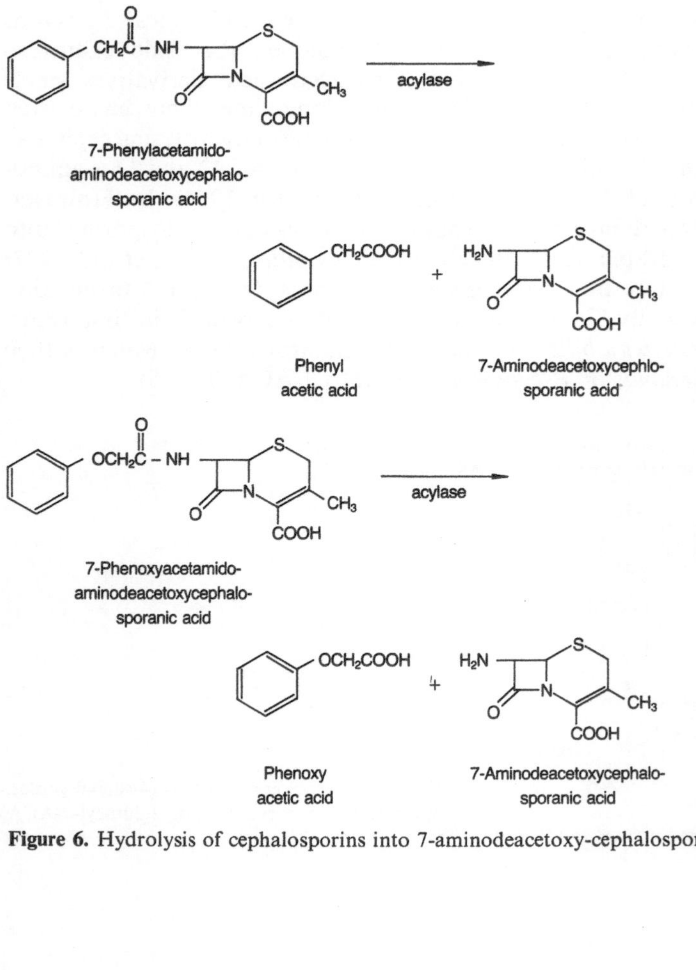

Figure 6. Hydrolysis of cephalosporins into 7-aminodeacetoxy-cephalosporanic acid (7-ADCA)

ACA acylase has been claimed also to be produced by *Bacillus*, *Arthrobacter* and *Alcaligenes* species (Takeda et al. 1977; Matsuda et al., 1978; Takeda and Matsuda, 1978; Inoue et al., 1979).

Immobilized enzyme or cell technology should soon be applied to this new type of acylase and might compete with conventional chemical methods for 7-ACA production which is the starting material for production of the medically important cephaloglycin and cephalothin.

Cephalosporins can also be produced from the precursored penicillin G or V by a series of chemical reaction that expand the five-membered thiazolidine ring of the penicillins into the six-membered dihydrothiazine ring of the cephalosporins. The obtained cephalosporins contain the penicillin G or V side chain which then can be removed chemically or enzymically by penicillin G or V acylase action. The obtained cephalosporin nucleus is named 7-aminodesacetoxycephalosporanic acid (7-ADCA) (Fig. 6). Acylation of 7-ADCA with D-phenylglycine produces the useful antibiotic cephalexin.

In contrast to enzymic acylation, chemical acylation procedures need blocked derivatives of the reactants. This fact, combined with the advantages of immobilised enzyme or cell technology, could favour the use of enzymic procedures in semi-synthetic cephalosporin synthesis.

So far, most studies on enzymic deacylation of cephalosporins are directed towards producing the 7-ADCA nucleus (Figure 6). The substrates for this reaction, 7-phenyl and 7-phenoxyacetamido-acetoxycephalosporanic acid are readily obtained by ring expansion of the precursored penicillins G or V and are easy substrates for penicillin G or V acylases. Several penicillin acylases have been described to readily deacylate 7-ADCA-derivatives and to reacylate 7-ADCA into cephalexin (Fig. 7).

Another interesting bioconversion would be to hydrolyse 7-ACA into 7-ADCA or to transform 7-ADCA into 7-ACA by coupling of an acetoxygroup. No published reports are available on such enzymatic or microbial reactions.

Immobilised enzymes as catalyst

Immobilised acylases have also been used for the deacylation or acylation of cephalosporin compounds. Extracellular penicillin G-acylases from *B. megaterium* and *P. rettgeri* were absorbed onto Celite, activated carbon, carboxymethylcellulose or an Amberlite ion-exchange resin (Toyo Jozo Company 1973a, b, 1975) and hydrolysed 7-phenylacetamido-ADCA. An immobilised *E. coli* enzyme performed the same reaction (Nys at al. 1980). The Celite-absorbed enzyme was selected for pilot-plant production of 7-ADCA, yielding 89.2% hydrolysis of the substrate. Upon addition of toluene to minimise bacterial contamination the immobilised system had a useful life of about 10 days. The partially purified penicillin-V-acylase from *E. aroideae* (Vandamme et al. 1971b) was entrapped in cellulose triacetate fibres at Glaxo Laboratories, UK. A fibre-packed column-hydrolysed 7-phenoxyacetamido-ADCA to 7-ADCA with 58% yield after 3 h of substrate circulation at 37 °C (Fleming et al. 1974). The immobilised *E. coli* acylase used by Marconi et al. (1975) to acylate 6-APA could also be applied for cephalexin synthesis from 7-ADCA and D-phenylglycinemethylester with a yield of 75%.

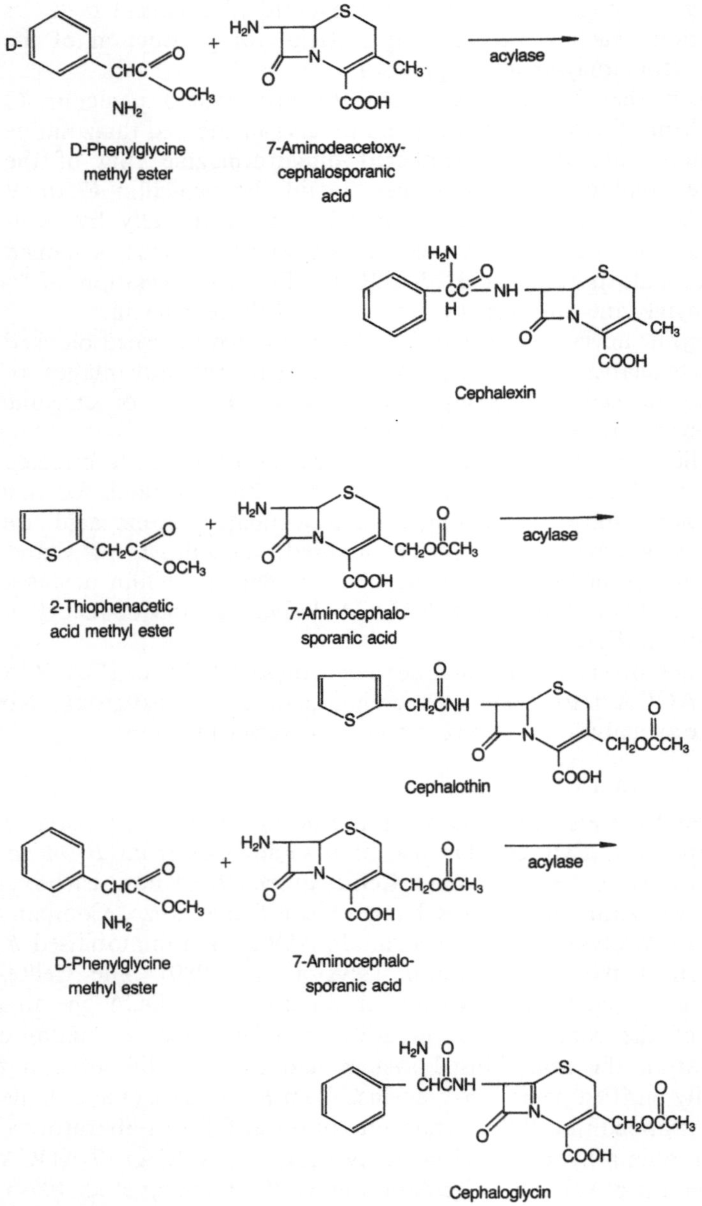

Figure 7. Synthesis of semi-synthetic cephalosporins from 7-ACA or 7-ADCA

The extracellular acylase from *B. megaterium* adsorbed on Celite, and a cell-free *Achromobacter* acylase, adsorbed onto hydroxyapatite, performed the same reaction with 80–85% yield (Abe et al. 1973; Toyo Yozo Company, 1974). The Celite-enzyme complex from *B. megaterium* has also been used to acylate 7-ACA with D-phenylglycinemethylester into cephaloglycin. Cephalothin was also obtained by acylating 7-ACA with 2-thiophene acetic acid methylester (Figure 7).

Immobilised cells as catalyst

Although several immobilised enzyme processes have been described to hydrolyse 7-phenyl- or 7-phenoxy-acetamido-ADCA into 7-ADCA (Toyo Jozo Company 1973a, b), only a few reports on such hydrolysis with immobilised cells have been published. *E. aroideae* cells entrapped in cellulose triacetate could perform this reaction (Fleming et al. 1974). Several immobilised cell systems for synthesising the clinically useful antibiotic cephalexin have been developed (Abe et al. 1973a, b; 1974) (Figure 7). Acetone-dried *Achromobacter*, *Beneckea hyperoptica*, *Alcaligenes faecalis* and *Flavobacterium aquatile* cells, adsorbed to DEAE, TEAE or carboxymethyl cellulose could synthesise cephalexin from 7-ADCA and D-phenylglycinemethylester. Highest yields (80–85%) were obtained with the *Beneckea* DEAE-cellulose complex. An immobilised cell system was described by Fukushima et al. (1976) for the preparation of 7-ACA: cells of *Pseudomonas* sp. or *Comamonas sp.*, entrapped in cellulose acetate capsules hydrolysed glutaryl-7-ACA into 7-ACA.

b) Cephalosporin acetylesterase bioconversions

Deacylated cephalosporins are useful intermediates in the synthesis of new cephalosporin derivatives (Abraham 1974). Indeed, the antibiotic properties of cephalosporins are determined not only by the side chain at the carbon-7 position, but also by the substituent at the carbon-3 position. After removal of the acetate group, many substituents can be added to the C-3 position of 7-ACA. Acetate can be chemically removed but yields are low, due to unwanted side reactions such as lactone formation and double bond migration. Enzymic hydrolysis is at the moment the only efficient method to prepare deacetyl-7-ACA (Fig. 8). The enzyme involved has been named acetylesterase. Such enzymes that hydrolyse acetate from the C-3 carbon of cephalosporins have been reported to occur in mammalian tissue (O'Callaghan et al. 1963), plant tissue (Jeffrey et al. 1961; Wick et al. 1971), in bacteria, actinomycetes and yeasts (Nishida et al. 1968; Konecny et al. 1980) and in fungi (Singh et al. 1980). Among a large number of fungi tested, only *Fusarium oxysporum* strains were found to produce cephalosporin acetylesterase (Singh et al. 1980). Spores and vegetative

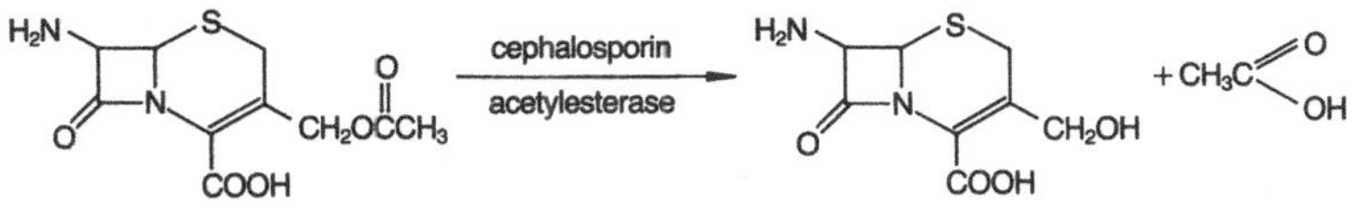

Figure 8. Conversion of 7-ACA into deacetyl-7-ACA by cephalosporin acetylesterase

cells transformed cephalothin, 7-phenylacetamido cephalosporanic acid (benzyl-cephalosporin) and phenoxymethylcephalosporin at pH 7,5–8,0 and 28 °C into the corresponding deacetyl-cephalosporins. Several processes that use cell suspensions or soluble enzyme preparations have been described to deacetylate cephalosporins. The production, isolation and purification of *Bacillus subtilis* NRRL-B-558 cephalosporin acetylesterase, splitting 7-ACA into acetate and deacetyl-7-ACA was described by Abbott and Fukuda (1975) and Abbott et al. (1976). The enzyme (mol. wt 190000) proved to be extremely stable in solution and to be very resistant to thermal inactivation.

The extracellular esterase was absorbed onto bentonite and used for multiple batch reactions to deacetylate 7-ACA solutions. During the reaction the enzyme dissociated from the bentonite particles. The complex could be partially stabilised by addition of aluminium hydroxide gel. The physical chacteristics of the enzyme made it also ideally suited to immobilisation by containment within an ultrafiltration device. With this technique, the enzyme was reused 20 times over an 11 day period; practically no enzyme inactivation occurred.

c) Bioconversion of cephalosporins and of 6-APA
  with α-amino acid ester hydrolase

Bacterial enzymes which are able to synthesise cephalosporins, starting from 7-ACA or 7-ADCA and an sppropriate α-amino acid ester have been described by Takahashi et al. (1972, 1973). Since the enzyme involved in these synthetic reactions utilises α-amino acid esters as acyl donors and transfers these acyl groups to both water (hydrolysis) and 7-amino-cephem compounds or 6-APA (transfer), it has been named 'α-amino acid ester hydrolase (Figure 9). As side-chain structures, α-D-phenylglycine methylester and analogues (glycine ethylester, D-alanine ethylester and the methylesters of D-α-cyclohexenylglycine, D-α-p-hydroxyphenylglycine and D-α-cyclohexylglycine) were coupled to 7-ACA or 7-ADCA at pH 6.0 by enzymic action of microorganisms belonging to the *Pseudomonadaceae*, including *P. melanogenum* IF012020, *P. maltophila* IF012690, *Xanthomonas oryzae* IF03995, *X. citri* 9F03855, *Acetobacter pasteurianus* ATCC 6033, *A. turbidans* ATCC 9325, *A. xylinus* IF03144, *Gluconobacter suboxydans* ATCC 621, *P. melanogenum* IF012020, *Mycoplana dimorpha* IF0A3212 and *Protaminobacter alvoflavus* IF0A3221. None of these bacteria displayed acylase activity towards benzyl- or phenoxy-methylpenicillin. The enzyme has been partially purified from *A. turbidans* ATCC 9325 and displayed a strong ester hydrolase activity with α-amino acid esters only. No activity was found with α-amino acid amides or N-(α-aminoacyl)-glycine, although cephalosporins and penicillins having an α-amino-acyl side chain, were also hydrolysed. The enzyme also displayed α-amino-acyltransferase activity in the presence of suitable acyl receptors such as 7-ADCA to produce cephalexin (Takahashi et al. 1974). Hydrolysis of the acyl-amide bond of cephalexin was also catalysed by the same enzyme. The molecular weight of the α-amino acid esterase from *Acetobacter* has been estimated to be 300000 (Kato et al. 1980a, b, c). The enzyme from *A. turbidans* ATCC 9325 has been immobilised to curdlan and used for continuous synthesis of cephalexin (Takahashi et al. 1977). The crude enzyme preparation obtained after sonication of washed cells and

CNBr-activated curdlan 13140 were mixed at pH 8 and 5 °C; after 20 h 93% of the enzyme activity was immobilised. For cephalexin synthesis, a solution (at pH 7.0) containing 7-ADCA (0.5%) and D-2-phenylglycine methylester hydrochloride (1.5%) was passed through a column packed with the immobilised enzyme. The reaction temperature was kept at 5 °C to avoid bacterial contamination and inactivation of the enzyme. At the start of the reaction, the conversion ratio of 7-ADCA: cephalexin was about 85% and even after 70 days of continuous reaction this ratio did not change substantially. The *X. citri* IF03835 enzyme was purified to homogeneity by ion-exchange chromatography and Sephadex G-100 gel filtration. The molecular weight was estimated to be 270000–280000. The enzyme is dissociable into four subunits of 72000. The kinetics of its catalytic properties were studied by Kato (1980a, b, c). The single enzyme catalysed both hydrolysis of α-amino acid esters (such as D-α-phenylglycine methylester) and cephalexin, but could also synthesise cephalexin from 7-ADCA and D-phenylglycine methylester in high yield (Fig. 9) (Rhee et al. 1980). 7-ACA and 6-APA can also be utilised as acyl-acceptors. The optimum pH for these reactions was 6.4 at an optimal temperature of 35 °C. The K_m for cephalexin deacylation was 2.99 mM (Takahashi et al. 1972; 1973; 1977).

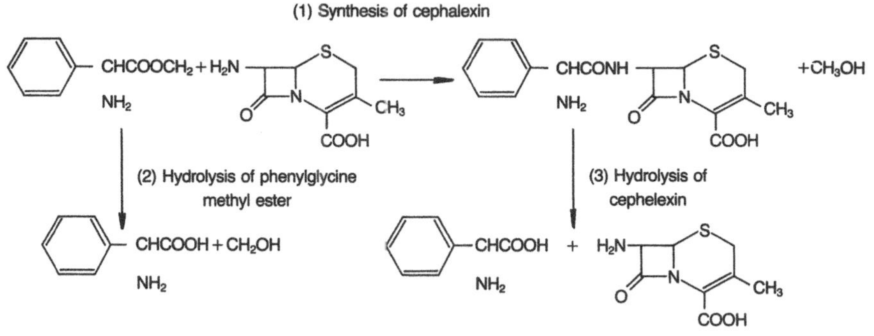

Figure 9. Reactions catalysed by α-amino acid ester hydrolase in the case of cephalexin synthesis

With washed cells, the conversion ratio of 7-ADCA to cephalexin was about 5% within 2 h. Amoxycillin could be synthesised from D-α-p-hydroxyphenylglycine methylester and 6-APA in a similar fashion, although the rate of synthesis was slower. A penicillinase-deficient mutant strain K24 was used here. Yields could be further improved by the addition of alcohols (sec-butanol) to the reaction mixture which lowered the rate of hydrolysis of the ester and increased that of the acylation of 6-APA. The conversion ratio of 6-APA to amoxycillin was about 93% (at pH 6.8 and 20 °C after 15 h of incubation) (Kato et al. 1980, a, b, c). Takahashi et al. (1977) also found that an *E. coli* IF013502 extract acylated at pH 7.5 and 25 °C, 7-amino-cephem-4-carboxylic acid esters (7-ADCA-methylsulphoxylethylester) with N-phenylacetyl-glycine or N-thienylacetylglycine to give the corresponding cephalosporin esters. These cephalosporin esters readily crystallised in the reaction mixture, allowing high yields to be obtained.

So far, enzymatic coupling of side chains to the penicillin or cephalosporin nucleus is rather inefficient and is as yet not competitive with chemical procedures; however, experiments based on the use of α-amino acid ester hydrolases rather than acylases look promising and use of immobilised enzyme or cell technology might improve efficiency.

V. Penicillin and Cephalosporin — Side Chain Preparation with Immobilised Bio-Catalysts

Microbial enzymes are not only used for bioconversion of penicillins and cephalosporins, but also in the manufacturing of certain side chain acids to be coupled to 6-APA or to the cephalosporin nucleus. Enzymic resolution processes have been developed for producing D(-)phenylglycine, used to prepare ampicillin, cephalexin and cephaloglycine and D(—)-p-hydroxyphenylglycine to prepare amoxycillin. Chemical procedures to resolve the racemic mixtures are also known, but enzymatic processes now attract industrial attention.

Immobilized *E. coli* penicillin G-acylase stereospecifically hydrolysed only the L(—)isomer of N-phenylacetyl-DL-hydroxyphenylglycine: partially purified enzyme was immobilised to an acrylic ester resin "Amberlite XAD7" and packed into a column reactor through which a 5% substrate solution was continuously flowed. The unhydrolysed D-derivative was recovered by solvent extraction and subsequently chemically hydrolysed into D(—)p-hydroxyphenylglycine (Savidge et al. 1974) (Fig. 10).

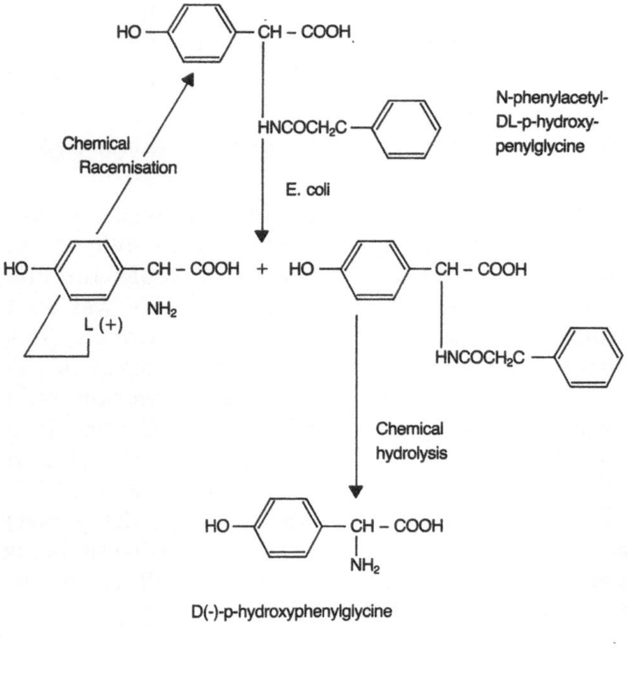

Figure 10. Enzymatic resolution of DL-p-hydroxyphenylglycine with *E. coli* penicillin G-acylase

A *Pseudomonas putida* amidase stereo-specifically hydrolysed L(+)phenyl-glycine amide in a racemic mixture, the remaining D(—)phenylglycine amide was then chemically hydrolysed into optically pure D(—)phenylglycine (Neilson, 1980).

A protease "subtilisin", covalently immobilised on methacrylate copolymers, stereospecifically hydrolysed 10% N-acetyl-DL-phenylglycine methylester to give N-acetyl-L(+)phenylglycine. The problem is that these enzymes are specific for the L(+)isomer which necessitates chemically hydrolysing the unaffected D(—)derivative into the desired compound. It would thus be advantageous if the D(—)isomer could serve as the enzyme-substrate.

D(—)amino acylases have indeed been detected in *Streptomyces olivaceus* and *S. tuirus* (Sugie and Suzuki, 1978, 1980). 2% N-acetyl-DL-phenylglycine was completely hydrolysed at pH 7.0 at 30 °C within 6 hours into optically pure D(—)acid: here the unaffected L(+)derivative has to be racemised before being reused in the process. A process initially developed by Snam Progetti in 1976 could avoid the above mentioned problems: a dihydroxy-pyrimidinase (or hydantoinase) from calf liver or from *Pseudomonas* strains was found to stereospecifically hydrolyse DL-p-(hydroxy)phenylhydantoin to a D(—)-carbamoyl derivative which yielded upon chemical hydrolysis the wanted D(—)p(hydroxy)phenylglycine; the remaining L(+)phenylhydantoin sponta-neously racemised such that in the end the DL-mixture was totally converted to the D(—)compound (Yamada et al. 1978). Further developments led to the isolation of strains (i.e. *Agrobacterium radiobacter* NRRL-B11291) with both D-hydantoinase as well as D(—)carbamoylase activity, thereby avoiding the need

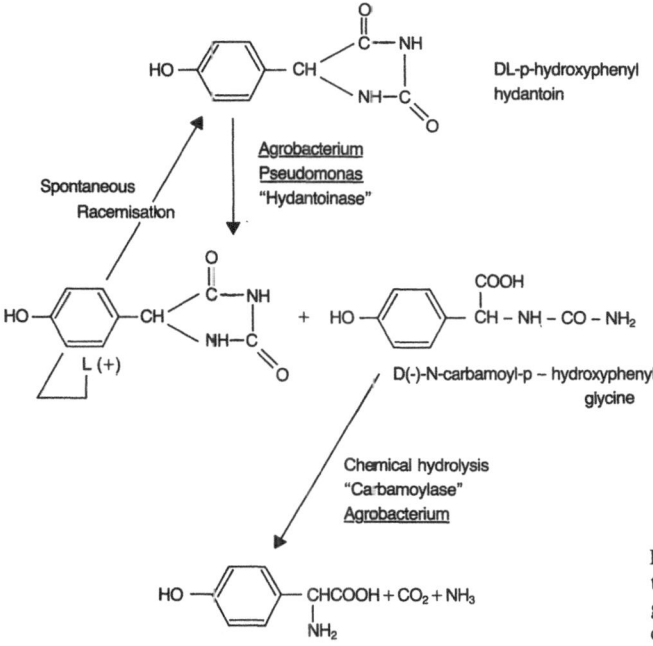

DL-p-hydroxyphenyl hydantoin

Agrobacterium
Pseudomonas
"Hydantoinase"

Spontaneous
Racemisation

L (+)

+ D(-)-N-carbamoyl-p – hydroxyphenyl glycine

Chemical hydrolysis
"Carbamoylase"
Agrobacterium

D(-)-p-hydroxyphenylglycine

Figure 11. Enzymatic resolu-tion of DL-p-hydroxyphenyl-glycine with dihydroxypyrimi-dinase and carbamoylase from *Agrobacterium*

for the complex chemical hydrolysis of the D(—)carbamoyl derivative (Olivieri et al. 1979): aromatic and heterocyclic hydantoins undergo rapid spontaneous racemization, therefore allowing the complete conversion of the racemic substrate to the D-amino acid in one single reactor. Resting cell suspensions could be used repetitively up to six times without appreciable loss of activity (Fig. 11).

It is obvious that these important bioreactions are candidates for immobilized enzyme or cell technology on an industrial scale.

VI. Bioconversion of other β-Lactam Antibiotics

Recently, several new natural β-lactams have been described including cephamycins, clavulanic acid, thienamycins, olivanic acids and nocardicins (Aoki and

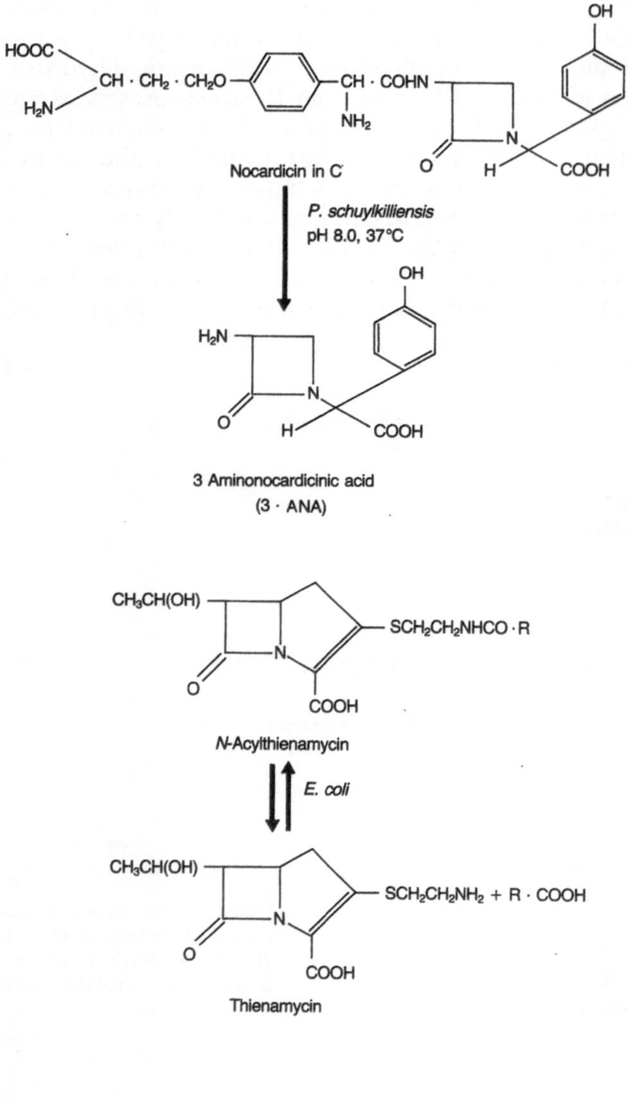

Figure 12. Bioconversions of Nocardicin C and Thienamycin

Okuhara 1980; Sebek 1980). Many of these natural compounds have been reported to be transformable by microbial cell of enzyme preparations (Fig. 12) (Okachi 1979). *Pseudomonas schuylkillensis* acylase hydrolysed the single β-lactam nocardicin C at H 8 and 37 °C into 3-amino-nocardicinic acid (2-ANA) (Komori et al. 1978). However, nocardicin A, the main component of the nocardicin complex produced by *Nocardia uniformis* was not affected by any of 1220 strains of bacteria, yeasts and fungi tested. The olivanic acid related PS-5 antibiotic was deacylated by *Pseudomonas* sp. 1158 cells, immobilised in polyacrylamide gel (Fukugawa et al. 1980) at pH 7.4 and 30 °C and by DEAE-Sephadex immobilised L- and D-amino acid acylases. An *E. coli* acylase has been reported to acylate thienamycin reversibly. So far, no practical use has been found or proposed for the products obtained.

Novel monocyclic β-lactam antibiotics, sulfazecin and isosulfazecin, have been detected in acidophilic bacteria, identified as *Pseudomonas acidophila* and *P. mesoacidophila* (Imada et al. 1981). These compounds are now classified as monobactams (Sykes et al. 1981) and similar compounds differing in side chain structure at the 3-amino group have been found in *Gluconobacter* sp., *Chromobacterium violaceum*, and *Agrobacterium radiobacter* (Fig. 13):

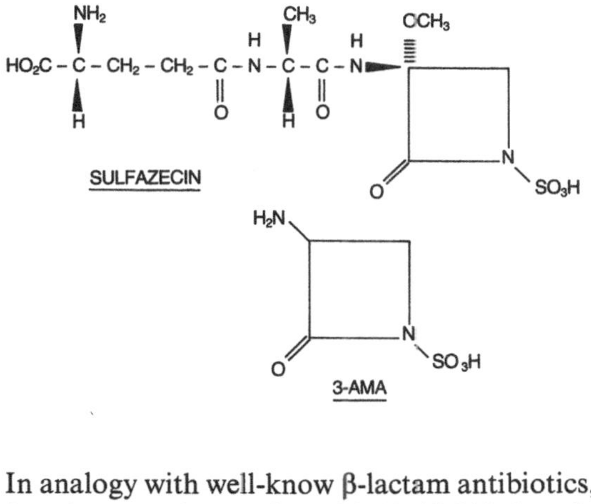

Figure 13. Structure of the monobactam, Sulfazecin and of its nucleus, 3-aminomonobactamic acid (3-AMA)

In analogy with well-know β-lactam antibiotics, the nucleus of the monobactams, 3-aminomonobactamic acid (3-AMA) has been prepared chemically, but it should also be possible in an enzymatic way; this in turn might lead to immobilised enzyme or cell technology to prepare semi-synthetic monobactams. Highly active β-lactamase stable derivatives have already been synthesized chemically (Sykes et al. 1981).

VII. Total Enzymic Synthesis of Peptide Antibiotics: Potential Field for Immobilized Enzyme Technology

A promising field for application of immobilised enzyme technology lies in the total *in vitro* enzymic synthesis of antibiotics.

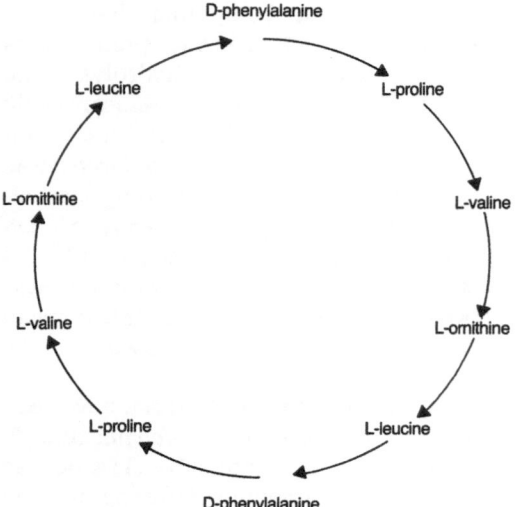

Figure 14. Schematic representation of the structure of Gramicidin S

Total enzymic synthesis *in vitro* has been studied particularly in the case of oligopeptide antibiotics (gramicidin S, tyrocidins, bacitracin, enniantin B).

Of the several systems described in the literature, the one that is best understood at the moment is the biosynthesis of gramicidin S, a cyclic decapeptide antibiotic produced by certain *B. brevis* strains (Fig. 14) (Demain et al. 1976; Kleinkauf and Koischwitz, 1980; Vandamme, 1981). The advantages of this model for the fundamental study of total enzymic synthesis are:

1. The number of enzymes in the biosynthesis pathway is small, i.e. two in number, GS synthetase 1 and 2, respectively.
2. The enzymes, substrates and cofactors involved have been characterised and the biosynthetic mechanism has been resolved in detail. Both enzymes have been purified to homogeneity. Phosphopantotheine is the cofactor, which is tightly bound to GS synthetase 2. The constituent amino acids L-Pro, L-Val, L-Orn, and L-Leu are activated by enzyme 2 while activation and racemisation of L-Phe to D-Phe is effected by enzyme 1. Enzyme 1 initiates the biosynthetic sequence while the peptide bonds (elongation and cyclisation) are catalysed by enzyme 2. The amino acid sequence in gramicidin S is hence determined by the unique location of specific subinits on the enzymes 1 and 2 (Froyshov et al. 1978). Synthesis of gramicidin S hence occurs independently and in a completely different way from protein biosynthesis (Vandamme 1981).

   The mechanism involved has been named "multi-enzyme thiotemplate" biosynthesis, and it requires — in addition to the two enzymes — only the building block amino acids or their analogues as precursors, ATP as an energy source, and Mg^{2+} ions.
3. The producing organism is a bacterium. Thus, it is more easily handled and grown than fungi and actinomycetes. Growth is faster and mutation

is simpler than with the filamentous microorganisms. Furthermore, much is known about the disruption of cells of *Bacillus* spp.
4. The antibiotic accumulates normally inside the cells.
5. Total organic synthesis of gramicidin S was achieved in 1957 and improved with the Merrifield solid-phase technique, but was never further developed.

The overall biosynthesis scheme can be simplified as:

$$10 \text{ ATP} + 10 \text{ amino acids} \xrightarrow[\text{Mg}^{++}]{\text{GS synthetase 1 and 2}} 1 \text{ GS} + 10 \text{ AMP} + 10 \text{ PPi} .$$

Using the two GS-synthetases isolated from high gramicidin S producing *B. brevis* ATCC 9999 cells gram quantities of gramicidin S have already been produced in vitro by the total enzymatic method (Demain et al. 1976; Hamilton et al. 1973; Demain and Wang 1976).

The successful operation of continuous enzyme reactors for such multistep synthetic reactions depends largely upon the effective regeneration of ATP with cheap energy sources. ATP regeneration, in view of total enzymic synthesis, has been investigated using immobilised enzymes and bacterial chromatophores (Gardner et al. 1974; Pace at al. 1976). The use of yeast enzymes (Asada et al. 1978), immobilised yeast cells (Asada et al. 1979) and of polyphosphates (Butler 1977) has also been proposed to continuously regenerate ATP. In future the two GS synthetases could be co-immobilised to construct a bioreactor, coupled to an enzymatic ATP-regeneration system which may allow for a continuous process to be developed. Furthermore, instead of using expensive amino acids as substrates cheap protein hydrolysates should be tested.

The relative simplicity of the gramicidin S system and its suitability for larger scale exploitation overrides the fact that it is at present not a commercially very important antibiotic and it should be considered as a model system. It is indeed the hope to apply the knowledge gained to other similar systems with high commercial value (tyrocidins, linear gramicidins, bacitracins). So far application of immobilised enzyme technology to the total enzymatic antibiotic synthesis-principle has not been unequivocally successfull.

VIII. Total Synthesis of Peptide Antibiotics with Immobilised Living Microbial Cells as an Alternative to Conventional Fermentation

a) Immobilised living cell technology versus conventional antibiotic fermentation

To circumvent the difficulties encountered at this moment with such complex total enzymic synthesis processes yet attempted using cellfree (immobilised) enzyme preparations, it was thought that the use of whole immobilised living cells could lead to much greater success in this respect. Generally, utilisation of immobilised whole living cells for total synthesis of fermentation or organic products is scarcely documented: laboratory-scale synthesis of α-amylase, glutamic acid, isoleucine, citric acid, hydrogen, ethanol, coenzyme A, ammonia and of a few antibiotics have recently been reported (Abbott 1976, 1977; Morikawa et al. 1979a, b; Chibata and Tosa 1977, 1981).

Again the vast potential of whole immobilised cells in the field of antibiotic production is only just now being fully recognised. This concept would be particularly valuable for those antibiotics which are or can be excreted into the culture medium. Only the production of penicillin G (Morikawa et al. 1979a), bacitracin (Morikawa et al. 1979b, Morikawa et al. 1980), cephamycin C (Freeman and Aharonowitz, 1981) and candicidin (Venkatasubramanian and Vieth, 1979) and nisin (Egorov et al. 1978), has so far been attempted with immobilised cells as catalysts. Compared to conventional fermentation, immobilised living cell systems can offer several important advantages: such as possibility of continuous operation (at high dilution rates with no wash-out danger) or plug-flow mode of action, 'column fermentation', reduction of non-productive growth phases, faster reaction rates possible at increased cell density, higher yields, easier rheological control and control of cell reproduction (Venkatasubramanian and Vieth, 1979; Morikawa et al. 1980). Other advantages have been outlined in section II.

b) Antibiotic production with immobilised cell technology

P. chrysogenum ATCC 12690 mycelium has been immobilised in polyacrylamide gel (PAA), collagen membranes and Ca-alginate beads: production of penicillin G from glucose with this catalyst was recently reported by Morikawa et al. (1979a). Mycelium was harvested from a conventional batch fermentation at its maximum rate of penicillin production. Collagen immobilisation resulted in low activity, while Ca-alginate entrapment yielded highest activity but yielded a gel too fragile to be used under reactor conditions to allow repeated use. With the 5% PAA gel-entrapped mycelium, penicillin was produced (0.77 U cm^{-3} h^{-1}) from glucose, $(NH_4)_2SO_4$, phosphate buffer (pH 7.0) and phenylacetate as precursor within 5 h but amounted only to 17% of that produced by washed mycelium; however, the half-life of immobilised mycelium activity was 6 days compared with 1 day for washed mycelium. The immobilised mycelium required oxygen for penicillin production. Adding 6-APA to the incubation mixture and more particularly the three penicillin constituent amino acids (L-α-amino adipic acid, L-cysteine and valine) instead of glucose, yielded higher penicillin titres. Repeated use of immobilised mycelium yielded an initial increase in penicillin formation (up to 200% after the third run), probably caused by growth of the mycelium within the gel and then a gradual decrease, whereas washed mycelium rapidly lost all activity. At least part of the immobilised mycelium remains viable during entrapment and reaction (Morikawa et al. 1979a). The total concentration and relative ratio of acrylamide and N,N'-methylene bisacrylamide (BIS) determine both the pore size of the interstitial space within which whole cells are entrapped and the physical properties of the complex. Increasing the content of BIS to 16% increased the synthetic activity of the immobilised mycelium and provided excellent mechanical rigidity.

Morikawa et al. (1979a) obtained evidence that the PAA gel polymerisation reagents individually inactivate the multi-enzyme systems in the mycelium. Such negative effects have also been encountered by Freeman and Aharonowitz (1981) when entrapping viable *Streptomyces* cells for cephamycin C production (Fig. 15). Indeed, the formation of PAA gels in the presence of viable cells

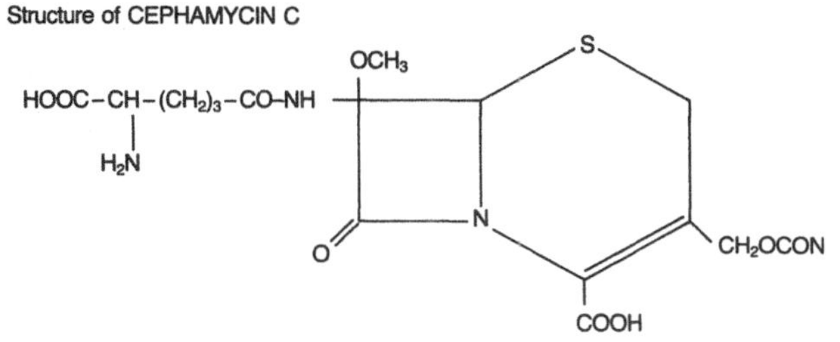

Structure of CEPHAMYCIN C

$$HOOC-CH-(CH_2)_3-CO-NH$$
$$H_2N$$

OCH₃   S

CH₂OCON

O

N

COOH

Figure 15. Structure of Cepha-mycin C

results in a considerable loss of biological activity, mainly due to toxicity of the monomers and heat evolution during polymerisation. Nevertheless, cross-linked PAA is one of the more common matrices used for entrapment of whole microbial cells. In order to retain the advantages of PAA and to avoid cell damage during preparative steps, Freeman and Aharonowitz (1981) recently proposed the use of preformed linear, water-soluble PAA chains (mol. wt 15 to 17×10^4), partially substituted with acylhydrazide groups. This pre-polymerised material was cross-linked in the presence of viable cells by stoichiometric amounts of bi- or multifunctional aldehydes or ketones (glyoxal, glutardialdehyde, poly-vinylalcohol). Use of cross-linking agents of various chain lengths (glyoxal, glutardialdehyde and periodate-oxidised polyvinyl alcohol) allowed control of gel compactness. This cross-linking reaction carried out in cold neutral physiological conditions resulted in gel-entrapped cells with mechanical properties similar to those of conventional PAA gels, but no cell damage occurred. *S. clavuligerus* cells from the early growth phase were entrapped by this pre-polymerisation PAA gel procedure and produced cephamycin C conti-nuously for 96 h with yields comparable to those of free resting cells (Freeman and Aharonowitz 1981; Morikawa et al. 1979a) obtained only 15% of the activity of washed *Penicillium* mycelium with the conventional PAA entrap-ment procedure. The concept of using preformed linear water-soluble PAA chains now deserves full attention and might improve the performance of entrapped living cell systems considerably.

The antibiotic bacitracin (Fig. 16) is also synthesised following the multi-enzyme thiotemplate mechanism which is involved in gramicidin S biosynthesis (Katz and Demain, 1977; Froyshov et al. 1978). While gramicidin S is

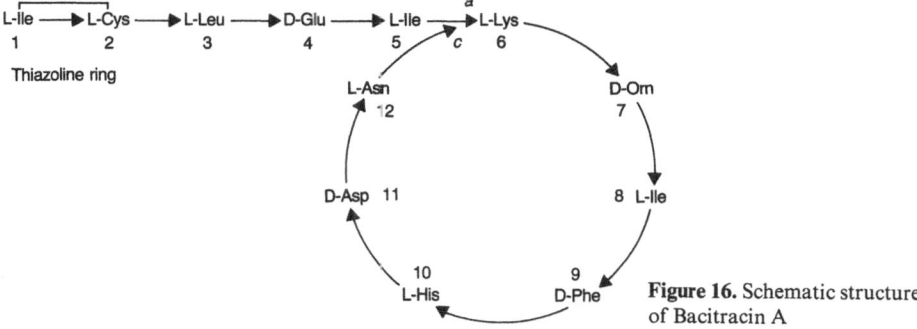

Figure 16. Schematic structure of Bacitracin A

retained in the cells of its producer, bacitracin is excreted in the medium; cell-free synthesis of bacitracin has also been studied in detail (Froyshov et al. 1978). These two factors have contributed to the selection of this process as a model for synthesis of a secondary metabolite and of an economically important antibiotic by immobilised living cells (Morikawa et al. 1979b; Morikawa et al. 1980). Whole cells of *Bacillus* KY 4515 were immobilised in PAA gel, and formation of the oligopeptide antibiotic bacitracin was studied under varying conditions. *Bacillus* cells were harvested from a batch fermentation when the rate of bacitracin production was maximal (40 U cm^{-3} h^{-1}). After immobilisation in 5% PAA gel, the catalyst was cut into small blocks (8—27 mm^3) and was added to a complex (starch-bouillon) fermentation medium and incubated at 30 °C for 4 h. The initial activity of the immobilised whole cells was only 20–25% of that of an equivalent amount of washed cells (rate of production 13–18 U cm^{-1} h^{-1}). Again polymerisation reagents (especially acrylamide and ammoniumpersulphate) and limited diffusion apparently lowered the antibiotic synthesis capacity. Successive use in 1% peptone or 0.5% meat extract as a reaction medium resulted in a gradual increase of the activity, reaching a steady state at 80–90% of the activity of freshly washed cells; this increase in activity seems to be caused by active growth of the cells in and on the gel, as could be observed with an electron microscope (Morikawa et al. 1979b). Bacitracin produced under anaerobic conditions was only 30% of that under aerobic conditions. The rate of bacitracin production increased with increasing amounts of cells in the gel, but upon repeated utilisation an equal level of antibiotic was ontained. Bacitracin production from its constituent amino acids and ATP did not occur with these immobilised cell preparations as catalysts, probably due to transport problems. The apparent half-life of this bacitracin synthesis catalyst was estimated to be at least 1 week. Continuous production of bacitracin has recently been achieved in an "immobilised whole-cell fermenter" (Morikawa et al. 1980). Maximum bacitracin levels were achieved after 1 day of fermentation after which a gradual decrease in productivity was observed. This was mainly caused by cell growth at the surface of the gel blocks: this prevented diffusion of oxygen and peptone into the gel towards the immobilised cells. Interval washing of the gel (flushing the 'chemostat' with saline for 2 h once each day) solved this problem. Under those operational conditions continuous bacitracin production remained high for 8 days. Continuous bacitracin production by this immobilised whole-cell fermenter displayed several advantages: bacitracin is produced from one simple nutrient, peptone; antibiotic recovery is easier since the culture liquid does not contain cells; productivity (even at high dilution rates) is higher than in conventional chemostat cultures.

Bruyneel and Vandamme (1982) immobilised the gramicidin S (GS) producer *B. brevis* ATCC 9999 in agar: cells were cultivated in high yielding yeast extract-peptone medium and harvested during exponential growth, washed cells were entrapped in agar beads: when suspended in a minimal medium based on L-fumarate as carbon source, up to 50 µg of GS/ml was formed within 24 hours whereas the free cells produced much less GS under the same reaction conditions. Optimisation studies of this system are in progress.

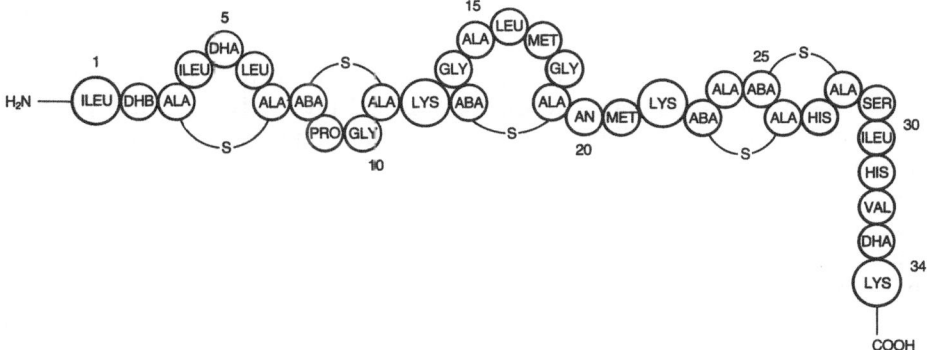

Figure 17. Schematic structure of Nisin

Nisin is an economically important polypeptide antibiotic produced by *Streptococcus lactis* (Hurst, 1981). Its structure is represented in Fig. 17. Its biosynthesis appears not to proceed according to the above mentioned multienzyme thiotemplate mechanism and rather follows protein synthesis mechanisms. It is active against gram-positive organisms including related streptococci: nisin resembles a bacteriocin rather than an antibiotic. Among antibiotics, nisin has the unique function of being exclusively used as a biological food preservative: it is non-toxic, and as a polypeptide, any residues remaining in food are digested. Chemistry, biosynthesis and applications of nisin have been reviewed recently by Hurst (1981). It is manufactured in several countries e.g. England (Aplin and Barrett Ltd.; Unigate Ltd.), Poland (Krakow Pharmac. Works) and the Sovjet Union on a large scale. In pH controlled (pH 6.0) batch fermentations at 20 to 30 °C without aeration, maximal nisin synthesis was related to maximal biomass formation. Normally, nisin is excreted into the culture medium but when cells are grown at pH 6.7, they retain the synthesized nisin. Media are based on corn extract, yeast autolysate and dairy waste. Maximal production can attain 2000 IU/ml or 50 µg/ml.

Nisin production by polyacrylamide gel-immobilised *Streptococcus lactis* cells was reported by Egorov et al. (1978), but few details of the process are available.

Another antibiotic so far described to be produced according to the immobilised cell principle is the polyene macrolide candicidin, produced by *Streptomyces griseus*. Immobilisation of the streptomycete in collagen led to an antibiotic production capacity from glucose of 14% of that obtained in the conventional batch fermentation process (Venkatasubramanian and Vieth, 1979).

c) Potentials and limits of immobilised living cells

The above described systems are all typical examples of synthesis by fixed viable cell systems of secondary metabolites, normally non-growth associated complex fermentation products. Now it seems already that the physiological state, culture and viability of the cells in the immobilised reactor are of primordial importance for effecting such complex multi-enzyme reactions (whereas in the case of single step reactions the viability of the cells is not mandatory). It might indeed seem necessary to control cell reproduction at a certain level within the immobilised matrix to obtain maximal productivity. In this perspective, new bioreactor configurations such as the "immobilised whole-cell fermenter" described by Morikawa et al. (1980) for bacitracin production, could be designed combining fermenter and immobilised-cell reactor characteritics, facilitiating synthesis of both growth-associated as well as non-growth-associated products. Also different cell species or types (spores) can be co-immobilised in such a fermentor and offer exciting perspectives for 'artificial pathway' or multi-step organic reactions. The same is true for systems where isolated enzymes, coenzymes or cellorganelles are combined with whole cells and co-immobilised (Hough and Lyons, 1972; Martin at al. 1976; Lowe, 1981; Takasaki, 1974; Kierstan and Bucke, 1977; D'Souza and Nadkarni, 1980). Combination of such biocatalytic potential indicates that enzyme and whole cell immobilisation can operate together successfully rather than compete with each other.

In contrast with bound mono-enzyme systems, optimal catalysis by multi-step enzyme complexes and by whole stabilised microbial cells is quite complex and many basic aspects are yet to be well understood. A comparison of immobilised cell systems reveals that very few immobilisation methods were used relative to the large number that have been employed with cell-free enzymes. Especially missing are methods for the attachment of cells to insoluble supports by covalent bounds. Cells covalently immobilised should be more stable to dissociation than ionically bound cells and less restricted by substrate and product diffusion than cells entrapped in a polymer, the usual technique so far. However, it is by now clear that immobilised cells can be considered as one of the tools of stereospecific and complex organic chemistry and bio-production in the near future. Furthermore, it has been and apparently still is a common penomenon in fermentation science and biotechnology that application and technique often preceded the fundamental understanding of the process. Indeed, several typical microbiological, biochemical, physiological and technical problems, inherent to this immobilised cell development, need to be examined further in order to arrive at an optimal performance of immobilised cell reactors; such as effect of physiological state, of viability, lysis or growth phenomena, quantification of growth and biomass, cell metabolism and maintenance energy at low to zero growth rate, cofactor utilisation and regeneration, microbial contamination problems and prevention of unwanted side reactions. Other important process design parameters such as enzyme (cell)loading factor, stability of the cell, catalyst packing density, oxygen transfer, mass transport and diffusion efficiencies and residence time distribution which determine overall cell reactor productivity, need further study (Venkatasubramanian and Vieth, 1979; Mason and Somerville, 1978; Carleysmith et al. 1980).

Presently, for most complex multi-enzyme processes, there is as yet no economic incentive to substitute well-established fermentation processes by immobilised cell reactor technology, though emphasis is clearly concentrated upon switching from free or immobilised single-enzyme reactions to immobilised (living) cell processes. Complex fermentations, such as the production of antibiotic and other secondary metabolite offer reat potential as future candidates for immobilised living cell technology on an industrial level.

Acknowledgements

The author gratefully acknowledges T. A. Savidge of Beecham Pharmaceuticals, UK, for providing unpublished data. The secretarial assistance of my wife Mireille was invaluable to deliver a complete and timely manuscript.

References

Abbott BJ (1976) Adv Appl Microbiol 20: 203
Abbott BJ (1977) In: Annual Reports on Fermentation Processes (Perlman D, Ed) Vol 1, Academic Press, New York, p 205
Abbott BJ, Fukuda DS (1975) Appl Microbiol 30: 413
Abbott BJ, Cerimele B, Fukuda DS (1976) Biotechnol Bioeng, 19: 1033
Abe J, Watanabe T, Yamaguchi T, Matsumoto K (1973) USA Patent 3761354
Abraham EP (1974) Biosynthesis and Enzymatic Hydrolysis of Penicillins and Cephalosporins, University of Tokyo Press, Tokyo, Japan
Abraham EP, Chain E (1940) Nature (London) 146: 837
Amin M, Zedan H, El-Tayeb O (1980a) Abstracts 6th International Fermentation Symposium, July 20–25 London, Ontario, Canada, p 19
Amin M, Zedan H, El-Tayeb O (1980b) Abstracts 6th Internat Ferment Symp July 20–25, London, Ontario, Canada, p 121
Aoki H, Okuhara M (1980) Ann Rev Microbiol 34: 159
Asada M, Nakanishi K, Matsuno R, Kariya Y, Kimuya A, Kamikubo T (1978) Agric Biol Chem 42: 1533
Asada M, Morimoto K, Nakahishi K, Matsuno R, Tanaka A, Kamikubo T, Kimura A (1979) Agric Biol Chem 43: 1773
Barker SA (1980) In: Economic Microbiology (Rose AH, Ed) Vol 5. Academic Press, New York and London, p 330
Beecham Group Ltd (1974) German Patent 2 356 630
Benveniste R, Davies J (1973) Ann Rev Biochem, 42: 471
Berdy J (1974) Adv Appl Microbiol 18: 309
Brandl E, Knauseder F (1974) Germ Offen 2: 503
Brodelius P (1978) In: Advances in Biochemical Engineering (Ghose TK, Fiechter A, Blakeborough N, Eds), Vol 10 Springer-Verlag, Berlin Heidelberg New York, p 75
Bruyneel B, Vandamme EJ (unpublished results)
Butler L (1977) Biotechnol Bioeng 19: 591
Carleysmith SW, Eames MBL, Lilly MD (1980) Biotechnol Bioeng 22: 957
Carleysmith SW, Lilly MD (1979) Biotechnol Bioeng 21: 1057
Carlsen F, Emborg C (1981) Biotechnol Letters, 3: 375
Charles M (1980) Abstracts VIth Int Symp July 20–25, London, Ontario, Canada, p 120
Chibata I, Tosa T (1977) Adv Appl Microbiol 22: 1
Chibata I, Tosa T (1981) Ann Rev Biophys Bioeng, 10: 197
Chibata I, Tosa T, Mori T, Matuo Y (1972) In: Fermentation Technology Today (Terui G, Ed), Soc Fermentation Technology, Osaka Japan, p 383
Demain AL, Piret JM, Friebel TOE, Vandamme EJ, Matteo CC (1976) In: Microbiology (Schlessinger D, Ed) American Society for Microbiology, Washington, DC, p 437
Demain AL, Walton RB, Newkirk JF, Miller IM (1963) Nature (London 199: 909
Demain AL, Wang DIC (1976) In: Second Internat Symposium on the Genetic of Industrial Microorganisms (McDonald KD Ed). Academic Press, New York, London, pp 115–128
Diers IV and Emborg C (1979) British patent, 2,021, 119

Dinelli D (1972) Process Biochemistry 7: 9

Drobnik J, Saudek V, Svec F, Kalal J, Vojtisek V, Barta M (1979) Biotechnol Bioeng 21: 1317

D'Souza SF, Nadkarni GB (1980) Biotechnol Bioeng 22: 2179

Egorov NS, Baranova IP and Kozlova YUI (1978) Antibiotiki USSR 23: 872–874

Ekstrom B, Lagerlof E, Nathorst-Westfelt L, Sjoberg B (1974) Svenska Farmaca Tidskrift 78: 531

Fleming ID, Turner MK, Napier EJ (1974) German Patent 2 422 374

Fosker GR, Hardy KD, Nayler JHC, Seggery P, Stover ER (1971) J Chem Soc 10: 1917

Freeman A, Aharonowitz Y (1981) Biotechnol Bioeng 23: 2747

Froyshov O, Zimmer TL, Laland SG (1978) In: International Review of Biochemistry; Amino Acid and Protein Biosynthesis II (Arnstein HRV, Ed), Vol 18, University Park Press, Baltimore, p 49

Fujii T, Hanamitsu K, Isumi R, Yamaguchi T, Watanabe T (1973) Japanese Patent 7 399 393

Fujii T, Matsumoto K, Watanabe T (1976) Process Biochem 11: 21

Fujii T, Shibuya T, Matsomoto K (1979) Proc Ann Mtg Ag Chem Soc Japan 1–4 April

Fukugawa Y, Kubo K, Ishikura T, Kouno KJ (1980) Antibiotics 33: 543

Fukushima M, Fujii T, Matsumoto K, Morishita M (1976) Japanese Patent 7 670 884

Gardner CR, Colton CK, Langer RS, Hamilton BK, Archer MC, Whitesides GM (1974) In: Enzyme Engineering (Pye EK, Wingard LB Eds), Vol 2, Plenum Press, New York, p 209

Goi H, Niwa T, Nofiri C, Miyado S, Seki M and Yamada Y (1978) Jpn Pat 53-94093

Hamilton BK, Montgomery JP, Wang DIC (1973) In: Enzyme Engineering (Pye EK, Wingard LB, Eds), Vol 2, Plenum Press, New York, p 153

Hough JS, Lyons TP (1972) Nature (London) 235: 389

Huper F (1973) German Patent 2 157 970

Hurst A (1981) Adv Appl Microbiol 27: 85–123

Inoue T, Matsuda K, Fukuo T and Kawate S (1979) Abstracts of Papers, Annual Meeting of Agricultural Chemical Society, April 1979, Japan, Tokyo, p 220

Ichikawa S, Murai Y, Yamamoto S, Shibuya Y, Fujii T, Komatsu K and Kodaira R (1981 a) Agr Biol Chem, 45: 2225–2229

Ichikawa S, Shibuya Y, Matsumoto K, Fujii T, Komatsu K and Kodaira R (1981 b) Agr Biol Chem, 45: 2231–2236

Imada A, Kitano K, Kintaka K, Muroi M and Asai M (1981) Nature (London) 289: 590–591

Jack TR, Zajic JE (1977) In: Advances in Biochemical Engineering (Fiechter A, Ed), Vol 5, Springer-Verlag Berlin Heidelberg New York, p 125

Jeffrey JDA, Abraham EP, Newton GGF (1961) Biochem J 91: 591

Kato K (1980) Agric Biol Chem 44: 1083

Kato K, Kawahara K, Takahashi T, Igarasi S (1980a) Agric Biol Chem 44: 821

Kato K, Kawahara K, Takahashi T, Kakinuma A (1980b) Agric Biol Chem 44: 1075

Kato K, Kawahara K, Takahashi T, Kakinuma A (1980c) Agric Biol Chem 44: 1069

Katz E, Demain AL (1977) Bact Rev 41: 449

Kierstan M, Bucke C (1977) Biotechnol Bioeng 19: 387

Klein J and Eng H (1979) Biotechnol Lett, 1: 171

Klein J, Hackel U, Wagner F (1979) In: Immobilized Microbial Cells (Venkatasubramanian K, Ed), ACS Symposium Series, Vol 106, p 101

Kleinkauf H, Koischwitz H (1980) In: Multi-functional Proteins (Bisswanger H, Schminke-Ott E, Eds); Wiley and Sons, Inc, p 217

Kluge M, Klein J and Wagner F (1982) Biotechnol Lett 4: 293

Komori T, Kunugita K, Nakahara K, Aoki H, Imanaka H (1978) Agric Biol Chem 42: 1439

Konecny J, Sieber M (1980) Biotechnol Bioeng 22: 2013

Lagerlof E, Nathorst-Westfelt L, Ekstrom B, Sjoberg B (1976) Meth Enzymol 44: 759

Lilly MD, Balasingham K, Warburton D, Dunnill P (1972) In: Fermentation Technology Today (Terui G, Ed), Society of Fermentation Technology, Kyoto, Japan, p 379

Lowe CR (1981) In: Topics in Enzyme and Fermentation Biotechnology (Wiseman A, Ed), Vol 5. Horwood-Wiley, p 13

Lowe DA, Romancik G, Elander RP (1981) Dev Ind Microb 22, p 163

Marconi W, Cecere F, Morisi F, Della Penna G, Rappuoli B (1973) J Antibiotics 26: 228

Marconi W, Bartoli F, Cecere F, Galli G, Morisi F (1975) Agric Biol Chem 39: 277

Martin CKA, Perlman D (1976) Abstracts of papers Fifth Internat Fermentation Symposium (Vth IFS) (Dellweg H, Ed), Berlin, p 297

Mason JR, Somerville MJ, Pirt SJ (1978) J Appl Chem Biotechnol 28: 770

Mayer H, Collins J, Wagner F (1979) In: Plasmids of Medical Environmental and Commercial Importance (Timmis KN, Pichler A, Eds), p 459

Mazzeo P, Romeo A (1972) J Chem Soc Perkin Trans I. 20: 2532

Mosbach K (1971) Scient Am 224: (3) 26

Mosbach K (1976) Methods in Enzymology Volume 44, Academic Press, New York, London

Matsuda K, Inoue T, Tanaka K, Fukuo T and Kawate S (1978) Abstracts of Papers, Annual Meeting of Agricultural Chemical Society, April 1978, Japan, Tokyo, p 111

Morikawa Y, Karube I and Suzuki S (1980a) Eur J Appl Microb 10: 23

Morikawa Y, Karube I, Suzuki S (1979a) Biotechnol Bioeng 21: 261

Morikawa Y, Ochiai K, Karube I, Suzuki S (1979b) Antimicrob Chemother 15: 126

Morikawa Y, Karube I, Suzuki S (1980b) Biotechnol Bioeng 22: 1015

Nara T, Okachi R, Kato F (1972) Abstracts of the Sixth Internat Symposium Kyoto, Japan, p 207

Neilson MH (1980) In: "13th International TNO Conference" (A Verbraeck, Ed), p 41–58

Nelson RP (1976) UK Patent 3 957 580

Nishida M, Yokota Y, Okui M, Mine Y, Matsubara T (1968) J Antibiotic 21: 165

Niwa T, Nojiri C, Goi H, Miyado S, Kai F, Seki M, Yamada Y and Niida T (1977) Jpn Pat 52-143289

Nys P, Satarova DE, Podshibhakina LV, Korchasin VB, Savitskaya EM (1980) Antibiotiki USSR 25: 803

O'Callaghan C, Muggleton PW (1963) Biochem J 89: 304

Okachi RJ (1979) Agric Chem Soc Japan 53: R 169

Okachi R, Kawamoto I, Yamamoto M, Takasawa S, Nara T (1973) Agric Biol Chem 37: 335

Olivieri R, Fascetti E, Angelini L and Degen L (1979) Enzyme Microb Technol 1: 201

Pace GW, Yang HK, Tannebaum SR, Archer MC (1976) Biotechnol Bioeng 18: 1413

Park JM, Choi CH, Seong BL and Han MH (1982) Biotechnol Bioeng 24: 1623

Perlman D (1977) ASM-News 43: 82

Poulsen PB (1981) Enzyme Microb Technol, 3: 271

Rhee DK, Lee SB, Rhee SJ, Ryu DDH, Hospoda J (1980) Biotechnol Bioeng 11: 1237

Rolinson GNJ (1979) Antimicrobial Chemother 5, Y

Ryu DH, Bruno CF, Lee EK (1972) Abstracts of the Sixth International Symposium Kyoto, Japan, p 53

Ryu DY, Bruno CF, Lee BK, Venkatasubramanian K (1972) In: Fermentation Technology Today (Terui G, Ed). Society of Fermentation Technology, Kyoto, Japan, p 307

Sato T, Tosa T, Chibata I (1976) Europ J Appl Microbiol 2: 153

Savidge TA, Cole M (1975) Meth Enzym 43: 705

Savidge TA, Powell LW and Lilly MD (1974) British Patent, 1, 357 317

Savidge T, Powell LW, Warren KB (1974) German Patent 2336829

Schneider WJ, Rohr M (1976) Biochim bioph Acta 452: 177

Sebek OK (1974) Lloydia, 37: 115

Sebek OK (1975) Acta microbiol hung 22: 381

Sebek OK (1980) In: Economic Microbiology (Rose AH, Ed), Vol 5, Academic Press, p 575

Sebek Ok, Perlman D (1971) Adv Appl Microbiol 14: 123

Self DA, Kay G, Lilly MD, Dunnill P (1969) Biotechn Bioeng 11: 337

Shaltiel S, Mizrahi R, Stupp Y, Sela M (1970) Europ J. Biochem 14: 509

Shaltiel S, Mizrahi R, Sela M (1971) Proc R Soc, Series B 179: 411

Shibuya Y, Matsumoto K and Fujii T (1981) Agr Biol Chem, 45: 1561–1567

Singh K, Seghal SN, Vezina D (1969) Appl Microbiol 17: 643

Singh K, Sun S, Rakhit S (1980) Europ J Appl Microb Biotechnol 9: 15

Skinner KJ (1975) Chem Eng News, August 18: 22

Sugie M and Suzuki H (1978) Agr Biol Chem 42: 107

Sugie M and Suzuki H (1980) Agr Biol Chem, 44: 1089

Svedas VK, Margolin AL, Berezin IV (1980) Microb Technol 2: 138

Svedas VK, Margolin AL, Borisov IL, Berezin IV (1980) Enzyme Microb Technol 2: 313

Sykes RB, Bonner DP, Bush K, Georgopapadaku NH and Well SS (1981) J Antimicrob Chemother, 8, (Supp E), 1–16

Szweczuk A, Ziomek E, Mordarski M, Siewinski M, Wieczorek JJ (1979) Biotechnol Bioeng 21: 1543

270   E. J. Vandamme

Takasaki Y (1974) Agric Biol Chem 38: 1081
Takeda H, Matsuda K (1978) Jpn Pat 53-86094
Takeda H, Matsumoto I, Matsuda K (1977- Kawakami Jpn Pat 52-128293
Toyo Jozo Company (1973) Belgian Patent 801044
Toyo Jozo Company (1973) Belgian Patent 1324159
Toyo Jozo Company (1975) Japanese Patent 7588694
Toyo Jozo Company (1974) UK Patent 1347665
Takahashi T, Kato K, Yamazaki Y, Isono M (1977) Jap J Antibiotics 30: 130
Takahashi T, Yamazaki Y, Kato K, Isono MJ (1972) Am Chem Soc 94: 4035
Takahashi T, Yamazaki Y, Kato K (1973) Abstracts of the Annual Meeting of the Agricultural Chemical Society of Japan, Tokyo p 287
Takahashi T, Yamazaki Y, Kato K (1974) Biochem J 137: 497
Vandamme EJ (1972) PhD Thesis, University of Ghent
Vandamme EJ (1976) Chem Ind 24: 1070
Vandamme EJ (1977) Adv Appl Microbiol 21: 89
Vandamme EJ (1980) In: Economic Microbiology (Rose AH, Ed), Vol 5, Academic Press, New York, London, p 467
Vandamme EJ (1981) In: Topics in Enzyme and Fermentation Biotechnology (Wiseman A, Ed), Vol 5, Horwood-Wiley, New York, London, p 185
Vandamme EJ (Ed) (1983, in press) Antibiotic Production by Fermentation Biotechnology Marcel Dekker Inc, New York
Vandamme EJ, Voets JP, Beyaert G (1971) Mededelingen van de Faculteit Landbouwwetenschappen 36: 577 (University of Ghent)
Vandamme EJ, Voets JP (1974) Adv Appl Microbiol 17: 311
Vandamme EJ, Voets JP Dhaese A (1971b) Annals Inst Pasteur, Paris 121: 435
Vezina C, Seghal S, Singh K (1968) Adv Appl Microbiol 10: 221
Venkatasubramanian K, Vieth WR (1979) In: Progress in Industrial Microbiology (Bull MJ, Ed), Vol 15, Elsevier Amsterdam, Oxford, New York, p 61
Walton RB (1964a) Developments in Industrial Micrology 5: 349
Walton RB (1964b) Science, 143: 1138
Warburton D, Balasingham K, Dunnill P, Lilly MD (1972) Biochim Biophys Acta 284: 278
Warburton D, Dunnill P, Lilly MD (1973) Biotechnol Bioeng 25: 13
Weisenburger HWO, Vanderhoeven MG (1970) Rec Trav chim Pays-Bas Belg 89: 1081
Wick WE, Wright WE, Kuder HV (1971) Appl Microbiol 21: 426
Wiseman AJ (1980) Chem Technol Biotechnol 30: 521
Yamada H, Takahashi S, Kil Y and Kumagai H (1978) J Ferment Technol, 56: 484-491

VI. Enzyme Reactors and Process Design

Reaction Technology of the Enzymatically Catalyzed Production of L-Alanine

E. Fiolitakis and C. Wandrey

Summary

Technological aspects and process characteristics of the homogeneous catalysis in a continuous flow enzyme-membrane-reactor have been discussed. High concentrations of the enlarged soluble form of coenzyme and of the two enzymes involved make it possible to achieve high conversion rates. The availability of membranes with high retention characteristics for both coenzyme and enzymes is the basis for the economical large scale application of such processes.

An engineering analysis is especially relevant in this situation in order to establish the optimal operating conditions (e.g. coenzyme- and enzyme concentrations) and the optimal reactor size. For large reactors the deactivation of enzymes and coenzyme was a dominant design parameter; for small reactors the retention of catalysts was dominant. This has been demonstrated on the L-Alanine production from pyruvate with NAD-regeneration with formiate.

Introduction

The enzymatic production of L-aminoacids proceeds usually by two routes which are demonstrated in Fig. 1. The first one, i.e. the resolution of N-acetyl-D,L-aminoacids, consists of a single reaction step but the D-substrate cannot be used without reracemization. The second one, i.e. the reductive amination of an α-ketoacid, yields an optically active aminoacid but it demands

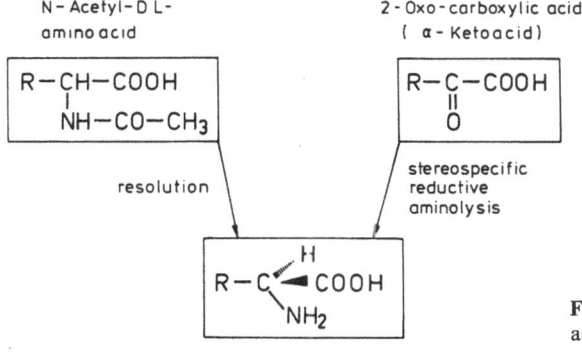

Figure 1. Enzymatic routes to L-amino-acids

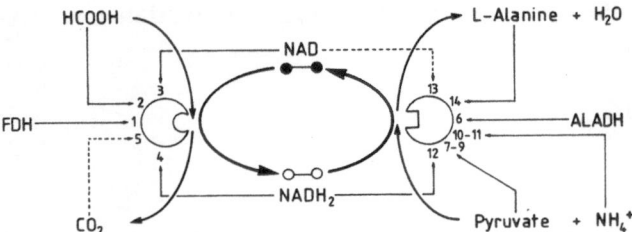

Figure 2. Reaction scheme of the multi-enzyme-system FDH-ALADH-NADH for the conversion of Pyruvate to L-Alanine with the coenzyme regeneration via formiate decomposition

two reaction-steps and the participation of two enzymes and one coenzyme. In the special case considered here the ALADH-catalyzed conversion of pyruvate to L-Alanine (see Fig. 2) is accomplished by the participation of NADH. The coenzyme regeneration takes place in the second reaction step, i.e. the FDH-catalyzed formiate decomposition.

The large scale application of a multi-enzyme-system, e.g. in the below described enzyme-membrane-reactor, depends essentially on its economical feasibility. Therefore, in the engineering analysis here reported we have concentrated the main point of interest on the question whether it is feasible to establish optimal operating points and an optimal reactor size in order to minimize the production costs.

System

The schematic representation of a continuous operated enzyme-membrane-reactor in Fig. 3 outlines the principles of the technological realization of the homogeneous multi-enzyme-catalysis (Wandrey et al., 1978). Besides, the advantages and disadvantages of this reactor type summarized in Table 1 the high catalyst costs have often the most significant contribution to the total production costs (Flaschel et al., 1978). Catalyst losses because of the incomplete retention of enzymes and especially of coenzyme, in spite of the fact that water soluble NADH-polymers can be synthesized in the meantime on an economical basis (Bückmann et al., 1981) are the main cause of catalyst supplementation in order to keep the activity as high as necessary. But also thermal deactivation of enzymes and coenzymes demands higher rates of catalyst supplementation.

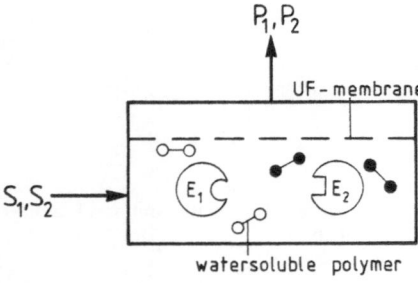

Figure 3. Schematic representation of a continuous flow enzyme-membrane-reactor

<table>
<tr><td colspan="2" align="center">Enzyme Membrane Reactor</td></tr>
<tr><td align="center">ADVANTAGES</td><td align="center">DISADVANTAGES</td></tr>
<tr>
<td>
- <u>HOMOGENOUS</u> CATALYSIS
  (NO ACTIVITY LOSS DURING FIXATION)
  (NO TRANSPORT LIMITATION)
  (NO CONTAMINATING CHEMICALS)

- <u>LOW</u> INVESTMENT PER UNIT ACTIVITY

- <u>HIGH</u> ACTIVITY PER VOLUME

- <u>CONSTANT</u> PRODUCTIVITY

- ULTRAFILTERED PRODUCT

- REACTOR EASY TO CLEAN,
  TO STERILIZE, TO CONTROL
</td>
<td>
- MEMBRANE COST

- ENZYME STABILITY IN SOLUTION

- BACKMIXING
</td>
</tr>
</table>

In the subsequent analysis we are goint out of a continuous supplementation of catalyst, which is necessary to keep the initial catalyst activity, although in the practice usually a periodical supplementation would be prefered.

Definition of the problem

Catalyst losses caused by incomplete retention

Due to the incomplete retention of catalysts by the membrane, their concentration in the reactor drops off with time if it is not compensated by supplement of fresh catalyst. Retention losses are as a rule small convective streams through the membrane and are characterized by the so-called retention parameter R, defined as follows:

$$(r_{RL})_i = c_i(1 - R) \frac{q}{V} = c_i \frac{(1 - R)}{\tau} \tag{1}$$

As you see from this definition a small catalyst stream $c_i(1 - R) q$ is lost through the membrane. Fig. 4 depicts the exponential decrease of the initial coenzyme concentration for two polymer enlarged coenzymes with molecular weights of 10.000 and 20.000 respectively.

Catalyst losses caused by thermal deactivation

We consider here thermal deactivation of catalysts expressed by the following first-order-rate kinetics

$$(r_D)_i = k_{Di} c_i . \tag{2}$$

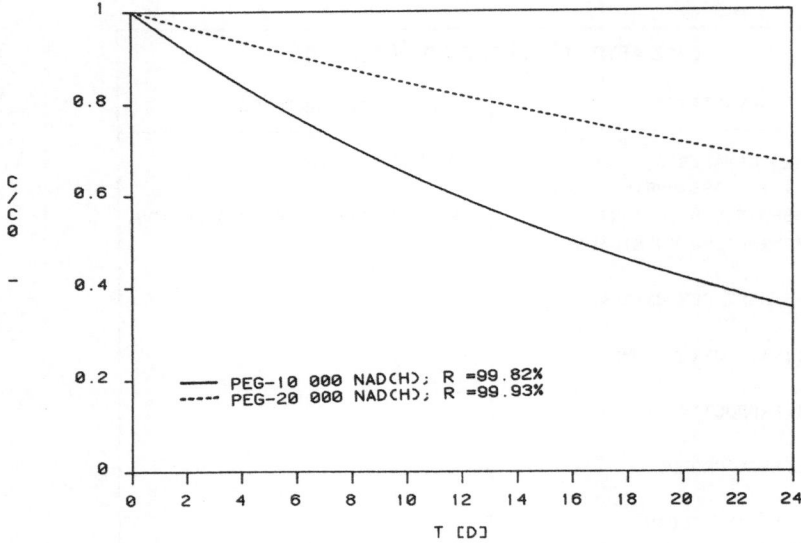

Figure 4. Losses due to incomplete retention of two water-soluble NAD-polymers with molecular weights 10 000 and 20 000

Thus the combined deactivation caused by retention losses and thermal deactivation follows an exponential decrease with time as depicted in Fig. 5.

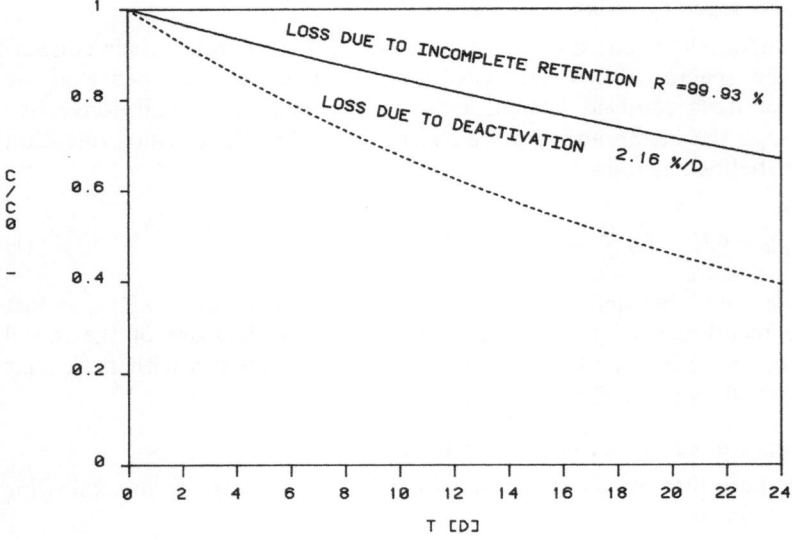

Figure 5. Loss of NADH-activity due to incomplete retention and thermal deactivation

Substrate costs

For complete conversion the substrate cost per mass unit of product is given by the following formula:

$$a_0 = \sum_{j=1}^{M} v_j f_{sj} M_{sj} / M_p \qquad (DM/kg \text{ Product}) \qquad (3)$$

which takes into account the M reaction steps (special case here $M = 2$) with the stoichiometric coefficients v_j and the substrate molecular weights M_{sj} with cost coefficients f_{sj} and the molecular weight of product M_p. For degree of conversion u which is lower than 100%, the following formula accounts for the substrate costs:

$$F_s = a_0 / u . \qquad (4)$$

Catalyst costs

The continuous supplement of catalyst necessary to compensate retention and deactivation losses can be estimated as the convective stream equal to the sum of both losses following Eq. (1) and (2):

$$\frac{(c_i)_F}{\tau} = (k_D)_i \, c_i + \frac{(1 - R) \, c_i}{\tau} \qquad (5)$$

where $(c_i)_F$ is the hypothetical catalyst concentration in the feed stream necessary to keep the initial activity. Following Eq. (5), the individual catalyst costs related to the production rate and the catalyst cost coefficients f_{ci} are introduced as follows:

$$F_{ci} = \frac{f_{ci}(c_i)_{F/\tau}}{(c_s)_F \, uM_{p/\tau}} = \frac{(k_D)_i \, c_i \tau + (1 - R) \, c_i}{(c_s)_F \, uM_p} f_{ci} \qquad (6)$$

Eq. (6) suggests that deactivation costs depend linearly on the product $(c_i\tau)$, retention costs depend linearly on c_i.

The following important assumption is made here: By changing the operating parameter τ the geometry of the membrane remains unchanged. It is obvious that for the same flow rate the retention losses differ for different membrane surface sizes.

Objectives

The objective function to be minimized here is the total product costs which are mainly composed of substrate costs and of catalysts costs:

$$F = F_s + \sum_{i=1}^{N} F_{ci} = a_0 / u + \frac{1}{(c_s)_F \, uM_p} \sum_{i=1}^{N} c_i f_{ci} [(1 - R) + (k_D)_i \, \tau] \qquad (7)$$

As a rule the degree of conversion at steady state u depends implicitly on the catalyst concentrations c_i and on the parameter τ. Thus it is preferred here to

reduce the optimization problem by one dimension and to find the optimal
catalyst concentrations at discrete values of the mean residence time τ. The
steady state value u can be obtained by different numerical procedures; it is
recommended here to use a numerical integration procedure until steady state
of the solution has been established.

Special systems and numerical values

In the special system considered here (see Fig. 2) a water soluble form of the
so-called PEG-NADH on polyethyleneglycole basis is used (Bückmann et al.,
1981). Excessive concentrations of ammonium formiate are used to guarantee
that the alanine production kinetic is ammonium-saturated. This is feasible
because of the low price of ammonium formiate. Here is suggested to have
$(NH_4COOH) = 2$. (Pyruvate) in the feed stream. Thus the maximal degree of
conversion of formiate amounts 0.50. Therefore in Eq. (3) the stoichiometric
coefficients should be selected to

$$v_1 = 2, \ v_2 = 1$$

and pyruvate or alanine are the key components.

In the specific application the objective function Eq. (7) is written in the
following form:

$$F = a_0 + \sum_{i=1}^{3} a_{Ri} c_i + \sum_{i=1}^{3} \tau a_{Di} c_i / u \tag{8}$$

For the numerical values of Table 2 the cost parameters listed on page 279
have been estimated:

Table 2. Numerical values relevant for the optimization of the multienzyme-
system ammonium formiate — Pyruvate — Alanine

(I) *Substrate and products*

	Feed concentrations	Molecular weights	Cost factors
Ammonium Formiate	400 mmol/l	63	2 DM/kg
Pyruvate	200 mmol/l	88	40 DM/kg
Alanine	0 mmol/l	89	—

(II) *Catalysts*

	Activity	Cost factors	Deactivation constants	Retention coefficient
FDH	2 U/mg	$0.35 \cdot 10^{-2}$ DM/U	0.05 (d^{-1})	99.9%
ALADH	500 U/mg	$0.50 \cdot 10^{-2}$ DM/U	0.05 (d^{-1})	99.9%
PEG-NAD	—	3.98 DM/mmol NAD	0.05 (d^{-1})	99.5%

Substrate:

$$a_0 \;=\; 41.39 \qquad DM \cdot (kg\ Ala)^{-1}$$

Retention:

$$a_{R1} \;=\; 1.12 \qquad\quad DM \cdot (kg\ Ala)^{-1} \cdot (mmol\ NAD)^{-1} \cdot 1$$
$$a_{R2} \;=\; 0.39 \qquad\quad DM \cdot (kg\ Ala)^{-1} \cdot (g\ FDH)^{-1} \cdot 1$$
$$a_{R3} \;=\; 140.45 \qquad DM \cdot (kg\ Ala)^{-1} \cdot (g\ ALADH)^{-1} \cdot 1$$

Deactivation:

$$a_{D1} \;=\; 7.75 \cdot 10^{-3}\ DM \cdot (kg\ Ala)^{-1} \cdot (mmol\ NAD)^{-1} \cdot 1 \cdot min^{-1}$$
$$a_{D2} \;=\; 1.37 \cdot 10^{-3}\ DM \cdot (kg\ Ala)^{-1} \cdot (g\ FDH)^{-1} \cdot 1 \cdot min^{-1}$$
$$a_{D3} \;=\; 4.88 \qquad\quad DM \cdot (kg\ Ala)^{-1} \cdot (g\ ALADH)^{-1} \cdot 1 \cdot min^{-1}$$

Material Balances

The following system of ordinary coupled differential equations accounts for the unsteady state material balances of the system:

Formiate:

$$\frac{dC_F}{dt} = \frac{C_{FF} - C_F}{\tau} - r_{FDH} \tag{9a}$$

Pyruvate:

$$\frac{dC_P}{dt} = \frac{C_{PF} - C_P}{\tau} - r_{ALADH} \tag{9b}$$

Alanine:

$$\frac{dC_{AL}}{dt} = \frac{-C_{AL}}{\tau} + r_{ALADH} \tag{9c}$$

Ammonium:

$$\frac{dC_{AM}}{dt} = \frac{C_{FF} - C_{AM}}{\tau} - r_{ALADH} \tag{9d}$$

NAD:

$$\frac{dC_N}{dt} = r_{ALADH} - r_{FDH} \tag{9e}$$

The kinetic expressions for r_{FDH} and r_{ALADH} and the numerical values of the parameters as stated in Table 3 (Wichmann 1981) suggest a strong non-linear dependence of the reaction rate on the total coenzyme concentration c_1 and a linear one on the enzyme concentrations c_2, c_3, a fact which is important for the structure of the optimal-cost-function. Eqs. (9a–e) cannot be solved analytically. They have been solved by numerical integration using a 5th or 6th order Runge-Kutta routine (Hull et al., 1976).

Table 3. Kinetics of the Alanine Production from Pyruvate with PEG-NAD-10,000 (Wichmann 1981):

$$r_{FDH} = c_2 V_{\max 1} \frac{c_F}{c_F + k_{m1}} \cdot \frac{c_N}{c_N + k_{m2} \left(1 + \dfrac{c_1 - c_N}{k_{m3}}\right)}$$

$$r_{ALADH} = c_3 V_{\max 2} \frac{c_p \left(1 + \dfrac{c_p}{k_{m4}}\right)}{k_{m5}\left(1 + \dfrac{c_{AL}}{k_{m6}}\right) + c_p\left(1 + \dfrac{c_p}{k_{m7}}\right)} \cdot \frac{c_1 - c_N}{c_1 - c_N + k_{m8}}$$

$$\times \frac{c_{AM}}{k_{m9} + c_{AM}\left(1 + \dfrac{c_{AM}}{k_{m10}}\right)}$$

Numerical values for the kinetic parameters:

V_{max1} = 2.60 (U $\cdot$ mg^{-1}) k_{m5} = 0.395 (mmol $\cdot$ l^{-1})

V_{max2} = 821 (U $\cdot$ mg^{-1}) k_{m6} = 1.24 (mmol $\cdot$ l^{-1})

k_{m1} = 269 (mmol $\cdot$ l^{-1}) k_{m7} = 14.7 (mmol $\cdot$ l^{-1})

Solution of the problem

The numerical solution of Eqs. (9 a–e) in order to determine the degree of conversion u at the steady state has been combined with a least squares minimization routine (Dennis et al., 1981) in order to find the optimal operating parameters c_1, c_2, c_3 at discrete values of the mean residence time τ. The computations have been performed on an IBM/370-168 computer. Because of the special character of the problem, an augmented model after Levenberg — Marquardt has been used in order to achieve effective convergence to the minimum.

Results and Discussion

The total product costs at different τ-values for the optimal operating catalyst concentrations is represented in Fig. 6. A well defined minimum at about 2.5 (h) can be observed. The optimal level of 71.6 DM/kg suggests that the process might become economically feasible. As depicted in Fig. 7, the catalyst costs follows the same qualitative dependence with a minimum also at 2.5 (h). On the other hand, the substrate costs decrease monotonously at increasing τ-values following the monotonous increase in the degree of conversion in Fig. 8. The small fluctuations of the optimization results are caused by numerical errors. The established optimal operating concentrations are depicted in Fig. 9. The enzyme concentrations decrease rapidly for increasing τ-values and the product $c_i\tau$ remains practically constant, as expected from the linear dependence of the reaction kinetics on the enzyme concentrations. The total coenzyme concentration decreases on the contrary very slowly for increasing τ-values as expected from the non-linear dependence of reaction kinetics on the coenzyme concentration. An analysis of the catalyst costs in retention and deactivation part as depicted in Fig. 10 suggests that at low τ-values the

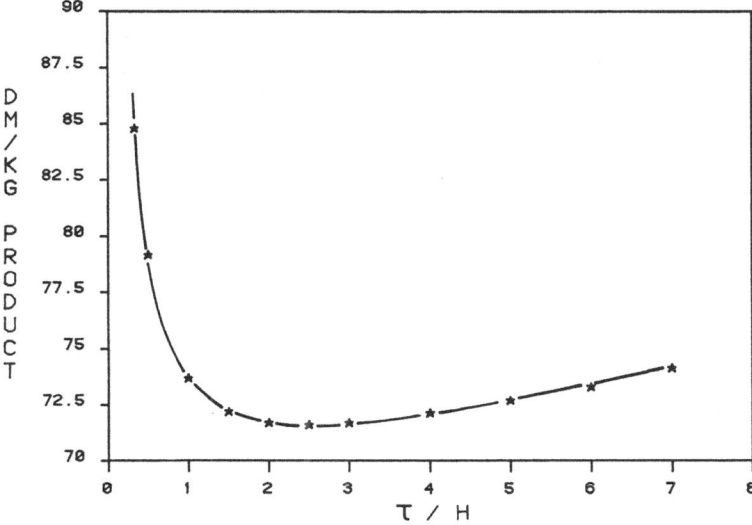

Figure 6. Total product cost as a function of mean residence time for the optimal catalysts concentrations

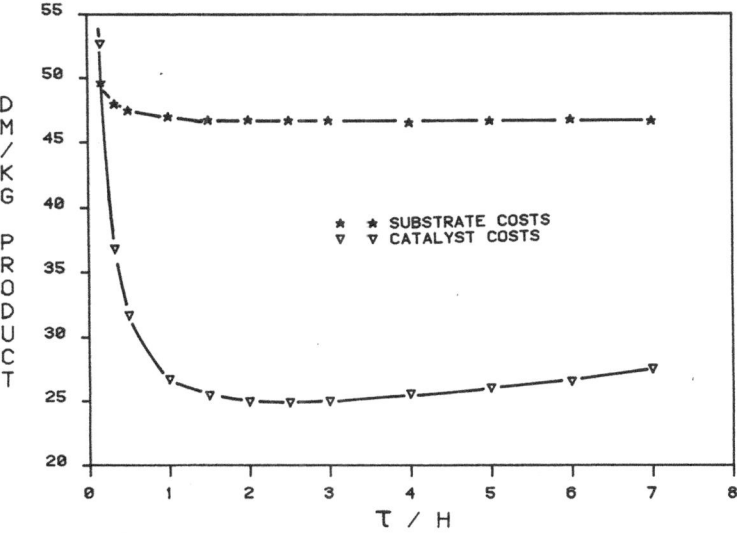

Figure 7. Individual product costs as a function of mean residence time for the optimal catalysts concentrations

retention losses become significant, whereas for large τ-values the deactivation ones are the most important. This is well in agreement with Eq. (7) and with optimal operating parameters depicted in Fig. 9. Thus for low τ-values the retention losses of enzymes are most important, for large τ-values the deactivation losses of coenzyme are most significant. At the optimum a turnover number of about 30.000 has been estimated.

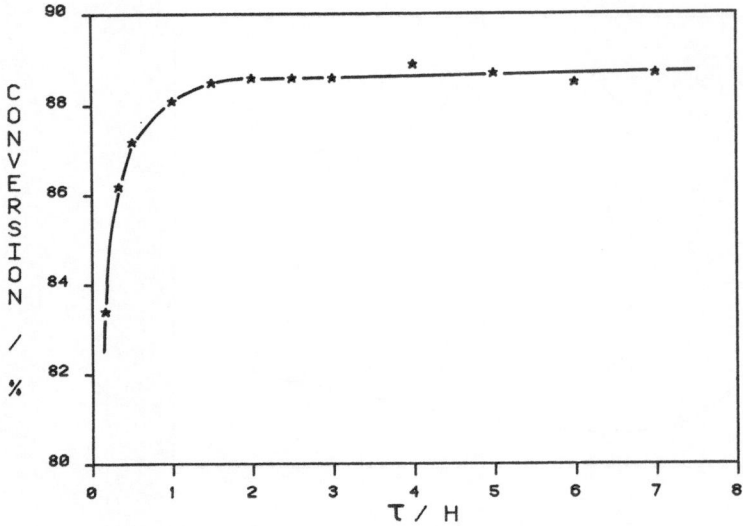

Figure 8. Optimal degree of conversion as a function of mean residence time

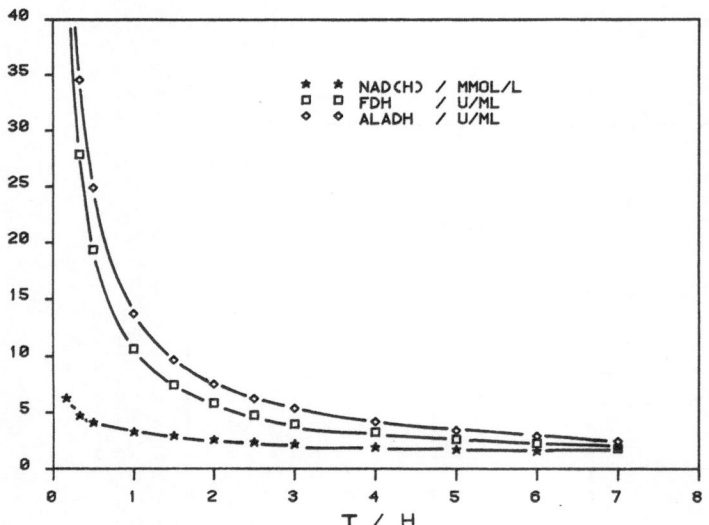

Figure 9. Optimal catalysts concentrations as a function of mean residence time

Conclusions

The optimization of the multi-enzyme-system outlined above depends on the special retention characteristics of membrane as well as on the stability of the coenzyme and of the enzymes. Because of the high substrate costs it should be attempted to achieve large degrees of conversion by selecting high catalyst

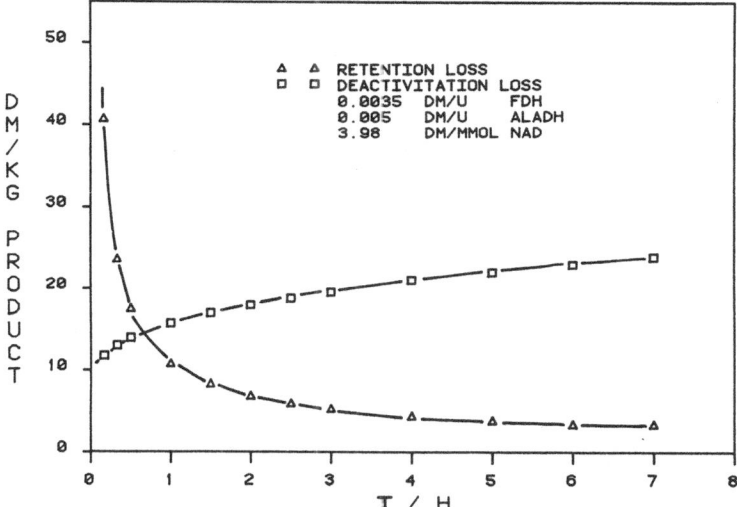

Figure 10. Individual catalysts costs as a function of the mean residence time for the optimal catalysts concentrations

concentrations. Catalyst costs will alter thus in a very significant way the optimum level. From the engineering point of view the above analysis could be interesting in determining a priori the optimal reactor size for multi-enzymatic-processes in membrane reactors.

Acknowledgement

The support of the computing centre KFA Jülich for the numerical computations is greatfully acknowledged.

List of Symbols

a_{Di}	deactivation cost factors (monetary units) $\cdot$ (product mass unit)$^{-1}$ $\cdot$ (catalyst concentration)$^{-1}$ $\cdot$ (time)$^{-1}$
a_0	substrate cost factor (monetary units) $\cdot$ (product mass unit)$^{-1}$
a_{Ri}	retention cost factors (monetary units) $\cdot$ (product mass units)$^{-1}$ $\cdot$ (catalyst concentration)$^{-1}$
c_{AL}	alanine concentration in the reactor (mmol/l)
c_{AM}	ammonium concentration in the reactor (mmol/l)
c_F, c_{FF}	formiate concentration in the reactor and in the feed (mmol/l)
c_N	NAD concentration in the reactor (mmol/l)
c_p, c_{pF}	pyruvate concentration in the reactor and in the feed (mmol/l)
c_1	total coenzyme concentration (mmol/l)
c_2	enzyme concentration (FDH) (g/l)
c_3	enzyme concentration (ALADH) (g/l)
$(c_s)_F$	feed concentration of the key-substrate (mmol/l)
F	objective function, product costs (monetary units) $\cdot$ (product mass unit)$^{-1}$
F_s	substrate costs (monetary units) $\cdot$ (product mass unit)$^{-1}$
F_c, F_{ci}	catalyst costs, total costs and individual costs (monetary units) $\cdot$ (product mass unit)$^{-1}$

f_{si}, f_{ci} substrate and catalyst cost factors
          (monetary units) $\cdot$ (mass unit)$^{-1}$
$(k_D)_i$ deactivation constant (time)$^{-1}$
M_{sj}, M_p molecular weights of substrates and of product (g/mol)
q   flow rate (m^3/h)
R   retention factor ($-$)
r_D rate of deactivation losses (concentration) $\cdot$ (time)$^{-1}$
r_{RL} rate of retention losses (concentration) $\cdot$ (time)$^{-1}$
r_{FDH} rate of the formiate decomposition (mmol $\cdot$ l^{-1} $\cdot$ min^{-1})
r_{ALADH} rate of the alanine production (mmol $\cdot$ l^{-1} $\cdot$ min^{-1})
u   degree of conversion ($-$)
V   reaction volume (m^3)
v_j stoichiometric coefficient ($-$)
τ mean residence time (min)

References

Bückmann AF, Kula MR, Wichmann R, Wandrey C (1981): An Efficient Synthesis of High Molecular Weight NAD(H)-Derivates Suitable for Continuous Operation with Coenzyme Depending Enzyme Systems, J Appl Biochem, p 301

Dennis JE, Gay DM, Welsch RE (1981): Algorithm 573, NL2SOL, An Adaptive Nonlinear Least Squares Algorithm, ACM Trans Math Software, Vol 7, p 369

Flaschel E, Wandrey C (1978) Economical Aspects of Continuous Operation with Biocatalysts. In: Broun GB, Manecke G, Wingard LB Enzyme Engineering, Vol 4. Plenum Pub Corp, New York, USA, p 83

Hull TE, Enright WH, Jackson KR (1976) User's Guide for DVERK, a Subroutine for Solving Non-Stiff ODE's, TR No 100, Department of Computer Sciences, Univ of Toronto, Canada

Wandrey C, Wichmann R, Bückmann AF, Kula M-R (1978) Immobilization of Biocatalysts using Ultrafiltration Technique. 1st European Congress on Biotechnology, 25.–29. 09. 1978, Interlaken, Schweiz, Preprints 2/44–2/47

Wichmann R (1981) Kontinuierliche enzymatische Synthese mit Coenzym-Regenerierung, Dissertation, Clausthal, Bericht Jül-Spez-119

Development of a Tubular Recycle Membrane Reactor for Continuous Operation with Soluble Enzymes

Erwin Flaschel, Eric Raetz, and Albert Renken

Summary

The performance of steady-state flow ultrafiltration membrane reactors can be considerably increased by improving their hydrodynamic characteristics.

A tubular recycle membrane reactor has been designed and constructed. This type of reactor approximates plug flow behaviour, uses few movable parts in addition to a small membrane area — compared with stagewise reactor designs. The performance of a pilot scale tubular recycle membrane reactor has been verified for the hydrolysis of lactose using lactase from *Aspergillus niger*. The operational stability of the lactase during a long-term operation of the reactor has been estimated. The substrate employed for this purpose has been a de-proteinized, fermented (lactic acid) whey — the base for a non-alcoholic drink.

Introduction

Steady-state flow ultrafiltration membrane reactors (UFMR) are usually designed as single stage plants (Flaschel et al. 1982a). This, in addition to the requirement for agitation of the reactor contents, either by pumping or by stirring, results in a behaviour similar to a continuous stirred tank reactor (CSTR). The agitation is required to avoid the enzyme being concentrated at the ultrafiltration membrane.

When an enzymatic process must be conducted at a high substrate conversion or the reaction suffers from severe product inhibition, reactors without backmixing can yield a several fold increase in catalyst utilization compared with those of the completely mixed type (Flaschel and Renken 1981). Backmixing in ultrafiltration membrane reactors can be suppressed by a stagewise design which avoids backmixing between separate stages. Since cascading of fully equipped UFMRs results in a multiplication of the investment costs, a cascade of conventional stirred tanks, in series with an ultrafiltration separation unit, has been preferred (Wandrey and Flaschel 1979). In such a cascade recycle membrane reactor the enzyme which is concentrated to some extent in the separation unit, is recycled to the first tank of the cascade. The degree of backmixing can be influenced by changing the number of stages. Stirred tanks are equipped with movable parts which require maintenance and increasing their number would multiply the investment costs. A more elegant method to obtain a low degree of backmixing has been proposed in which the cascade of stirred tanks is replaced by a tubular reactor (Flaschel et al. 1982b).

The design of such a tubular recycle membrane reactor (TRMR) will be discussed. Experience of TRMR operation has been obtained with lactose hydrolysis by means of the lactase from *A. niger*. Some results will be presented concerning enzyme stability and the adjustment of the main operational variables in order to obtain optimum enzyme utilization.

Materials and Methods

The lactase employed was purchased from Rapidase, Seclin, F (Lactase AN, 200 ONPG-kU/g) and was employed without further purification.

Edible-grade lactose (Lactose Edible Extra) was obtained from Coöperative Condensfabriek Friesland, Leeuwarden, NL.

The pH of lactose solutions was adjusted to 3.5 by the addition of dilute hydrochloric acid.

A lactic acid-fermented whey concentrate which had been de-proteinized by heat treatment was placed at our disposal by a company producing non-alcoholic drinks based on whey. The concentrate was diluted with water to a concentration equivalent to the original fermented whey. This diluted concentrate (pH 3.55) contained lactose (97.5 mol/m^3), galactose (19.4 mol/m^3) and traces of glucose (0.3 mol/m^3) and was stabilized with sodium benzoate (0,17 kg/m^3).

Glucose was analyzed by means of a Glucose Analyzer (Model 23, Yellow Springs Instruments, Ohio, USA).

Galactose was analyzed photometrically (Boehringer Mannheim, Cham, CH, art. No. 176 303).

Laktose was analyzed via hydrolysis to glucose as follows: One millilitre of sample was mixed with 5 ml of a 10 mM KCl-solution. The pH was adjusted to 6.5. Twenty microlitres of Maxilact Lx 5000 (Gist-Brocades, Delft, NL) was added to the mixture and followed by incubation for 30 min at 30 °C. The glucose concentration was analyzed before and after incubation.

The tubular recycle membrane reactor

A schematic flow sheet of the tubular recycle membrane reactor (TRMR) is given in Fig. 1. The enzyme is usually present in the tubular reactor as well as in the separation unit and is preferentially kept in the soluble state. The substrate solution flows through the tubular reactor and is pumped to the separation unit. This pump forces a product stream equivalent to the feed flow rate across the membrane. The other portion of the flow entering the separation unit is fed back to the entrance of the tubular reactor. The separation unit is equipped with a pump in a closed loop with an ultrafiltration device to avoid concentration of the enzyme at the membrane.

The three main variables which govern the performance of the TRMR are:
— the degree of backmixing in the tubular reactor
— the recycle ratio (f_{rec}), i.e. the flow rate of back-feeding divided by the flow rate of substrate feeding

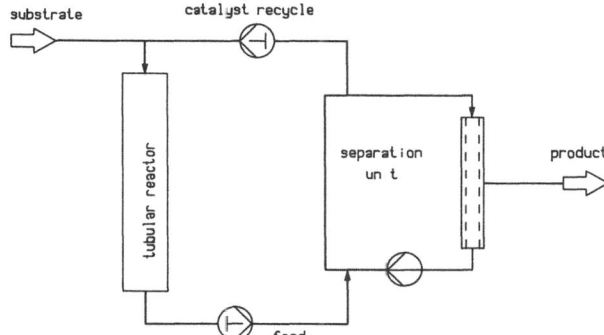

Figure 1. Schematic flow sheet of a tubular recycle membrane reactor

— the volume ratio (f_v), i.e. the volume of the separation unit divided by the volume of the whole TRMR.

These variables determine the over-all degree of backmixing and the catalyst repartition between the tubular reactor and the separation unit.

Since the TRMR will be operated at moderate enzyme concentrations, the usual mean residence time of the substrate will be situated in the range of 30 to 60 min. Such long reaction times imply that the fluid velocity in the tubular reactor will be low, approximately 0.25 to 2.5 mm/s. At such low linear fluid velocities the flow profile in normal columns is disturbed by free convection which gives rise to unpredictable axial backmixing. Since this effect increases with increasing column diameter, columns devoid of internals are not recommended for application as tubular reactors for pilot and production scale TRMRs. Special packings or internals have to be used to avoid excessive backmixing. In the range of very low linear fluid velocities, motionless mixers (type SMX, Sulzer AG, Winterthur, CH) have proven to be very effective in suppressing backmixing as well as equalizing radial flow profiles.

The degree of backmixing has been determined in a column with a diameter of 4 cm and a length of 1.6 m (sizes planned for a pilot scale TRMR). For these measurements the tubular reactor of 1.6 m in length had been provided with an entry section of 0.6 m and an exit section of at least 0.3 m which were likewise filled with the SMX mixer. For tracer experiments either a photometric (Nguyen-Khac-Tien 1982) or a conductimetric analysing method has been employed. For the photometric analysis, Patentblue was applied as tracer, for conductimetric analysis diluted HCl has been applied.

A main current of water was mixed with a secondary one of low flow rate in the entry section of the column. The secondary flow was doted with tracer by means of an HPLC injection valve. The concentration-time profile of the tracer was measured twice — at the entrance of the tubular reactor and at the exit, both 1.6 m apart.

The mean residence time and the Bodenstein number have been estimated by the least squares method according to Marquardt (Marquardt 1963). The exit signal has been calculated by convolution of the entry signal applying the dispersion model for open tubular reactors and large dispersion (Levenspiel and Smith 1957). The objective function was defined as the sum of squares of the deviations between the calculated and the experimental exit tracer response.

The results are gathered in Fig. 2. Since the Bodenstein number depends on the length of the reactor, it is given here per unit reactor length. This quantity is equal to the quotient of the mean linear fluid velocity ($\bar{u}$) and the axial dispersion coefficient (D_{ax}). The Reynolds number is defined by the effective mean linear fluid velocity ($\varepsilon_{SMX} = 0.905$) times the reactor diameter of the empty column divided by the kinematic viscosity of the liquid. It must be pointed out that the Reynolds number is only given for convenience and that an extrapolation of the relation shown in Fig. 2 to reactors of other radial dimensions is not valid.

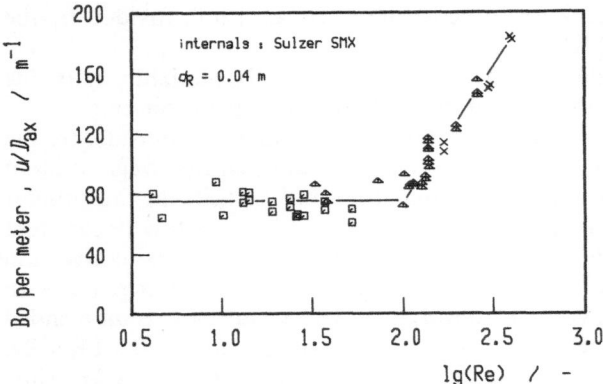

Figure 2. Axial backmixing in a tubular reactor equipped with internals of the type SMX
□ photometric analysis, water, 20 °C (Nguyen-Khac-Tien 1982)
△ conductimetric analysis, water, 30 °C
× conductimetric analysis, water, 50 °C

The relation in Fig. 2 shows that for Reynolds numbers less than 100 a Bodenstein number of 70–80 per meter reactor length results. This is the range of flow rates which would be applied in TRMR operation. Therefore, a reactor of 1.6 m length should behave essentially as an ideal plug flow tubular reactor (PFTR). The distribution of the enzyme and the mean residence times of the reaction mixture in the tubular reactor and the separation unit depend on the volume ratio ($f_V = V_S/(V_R + V_S)$) and the recycle ratio ($f_{rec} = \dot{V}_{rec}/\dot{V}_{feed}$) (Flaschel et al. 1982a). Lowering the volume ratio will increase the reactor performance since the enzyme distribution is changed in favour of the tubular reactor, where the activity is more efficiently utilized than in the separation unit. The recycle ratio has a more complex influence which can be explained best by the extreme situation. When no enzyme is recycled from the separation unit to the entrance of the tubular reactor, the catalyst will be concentrated in the separation unit. This would result in a single continuous stirred tank reactor (CSTR) of small volume. When the recycle flow rate

($\dot{V}_{rec}$) exceeds the feed rate ($\dot{V}_{feed}$) by several orders of magnitude, a CSTR of large volume would result. Since in both cases the behaviour of the tubular reactor would be completely destroyed, the optimum TRMR performance should be found at moderate recycle rates.

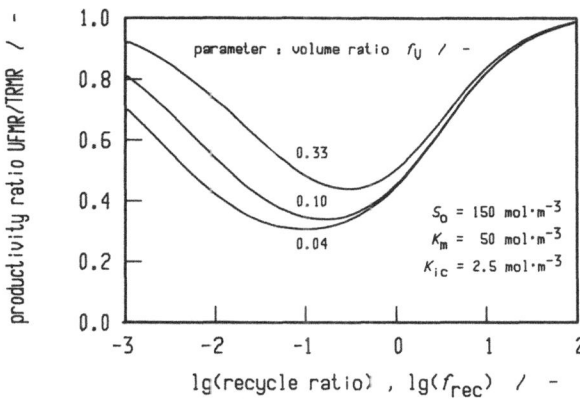

Figure 3. Influence of the recycle ratio on the productivity of a tubular recycle membrane reactor (X = 0.8)

The influence of both the recycle ratio and the volume ratio are shown in Fig 3 for an enzymatic reaction exhibiting Michaelis-Menten type kinetics with competitive product inhibition. The TRMR performance is compared to the performance of a simple UFMR under equal conditions. The catalyst utilization is a factor of 2 to 3-fold better with the TRMR as compared with a simple single stage UFMR. The optimum recycle ratio is shifted to lower values when the volume ratio decreases because the enzyme concentration in the separation unit increases.

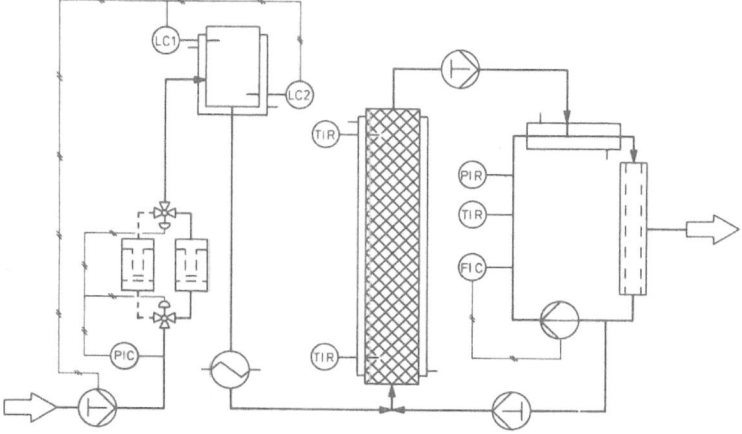

Figure 4. Flow sheet of the pilot scale tubular recycle membrane reactor

The pilot scale TRMR which has been constructed is shown in Fig. 4. The basic design data are given in Table 1. The substrate solution is periodically pumped through sterilizing filters to a storage tank. The solution passes a heat exchanger and the jacketed tubular reactor, which is equipped with motionless mixers of the type SMX, by gravity. A reciprocating pump conveys it to the separation unit where a part of the solution leaves the installation across the ultrafiltration membrane. Another part is recycled by means of a rotary piston pump to the entrance of the tubular reactor. The TRMR has a maximum capacity of ca. 150 l/d, limited by the feeding pump.

Table 1. Main design parameters and equipment of the pilot scale tubular recycle membrane reactor

Tubular reactor:		
length	l_R/m	1.6
diameter	d_R/m	0.04
volume	V_R/l	1.9
Separation unit:		
membrane area	A_m/m^2	1.4
volume	V_S/l	0.95
volume ratio	$f_V/-$	0.33
Special equipment:		
membrane module:	Romicon HF-15-43-PM 10	
membrane cut-off:	10000 g/mol	
recirculating pump:	centrifugal, speed controlled	
internals:	Sulzer type SMX	
sterilizing filters:	Gelman cartridge No. 12117 (0.2 μm)	

Experimental Verification of the Reactor Performance

The performance of the tubular recycle membrane reactor has been verified by varying the recycle ratio (f_{rec}) and measuring the substrate conversion after the tubular reactor in addition to the final conversion.

The reaction conditions for two series of tests are reported in Table 2. The TRMR has been sterilized by the passage of an aqueous solution of 2.5% formalin followed by the lactose solution. When the formalin was completely displaced, the enzyme has been injected into the separation unit in the form of an aqueous solution. The TRMR was then operated for 3 days to avoid measurements during the very first period after enzyme injection when the rate of deactivation is considerably higher than later. For this period the TRMR was operated as a simple single stage ultrafiltration membrane reactor (UFMR) by shutting down completely the recycling of enzyme. After three days, the substrate conversion has been determined and used to re-calculate the effective initial enzyme activity ($\eta = m_E/m_{EO}$). This calculation and further simulations are carried out by combining the mass balance of the TRMR (Flaschel et al. 1982a) with the kinetics of the lactase from *A. niger* (Flaschel et al. 1982c).

Table 2. Reaction conditions for the verification of the TRMR performance

Substrate	edible grade lactose	
Enzyme	lactase from *A. niger*	
pH	3.5	
S_o	150 mol/m^3	
Temperatures:		
tubular reactor	50 °C	
separation unit	46 °C	

Quantity	1. test	2. test
m_{Eo}/g	8	24
$\eta(3d)/-$	0.46	0.64
$\dot{V}_{feed}/l/h$	1.12	1.53
$\dot{V}_{loop}/l/h$	120	120

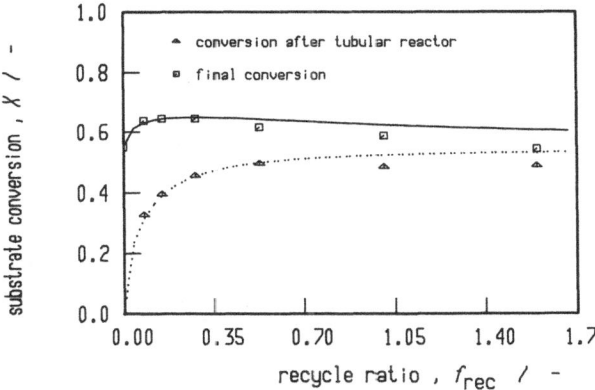

Figure 5. Verification of the performance of the tubular recycle membrane reactor at moderate lactose conversion

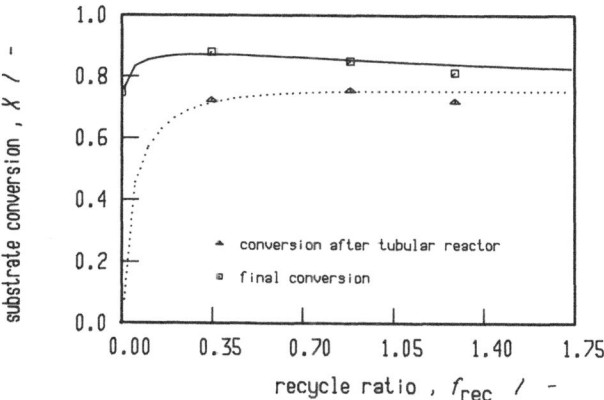

Figure 6. Verification of the performance of the tubular recycle membrane reactor at high lactose conversion

The recycle ratio was then varied step by step. After each variation the TRMR was operated for about one day in order to stabilize it before the conversion after the tubular reactor and the final conversion have been analysed. The expected substrate conversions were calculated on the basis of the corrected initial amount of enzyme.

The results of the two series of tests are shown in Fig. 5 and 6. The agreement of the experimental data with the model calculations is only slightly impaired by the deactivation of the enzyme during the measurement period. The optimum recycle flow rate is found to correspond to 30% of the feed flow rate with a rather flat optimum.

Operational Stability of Lactase from Aspergillus niger

The lactase from *A. niger* is a relatively expensive enzyme which suffers, in addition, from severe product inhibition (Flaschel et al. 1982c). This enzyme would therefore be preferentially employed in cases where a high temperature stability is needed at low pH and where the enzyme-dependent costs are not so critical as in pure whey treatment. Consequently an unusual substrate has been employed: a lactic acid fermented whey which had been de-proteinized by heat treatment. This liquor is normally employed as a base for non-alcoholic drinks. The aim has been to reduce the lactose content of this fermented whey by a factor of 4. Thus, the TRMR has been operated for 20 d to observe the stability of the lactase in the presence of this unusual substrate, whose composition is given in 'Materials and Methods'.

The operating conditions are specified in Table 3. The reactor has been sterilized by means of a 2.5% formalin solution before the fermented whey has been applied. When the fermented whey had displaced the formalin, an initial amount of lactase (m_{E0}) was injected into the separation unit in the form of an aqueous solution. The substrate conversion has been analysed and fresh lactase added when the conversion tended to fall below 70%. The amount of lactase necessary to reestablish a conversion of 80% was calculated before each new enzyme injection. For calculations it had to be taken into account that the maximum reaction rate of the lactase decreases to 85% of the value for pure lactose, in addition to the inhibition by the galactose present.

Table 3. Operating conditions for the lactose hydrolysis in fermented whey

mean conversion	$\bar{X}/-$	0.75
pH	$/-$	3.55
Temperatures:		
tubular reactor	$t_c/°C$	50
separation unit	$t_c/°C$	46
feed flow rate	$\dot{V}_{feed}/l/h$	2.5
loop flow rate	$\dot{V}_{loop}/l/h$	80
recycle ratio	$f_{rec}/-$	0.32
pressure drop	$\Delta p/bar$	0.3
mass of lactase introduced:		
initially	m_{E0}/g	25
at t = 8 d	m_{E1}/g	7.1
at t = 13 d	m_{E2}/g	8.8

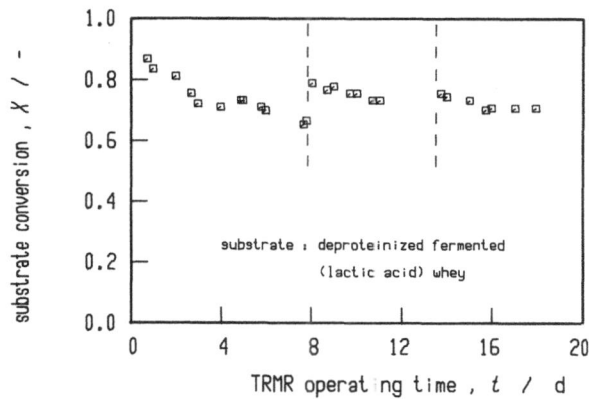

Figure 7. Operational stability of lactase from *A. niger* in the tubular recycle membrane reactor

The original data are shown in Fig. 7 in the form of substrate conversion as a function of operating time. The vertical broken lines indicate the times at which fresh lactase had been added.

Figure 8 shows the same data, only the substrate conversion has been transformed to the actual residual enzyme activity. If the first steep decline is neglected, a deactivation rate of 5.3 %/d relative to the initial mass of lactase (m_{E0}) is estimated. This corresponds to a lactase consumption of 20 g/m^3 of fermented whey.

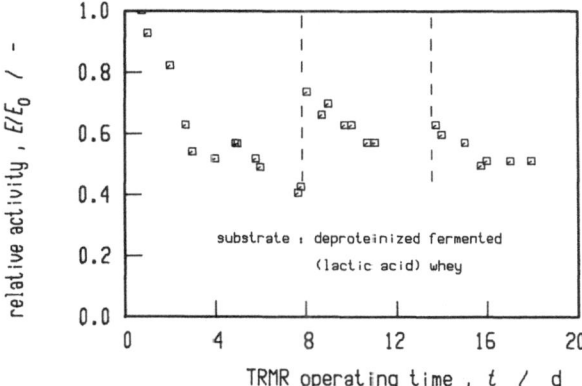

Figure 8. Operational stability of lactase from *A. niger* in the tubular recycle membrane reactor

Conclusion

The performance of a tubular recycle membrane reactor has been verified by means of lactose hydrolysis with the lactase from *Aspergillus niger*. The accordance of the simulations and the experimental results have shown that the tubular reactor equipped with special internals can be modelled as an ideal plug flow reactor.

The design of the tubular recycle membrane reactor is rather simple, needing only an additional control for the recycling of enzyme compared to single stage ultrafiltration membrane reactors. The maintenance requirement is lowered considerably compared to a stagewise design because the use of a tubular reactor avoids the installation of additional movable parts as well as excessive membrane areas.

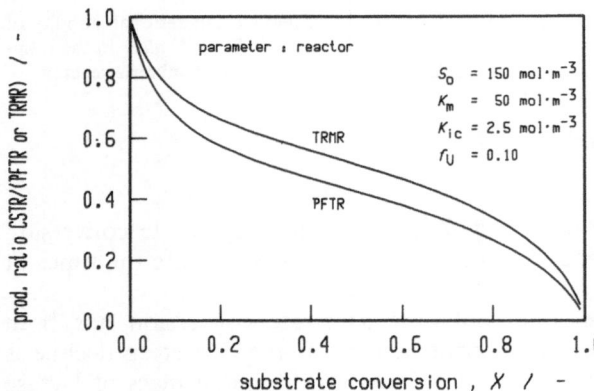

Figure 9. Comparison of the performance of a tubular recycle membrane reactor with an ideal plug flow reactor, both referred to a continuous stirred tank reactor

To illustrate the performance of the tubular recycle membrane reactor (TRMR), Fig. 9 shows the utilization of the enzymatic activity for this reactor in comparison to the best case, the ideal plug flow tubular reactor (PFTR) — both referred to the productivity of a simple single stage membrane reactor. For the same kinetics as taken for Fig. 3 and for a moderate volume ratio (f_v), the tubular recycle membrane reactor shows a performance nearly as favourable as an ideal plug flow reactor.

Acknowledgement

The support of the Board of the Swiss Federal Institutes of Technology is gratefully acknowledged.

Symbols

A	m²	area
Bo	—	Bodenstein number ($\bar{u} \cdot 1/D_{ax}$)
d	m	diameter
D_{ax}	m²/s	axial dispersion coefficient
E	kg/m³	enzyme concentration
f_{rec}	—	recycle ratio ($\dot{V}_{rec}/\dot{V}_{feed}$)
f_v	—	volume ratio ($V_S/(V_S + V_R)$)
K_m	mol/m³	Michaelis-Menten constant
K_{ic}	mol/m³	constant of competitive inhibition
l	m	length
m	g	mass

p	bar	pressure
Re	—	Reynolds number ($\bar{u} \cdot d/\nu$)
s	mol/m^3	substrate concentration
t	specific	time
t_c	°C	temperature
u	m/s	linear velocity
V	l	volume
$\dot{V}$	l/h	volumetric flow rate
X	—	substrate conversion

Greek

ε	—	voidage
η	—	effectiveness
ν	m^2/s	kinematic viscosity

Subscripts

o	initial
E	enzyme
feed	referred to feed
loop	referred to internal recirculation in the separation unit
m	membrane (except K_m)
R	tubular reactor
rec	referred to recycling of enzyme
S	separation unit
SMX	referred to internals SMX

Subscripts

—	mean

References

Flaschel E, Renken A (1981) Développement des réacteurs continus pour la catalyse enzymatique. Swiss Chem 3(9):102–107

Flaschel E, Wandrey C, Kula M-R (1983)Ultrafiltration for the Separation of Biocatalysts. Adv Biochem Eng 26:(in press)

FlaschelE, Raetz E, Renken A (1982a) A New Approach to Membrane Reactor Design and Operation. Enzyme Engineering 6:313–314

Flaschel E, Raetz E, Renken A (1982b) The Kinetics of Lactose Hydrolysis for the β-Galactosidase from *Aspergillus niger*. Biotechnol Bioeng 24:2499–2518

Levenspiel O, Smith WK (1957) Notes on the Diffusion-type Model for the Longitudinal Mixing of Fluids in Flow. Chem Eng Sci 6:227–233

Marquardt DW (1963) An Algorithm for Least-Squares Estimation of Nonlinear Parameters. J Soc Indust Appl Math 11(2):431–441

Nguyen-Khac-Tien (1982) Mélangeur statique comme réacteur tubulaire de polymérisation. Doctoral thesis No 445, ETH-Lausanne, Switzerland

Wandrey C, Flaschel E (1979) Process Development and Economic Aspects in Enzyme Engineering. Acylase L-Methionine System. Adv Biochem Eng 12:147–218

Kinetically VS. Equilibrium-Controlled Synthesis of C—N Bonds in β-Lactams and Peptides with Free and Immobilized Biocatalysts

Volker Kasche, Boris Galunsky, Uschi Haufler, and Roswitha Zöllner

Summary

Hydrolytic enzymes (proteases, penicillin amidase) can be used also for synthetic purposes. The interest to use these enzymes for synthetic purposes as the semisynthesis of β-lactam antibiotics and peptides has increased recently.

This enzyme catalyzed synthesis of C—N (peptide) bonds can be performed under equilibrium of kinetically controlled conditions.

The latter is studied here for transpeptidations or transfer of β-lactam side chains to 6-amino penicillanic acid using activated side chains (esters, amides). In this enzyme catalyzed reaction a transient maximum concentration *(kinetically controlled maximum)* much larger than the final equilibrium concentration of the desired product (peptide, β-lactam) can be obtained.

The yield at this maximum depends on the ratio of the deacylation rate constants of the acyl-enzyme with the nucleophiles and H_2O. The particle size-, ionic-strength-, pH- and nucleophile structure-dependence of this ratio has been experimentally studied for free and immobilised biocatalysts. The data were found to correlate well with theoretical predictions.

The kinetically controlled maximum is compared with the equilibrium controlled synthesis as an endpoint for enzyme catalyzed semisynthetic procedures.

Abbreviations

Ac—Tyr, Ac—Tyr—O—Et, Ac—Tyr—O—Me = N-acetyl-L-tyrosine (ethyl- or methylester); Ac—Tyr—Tris = N-acetyl-L-Tyrosine-Tris; 7-ACA = 7-aminocephalosporanic acid; 7-ADCA = 7-aminodesacetoxycephalosporanic acid; AMP = ampicillin; 6-APA = 6-aminopenicillanic acid; BP = benzyl-penicillin; CT = α-chymotrypsin (E.C. 3.4.21.1); Lys—O—Me = L-lysine-methyl-ester; PA = penicillin amidase (E.C. 3.5.1.11); PAA = phenyl-acetic-acid; PAG = phenyl-acetyl-glycine; PG = D-phenyl-glycine; PG—O—Me = D-phenyl-glycine-methyl-ester; Tris = tris(hydroxymethyl)methylamine.

Introduction

Enzymes catalyze a reaction in both directions but they cannot influence the thermodynamic equilibrium of this reaction. This truism was first experimentally demonstrated for enzymes associated with hydrolytic (catabolic) processes more than 50 years ago (1, 2). When the catalytic function of an enzyme is discussed the catalysis in one direction still receives the main interest. Hydrolases are normally associated with catabolic (hydrolytic) reactions and not with the reverse anabolic reactions that are part of bio-synthetic pathways or of potential interest in enzyme technology. *In vivo* different enzymes are often used to catalyze the catabolic and the reverse anabolic process (protein and polysaccharide hydrolysis and synthesis). This is, however, not generally the case. In polynucleotide metabolism the same enzymes can be involved in both catabolic and anabolic reactions (3).

In biotechnology enzymes are now mainly used to catalyze hydrolytic reactions. Recently new methods have been introduced to increase the enzyme production in single cells (genetic engineering) and improved purification methods are developed. It is therefore expected that the potential use of enzymes as catalysts for synthetic reactions — now performed by chemical methods — will receive an increased attention.

This is reflected in some recent original papers and reviews on β-lactam (4–6) and peptide (7, 8) semisynthesis using hydrolases. In these reactions the enzymes act as transferases, where a group A is transferred from B to C in an enzyme catalyzed process as shown in Fig. 1 where both AB and AC are substrates for the enzyme. Before the yield of AC at the equilibrium that cannot be influenced by the enzyme is reached it is under kinetic control, i.e. also influenced by the properties of the enzyme. Whether higher yields than at equilibrium can be obtained in the latter case is investigated here.

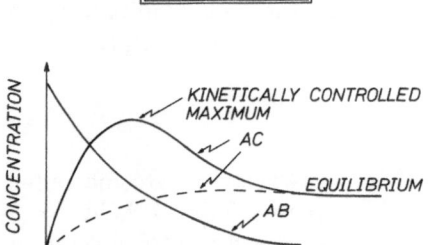

Figure 1. Kinetic and equilibrium control in enzyme catalyzed semisynthesis. Time dependence of the concentration of AB and AC after mixing AB, C and enzyme catalyzing the group transfer reaction shown on top

Kinetic or equilibrium control of enzyme catalyzed semisynthesis

In most applications of enzymes to synthesize compounds AC as shown in Fig. 1, the reaction is carried out to the thermodynamic equilibrium. Then the yield of AC can only be influenced by changing the parameters (solvent, pH, temperature) that can perturb the equilibrium. This cannot be perturbed by the properties of the enzyme used as a biocatalyst. In some early studies on β-lactam (9) and peptide (10) semisynthesis a transient maximum in AC larger than the final equilibrium content was observed (Fig. 1). This indicated that the yield in AC may be under transient kinetic control. The factors influencing the existence and magnitude of such *kinetically controlled maxima* in the enzyme catalyzed semisynthesis of β-lactams have been analyzed in detail recently (6). This analysis also applies for enzyme catalyzed transpeptidations. The important reactive intermediate is the (covalent) acylenzyme E—A that can be deacylated in competing reactions by C, H_2O and other nucleophiles in the reaction mixture (Fig. 2). When a kinetically controlled maximum in AC exists, the following relation can be derived for the fraction of C, $[C]_0$ initially present transferred to A at this maximum (6)

$$\frac{[AC]_{max}}{[C]_0} = \frac{\alpha}{1 + \alpha} \tag{1}$$

where

$$\alpha = \frac{k'_3}{\underline{k}_3[H_2O]} \cdot \frac{[AB]}{1 + \frac{k'_3[B]}{\underline{k}_3[H_2O]}} \cdot \frac{\left(\frac{k_{cat}}{K_M}\right)_{AB}}{\left(\frac{k_{cat}}{K_M}\right)_{AC}} \tag{2}$$

and [B] and [AB] are the concentrations of B and AB when $[AC] = [AC]_{max}$.

From Eq. (1) and (2) follows that the magnitude of the maximum is independent of the total enzyme content. The time when it occurs after mixing AB, C and enzyme is, however, dependent on the total enzyme content. The magnitude

ENZYME CATALYZED SEMISYNTHESIS OF β-LACTAMS AND PEPTIDES
BY NUCLEOPHILIC ATTACK ON ACYL-ENZYME E-A

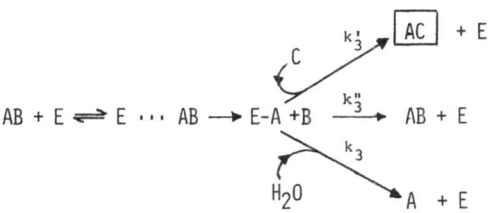

β-LACTAMS		PEPTIDES
E	PENICILLIN AMIDASE	PROTEASES
AB	PHENYL-ACETYL-GLYCINE	PEPTIDE (AMINO ACID) ESTERS
	PHENYL-GLYCINE-ESTERS	PEPTIDE (AMINO ACID) AMIDES
B	ALCOHOL; GLYCINE	ALCOHOL, AMINO ACID, NH_3
C	6-APA; 7-ACA, 7-ADCA	AMINO ACID (AMIDE, ESTER), TRIS, R-OH
AC	PENICILLINS; CEPHALOSPORINS	DIPEPTIDES
A	PHENYL ACETIC ACID	PEPTIDE (AMINO ACID)

Figure 2. Reaction mechanism substrates and products in the enzyme catalyzed semisynthesis of β-lactams and peptides

of the maximum is contrary to the AC-content at the final thermodynamic equilibrium dependent on properties of the enzyme and increases with:

(i) the ratio of the deacylation rate constants (k_3'/k_3),

(ii) the ratio $(k_{cat}/K_M)_{AB}/k_{cat}/K_M)_{AC}$, i.e. the difference in the specificities of the two substrates AB and AC.

Besides this, the magnitude of the maximum is also influenced by the H_2O-content and the initial amount of the substrates AB and C.

From this follows that for an enzyme catalyzed synthesis as shown in Fig. 1 the kinetically controlled maximum in AC may, if it exists, be an alternative to the equilibrium as an end point of the reaction. Then higher yields in the desired product are obtained in shorter time. The reaction must then be stopped at the maximum by removing the catalyst, for this immobilized biocatalysts (enzymes, cells) are more convenient to use than free enzymes.

Due to the immobilization new yield-controlling factors are introduced besides those given in (i) and (ii) above that also may be influenced by the immobilization. These yield-controlling factors are summarized in Table 1.

We are studying these for the semisynthesis of β-lactams and peptides with free and immobilized hydrolyses. Some results relevant in this context will be presented and analyzed here.

Table 1. Effects of immobilization and substrate on the ratio of the deacylation rates ($k_3'[C]/k_3[H_2O]$) and maximum yield of the product AC for the semi-synthesis reaction given in Fig. 2

	GENERAL EFFECTS DUE TO BIOCATALYST IMMOBILIZATION		SUBSTRATE SPECIFIC EFFECTS	
			AB	C
	EXTERNAL AND INTERNAL MASS TRANSFER LIMITATION	PERTURBATION OF INTRINSIC RATE CONSTANTS	LEAVING GROUP (B) EFFECTS	SUBSTRATE CHARGE - PARTITION IN MATRIX - REPULSION/ATTRACTION BY ACTIVE SITE pK OF NUCLEOPHILIC GROUP
THE RATIO $\left(\dfrac{k_3' [C]}{k_3 [H_2O]}\right)$ AND MAXIMUM YIELD OF $\boxed{AC}$ IN KINETICALLY CONTROLLED SYNTHESIS	IS REDUCED	INCREASES OR DECREASES (DEPENDING ON MATRIX CHARGE	INCREASES OR DECREASES	INCREASES OR DECREASES

Experimental procedure

Materials:

Amino acids, amino acid derivatives (Ser—NH$_2$, Val—NH$_2$, Lys—O—Me, Ac—Tyr), PAA and PG were obtained from Sigma. Ac—Tyr—O—Et was purchased from Serva. Samples of PG—O—Me, 6-APA, BP, AMP, PA ((E.C. 3.5.1.11) free and immobilized in Eupergit C) were kindly provided by Dr. K. Sauber (Hoechst) and Dr. D. Krämer (Röhm). Immobilized proteinase K (E.C. 3.4.21.14) was obtained from Merck, and C.T. (E.C. 3.4.21.1) from Worthington. PA and CT were immobilized to LiChrospher-500 NH$_2$ (kindly provided by E. Merck) and Sepharose C-2B and DEAE-Sephadex as described in (6, 11). *E. coli* with PA-activity immobilized in agar (average particle diameter

∅ 2 mm) was a gift from the Institute of Microbiology, Chinese Academy of Sciences, Beijing. All other chemicals were analytical grade.

Enzyme catalyzed semisynthesis

Biocatalyst, substrate and nucleophile were mixed in buffer. The reaction mixture in a scintillation vial was suspended using a rotary mixer. At different times after mixing samples were withdrawn, diluted, filtered and analyzed by HPLC. The reaction was carried out at constant temperature. The pH was kept constant by adding NaOH when necessary.

HPLC-analysis

The products and reactants were identified and analyzed by HPLC using a Spectra Physics SP 8700 solvent delivery system with a SP 8400 detector and a 250×4.6 mm RP-18 (10 µm) column (Knauer).

In β-lactam semisynthesis the conditions for the analysis were: flow rate 2 ml/min, gradient from 10% MeOH (w/w) in $0.067\ M$ KH_2PO_4 (pH = 4.7) to 50% MeOH in $0.067\ M$ KH_2PO_4 (pH = 4.7) in 5 min, column temperature 50 °C. In peptide semisynthesis the conditions were: flow rate 2 ml/min, 100% H_2O 4 min, gradient from 100% H_2O to 100% MeOH in 6 min, 100% MeOH 5 min. Column temperature 25 °C.

The amount of products and reactants was determined from calibration curves determined from analysis of stock solutions of known content and mass balances (for products in the peptide semisynthesis).

Results

Effects of mass transfer limitations

In Fig. 3 the time dependence of the semisynthesis of benzylpenicillin with immobilized *E. coli* cells is shown. The kinetically controlled maximum but not the final equilibrium content depends on the particle size. The smaller

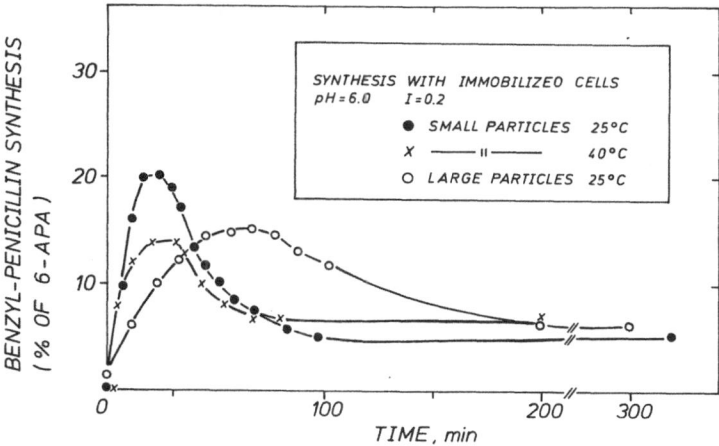

Figure 3. Time dependence of the benzyl-penicillin concentration after adding the same amount of immobilized *E. coli* cells to 10 mM 6-APA and 10 mM PAG in phosphate buffer of pH = 6.0 (I = 0.2). The particle size of the immobilized cells was reduced with a factor of > 10 by gentle grinding

the particle size the larger is the maximum that is also obtained after a shorter time. The latter indicates that the effective biocatalyst activity is reduced with the particle size. From Eq. (1) and (2) this should not give a reduction in the maximum as this does not depend on the amount of the biocatalyst.

Due to mass transfer limitations, the effective concentration of 6-APA in the biocatalyst particle decreases with increasing particle size. As the H_2O-content does not decrease, less acyl-enzyme is deacylated by 6-APA and the maximum in BP is reduced with increasing particle size.

This mass transfer limitation effect on the kinetically controlled maximum can be made negligible when the particle size is reduced or by increasing the substrate contents to values $\gg \underline{K}_M$ for the different substrates (11, 12).

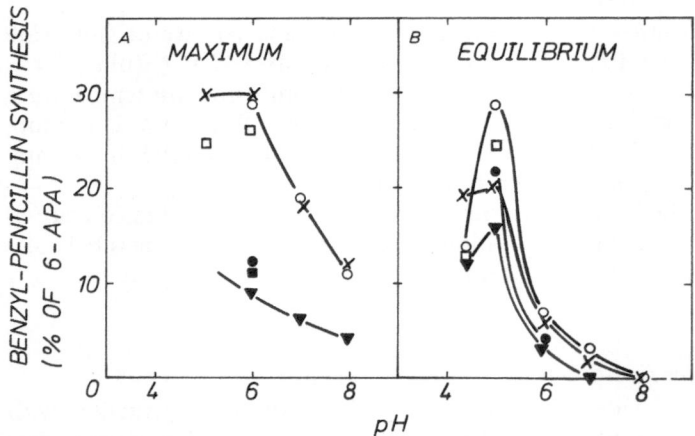

Figure 4. The pH-dependence of the concentration of benzylpenicillin (as % of initial 6-APA content) at the kinetically controlled maximum and the final equilibrium after adding free and immobilized penicillin amidase (PA) to 10 mM 6-APA and 10 mM PAG ($I_{buffer} = 0.2$, 25 °C) ○ free PA; ■ immobilized *E. coli* cells; PA immobilized in charged (×) and uncharged (△) LiChrosper; ● PA immobilized in Eupergit C

Effects on intrinsic rate constants

The ionic strength dependence of the kinetically controlled maximum and the final equilibrium in the BP-semisynthesis with free and immobilized PA is given in Fig. 4. Below pH = 5 and pH = 4.5 no maxima could be observed with free and immobilized PA, respectively. This indicates that these pH-values are outside the range of existence for kinetically controlled maxima. The maximum is more markedly influenced by the ionic strength than the final equilibrium content. From Eq. (1) and (2) follows that the maximum increases with increasing ratio of the deacylation rate constants ($\underline{k}'_3/\underline{k}_3$). This ratio was determined from the ratio of the inital rates of the formation of BP and PAA (or AMP and PG for ampicillin semisynthesis) (13). The results are given in Table 2. They show that the ionic strength dependence on the ratio $\underline{k}'_3/\underline{k}_3$ is larger for the immobilized than for the free enzyme in the case of matrices with stationary charges. The ionic strength influences the ratio $\underline{k}'_3/\underline{k}_3$ due to:

Table 2. Ionic strength- and pH-dependence of the ratio of the apparent deacylation rate constants k_3' (with 6-amino-penicillanic acid) and k_3 (with H_2O) for the deacylation of free and immobilized phenyl-acetyl- or phenyl-glycyl-penicillin amidase (25 °C). The acyl enzyme was formed using PAG (benzylpenicillin semisynthesis) and PG-O-Me (ampicillin semisynthesis)

Biocatalyst	Charge on support	pH	Ionic strength	$\dfrac{k_3'}{k_3}$ + app
Benzylpenicillin semisynthesis E-A = Phenyl-acetyl-penicillin amidase				
Free enzyme		4.4	0.2	1100
		5.0	0.2	3300
		6.0	0.2	4900
		6.0	1.0	2300
		7.0	7.0	4500
Enzyme	0	6.0	0.2	4300
immobilized	0	6.0	1.0	1700
in LiChrospher	0	6.0	1.0	1700
	+	4.4	0.2	1600
	+	5.0	0.2	2200
	+	6.0	0.2	5400
	+	6.0	1.0	1600
	+	7.0	0.2	4800
Immobilized cells (particle diameter 3 mm)		6.0	0.2	2800
Ampicillin semisynthesis E-A = Phenyl-glycyl-penicillin-amidase				
Free enzyme		6.0	0.2	1000
		7.0	0.2	1500
		8.0	0.2	800
Enzyme	+	6.0	0.2	1100
immobilized in	+	7.0	0.2	1700
Eupergit C	+	8.0	0.2	1000

(i)   the ionic strength dependence of the pK_a for amino group on 6-APA, it increases with ionic strength;

(ii) partitioning of the deacetylating (negatively charged) form of 6-APA in the electric double layer around the charged enzyme support and in the bulk solution;

(iii) the ionic strength dependence of the rate constant for reactions between ions (6-APA and E—A) and H_2O and the ion E—A (14);

(iv) the ionic strength effect on possible nucleophile association with E—A;

(v)   the stabilization of the tetrahedral intermediate, in the deacylation process, by the electric field at the active site that is influenced by the ionic strength.

(ii) and (v) have no influence when a free enzyme or an enzyme on an uncharged support is used as a biocatalyst. From this follows that the observed ratio (k_3'/k_3) and the kinetically controlled maximum is a complex function of ionic strength. For immobilized biocatalysts we observed that several factors can be changed to increase the yields at the kinetically controlled maximum. For a charged nucleophile as 6-APA the partitioning and the effect on the rate constants (iii) should have the largest influence on the observed ionic strength dependence at pH > pK_a for the amino group (<5).

Charged matrices may also influence the yield at the kinetically controlled maximum when uncharged molecules are used as nucleophiles. This is shown in Fig. 5 where the concentration of Ac—Tyr—O—Me and Ac—Tyr—Tris formed the deacylation of Ac—Tyr—CT with MeOH and Tris is given as a function of time using CT immobilized in DEAE-Sepharose as a biocatalyst. The yield of Ac—Tyr—O—Me is hardly ionic strength dependent whereas the yield in Ac—Tyr—Tris shows a marked pH-dependence.

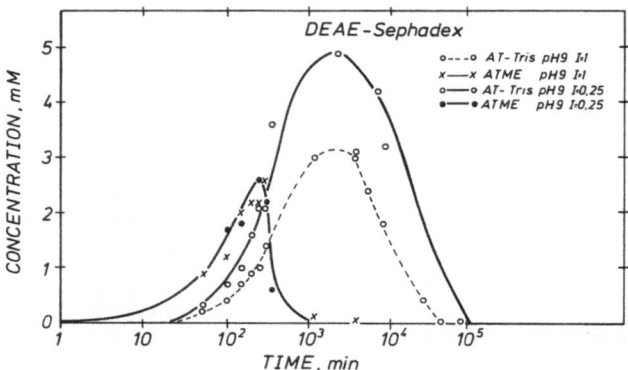

Figure 5. The concentration of Ac-Tyr-Tris and Ac-Tyr-O-Me as a function of time after addition of CT immobilized in DEAE-sephadex to 10 mM Ac-Tyr-O-Et in buffers of pH = 9.0 of different ionic strength (25 °C).
Ac-Tyr-Tris: O———O I = 1.0; O—O I = 0.25.
Ac-Tyr-O-Me: ×—× I = 1.0; ●—● I = 0.25

These results show that uncharged Tris is a good nucleophile for the nucleophilic deacylation of acyl-enzymes. It must therefore not be used in buffers used in enzyme catalyzed β-lactam or peptide semisynthesis (15).

In Fig. 5 is also shown that the final equilibrium concentration of the products formed by the deacylation of the acyl-enzyme is practically zero. The yield at the kinetically controlled maximum is >50% in terms of the acyl-concentration. Thus enzyme-catalyzed semisynthesis to the kinetically controlled maximum can also give considerably product yields even in the case where the yield of these products at the thermodynamic equilibrium is negligible. This can also be observed in the semisynthesis of β-lactams as shown for the case of ampicillin semisynthesis given in Fig. 6.

Effects of nucleophile structure

In the Table 3 the ratio of the deacylation rate constants and the yield at the kinetically controlled maximum — when it exists — are given for dipeptide semisynthesis with the endopeptidases CT and proteinase K immobilized in

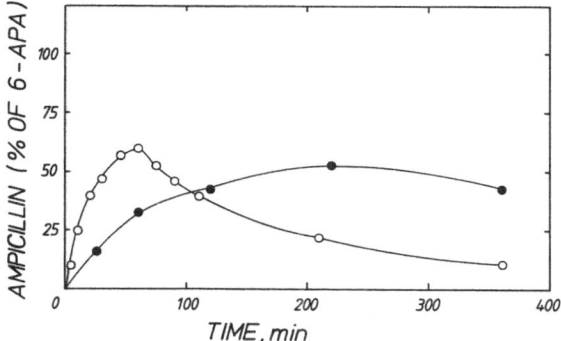

Figure 6. The concentration of ampicillin as a function of time after addition of penicillin amidase immobilized in Eupergit C to 93 mM 6-APA and 300 mM PG-O-Me in buffers of different pH at 25 °C. ○ pH = 7.0, ● pH = 6.0

uncharged matrices. A marked dependence on the charge on the nucleophile is observed. With protected (uncharged) carboxyl groups much larger values are observed than with charged carboxyl groups. This has also been observed qualitatively in peptide semisynthesis with other endo- (16) and exo-peptidases (17), but was not analyzed further in these studies. For exopeptidases the differences are not as large as observed for endo-peptidases. Sepharose may have some stationary negative charges that may result in a partitioning of the nucleophile that (18) perturbs the ration and the maximum yield as observed in the case of β-lactam semisynthesis. This nucleophile partitioning should, however, be of marginal importance for the order of magnitude differences given in Table 3. Most endopeptidases have a carboxyl group adjacent to or in the active

Table 3. Ratio of the deacylation rate constants for the deacylation of Ac-Tyr-CT or Ac-Tyr-Proteinase K with different nucleophiles (k_3') and water (k_3) in carbonate buffers (I = 0.2) of different pH at 25 °C. The charge on the nucleophile is given for the reactive form (one uncharged amino group)

Immobilized enzyme	Nucleophile	Charge	pH	$\dfrac{k_3'}{k_3}$
α-chymotrypsin	Val-NH$_2$	0	9	2200[a]
in	Val-NH$_2$	0	10	2000
Sepharose	Val-OH	—	10	1
	Ser-NH$_2$	0	9	450[a]
	Ser-NH$_2$	0	10	650
	Ser-OH	—	10	1
	Lys-O-Me	+	9	300
	Lys-O-Me	+	10	650
	Lys-OH	0	10	100[a]
Proteinase K	Lys-OH	0	10	45
in SAH-Cellulose				

[a] Kinetically controlled maximum has been observed

site, whereas exopeptidases do not have such groups near or in the active site (19). For endopeptidases this negatively charged group should repel negatively charged nucleophiles that may deacylate the acyl-enzyme. This can explain the order of magnitude difference observed in the ratio and maximum yield observed for the same nucleophile with and without protected carboxyl group.

This also explains why endopeptidases are poor exopeptidases. The unprotected carboxyl-endgroup is repelled by the active site. The observed dependence on pH for the same nucleophile is due to the pK_a of the amino group. The difference between the amino-acids used as nucleophiles probably reflects the sequence specificity of the enzymes (7, 8).

The data in Table 3 also show that the deacylation rate for the same acyl and different enzymes is enzyme specific. This may also be used as a variable to optimize enzyme catalyzed peptide synthesis. For this aim more comparative and quantitative studies on different proteolytic enzymes are required.

This effect due to nucleophile-active site interactions may be perturbed when the enzyme is immobilized in a charged matrix as outlined above. Thus by changing the properties of the matrix, solvent and the charge on the nucleophile using free and immobilized proteinases, fundamental data on the nature of the nucleophile-active site interactions can also be obtained in studies on enzyme catalyzed semisynthesis as outlined here.

Discussion

Kinetically controlled maxima in product concentrations that are larger than the final equilibrium content have been observed in the enzyme catalyzed semi-synthesis of β-lactams and peptides using free and immobilized biocatalysts. The yield at this maximum can be influenced by several factors when the biocatalyst is immobilized. These are given in Table 1. In this study the effect of mass transfer limitations, ionic strength dependence on intrinsic rate constants and nucleophile charge and structure have been demonstrated and analyzed. These results can be used to optimize the product yields at the kinetically con-trolled maximum under conditions that are used in technical systems. In these substrate concentrations up to 1 M are used. As the substrates are ions, ionic strength and substrate charge effects will influence the product yield markedly. Table 1, Eq. (1) and (2) can be used to design rational optimization procedures. For a positively charged matrix as biocatalyst support the substrate partitioning effect should as observed in the β-lactam synthesis increase the yield of product with a negatively charged nucleophile, as the ratio $(k_3'/k_3)_{app}$ will increase as compared to an uncharged matrix or the free enzyme. At increasing ionic strength this ratio is expected to decrease when the effective charge on the enzyme is positive. This has also been observed (Table 2) and may limit the yield at high substrate concentrations.

For peptide semisynthesis with endopeptidases, increasing ionic strength should increase the yield of semisynthesis with nucleophiles whose carboxyl groups are unprotected. This has not yet been observed. Immobilized biocatalysts should be favourable to use in enzyme catalyzed semisynthesis independent of whether the equilibrium or kinetically controlled maximum is used as an end point for the process. The equilibrium is not perturbed by the immobilization whereas the product yield at the kinetically controlled maximum, as shown here, may be influenced when the biocatalyst is immobilized. As compared to the equilibrium controlled synthesis the kinetically controlled synthesis requires more expensive (energy rich) substrates. The shorter reaction times and the higher yield may counterbalance this disadvantage. Thus the *kinetically controlled maximum* may eventually become an interesting end point for enzyme catalyzed synthetic processes. A more complete evaluation does, however, require further studies on the yield-controlling factors. In these leaving group and solvent effects, as already analyzed for the equilibrium controlled synthesis (20, 21) must be included.

Besides being of interest for a rational evaluation of the application of (immobilized) biocatalysts for synthetic purposes, such studies should also be of more fundamental interest. The kinetic data on the interaction between the acyl-enzyme and different nucleophiles can also be used for detailed analysis of structure-function relationship for the enzymes used here.

Acknowledgements

This work has been supported by the Deutsche Forschungsgemeinschaft, Fonds der Chemischen Industrie, and Dr. O. Röhm Gedächtnisstiftung.

References

1. Henriques V, Gjaldbaek IK (1912) Untersuchungen über die Plasteinbildung. II Mitteilung, Z physiol Chem 81:439–457
2. Bergmann M, Fruton JS (1938) Some synthetic and hydrolytic experiments with chymotrypsin, J Biol Chem 124:321–329
3. Kornberg A (1980) DNA-Replication. Freemann, San Francisco
4. Svedas VK, Margolin AL, Borisov IL, Berezin IV (1980) Kinetics of the enzymatic synthesis of benzylpenicillin. Enzyme Microb Technol 2:313–317
5. Konecny J (1981) Kinetics and mechanism of acyl transfer by penicillin amidase. Biotechnol Letters 3:507–512
6. Kasche V, Galansky B (1982) Ionic strength and pH-effect in the kinetically controlled synthesis of benzylpenicillin by nucleophilic deacylation of free and immobilized phenyl-acetyl-penicillin-amidase with 6-aminopenicillanic acid. Biochem Biophys Res Commun 104:1215–1222
7. Fruton JS (1982) Protein-catalyzed synthesis of peptide bonds. Adv Enzymol Relat Areas Mol Biol 53:239–306
8. Jakubke H-D, Kuhl P (1982) Proteasen als Biokatalysatoren für die Peptidsynthese. Die Pharmazie 37:89–106
9. Cole M (1969) Factors affecting the synthesis of ampicillin and hydroxypenicillins by the cell bound penicillinacylase of *Escherichia coli*. Biochem J 115:757–769
10. Goldberg MI, Fruton JS (1969) Beef liver esterase as a catalyst of acyl transfer to amino acid esters. Biochemistry 8:86–97
11. Kasche V, Buchholz K (1979) Mass transfer influence on effectiveness. In: Dechema Monographs, Vol 84, Characterization of immobilized biocatalysts. Verlag Chemie, Weinheim p 208
12. Kasche V (in press) Correlation of experimental and theoretical data for artificial and natural systems with immobilized biocatalysts. Enzyme Microb Technol
13. Bender ML, Clement GE, Gunter CR, Kézdy FJ (1964) The kinetics of α-chymotrypsin reactions in the presence of added nucleophiles. J Am Chem Soc, 86:3697–3703
14. Wiberg KB (1968) Physical organic chemistry. John Wiley & Sons, New York, p 390–395

15. Kasche V, Zöllner R (1982) Tris(hydroxymethyl)-methylamine is acylated when it reacts with acylchymotrypsin. Hoppe-Seyler's Z Physio Chem 365:531–534
16. Morihara K, Oka T (1977) α-Chymotrypsin as the Catalyst for Peptide Synthesis. Biochem J 163:531–542
17. Widmer F, Breddam K, Johansen JT (1981) Influence of the structure of amine components on carboxypeptidase Y catalyzed amide bond formation. Carlsberg Res Commun 46:97–106
18. Kasche V (1973) Effects of the microenvironment on the specific interaction between α-chymotrypsin and immobilized soybean trypsin inhibitor. Studia Biophys 35:45–56
19. Dugas H, Penney C (1981) Bioorganic Chemistry. Springer, Berlin, Heidelberg, New York, Chapt 4
20. Homandberg GA, Mattis JA, Laskowski Jr, M (1978) Synthesis of peptide bonds by proteinases. Addition of organic cosolvents shifts peptide bond equilibria toward synthesis. Biochemistry 17:5220–5227
21. Martinek K, Semenov AN, Berezin IV (1981) Enzymatic synthesis in biphasic aqueous-organic systems. Biochim Biophys Acta 658:76–89

Kinetics and Thermodynamics of Reactions Catalyzed by Penicillin Acylase — Type Enzymes

Jan Konecny

Summary

The thermodynamics and mechanism of reactions catalyzed by penicillin acylase-type enzymes, as seen at present, are outlined. Most if not all that is known about these proteins is consistent with the view that they are hydrolases of both esters and amides of α-aminoacids like phenyglycine and, in some cases, also of structurally related compounds like phenylacetic acid. They apparently react, like chymotrypsin, via an acyl-enzyme intermediate which then transfers the acyl group to water, amines or alcohols as receptors.

Three short communications describing the hydrolysis of penicillin G to 6-amino-penicillanic acid (6-APA) by *E. coli* [1, 2] and by enzymes from other organisms [3] appeared nearly simultaneously over thirty years ago. They contain significant observations about the reversibility of the reactions at low pH [1–3], the enzymatic acylation of 6-APA by N-phenylacetamidoglycine as an "activated form" of phenyl acetic acid [2], and also the observation that the specificity of the extracellular acylase from *Alcaligenes faecalis* for the hydrolysis of different penicillins approximates that for the reverse reaction [3]*. Since then thousands of organisms have been screened as potential producers of enzymes for such applications and industrial processes for such enzymatic reactions as the hydrolysis of penicillins G and V and the acylation of deacetoxycephalosporanic acid to cephalexin have been realized. Although literature on the subject is extensive [4–8], investigations relating to the kinetics, mechanism and thermodynamics of the reactions are few and mainly of recent origin.

This paper outlines the evidence that reactions catalyzed by the well characterized enzymes from *E. coli*, *B. megaterium*, *Acetobacter turbidans* and *Xanthomonas* citri are mechanistically related to those of proteases like chymotrypsin, an analogy first noted by Kato [9] in connection with his work with the *X. citri* protein. Mathematical treatments of the equilibria [10] and kinetics [11, 12] published recently and analyses of the available kinetic data [9, 13, 14] have been omitted. Instead, some reflections on the development of these ideas

---

* This paper also contains the observation that enzymes with a preference for penicillin V are widely distributed among actinomycetes and fungi, while activities for penicillin G had only been detected among bacteria. A penicillin V hydrolyzing bacterium has been found by researches at NOVO only recently

whose germ is contained in the three earliest publications, and also some references illustrating the scope and nature of the Japanese effort in this area have been included.

Equilibria

For the reactions of the electrically neutral species

$$C_6H_5CH_2CO_2H \qquad\qquad + RNH_2 \rightleftharpoons C_6H_5CH_2CONHR + H_2O \qquad\qquad (1)$$

$$C_6H_5CH_2CO_2C_2H_5 \qquad + RNH_2 \rightleftharpoons C_6H_5CH_2CONHR + C_2H_5OH \qquad (2)$$

$$C_6H_5CH_2CONHCH_2CO_2H + RNH_2 \rightleftharpoons C_6H_5CH_2CONHR + H_2NCH_2CO_2H \qquad (3)$$

(where RNH_2 = 6-APA) the estimated changes of the standard free energy ΔG_c^0 in water at 25 °C are -4.8, -6.4 and $+1.4$ Kcal respectively [10]. In calculating the actual changes ($\Delta G_c^{0\prime}$) which determine the equilibrium yields of the antibiotic it is necessary to take the dissociation of $C_6H_5CH_2CO_2H$ (pK = 4.2), of RNH_3 (pK = 4.6) and of the amino- and carboxygroups of glycine and N-phenylacetylglycine into account. Plots of $\Delta G_c^{0\prime}$ against pH then show that the values at pH 3 are about -4 Kcal for all three reactions and then diverge, the curves for reactions (1) and (3) being surprisingly about the same [10]. These two lines pass through a shallow minimum at pH 4.5 and then increase steeply, equilibrium for the acylation becoming increasingly unfavourable. For acylation by the ester, $\Delta G_c^{0\prime}$ approaches the optimal value of -6 Kcal at pH 5.5 and then remains constant in more alkaline solutions. The thermodynamic parameters for the synthesis of most other penicillin and cephalosporin antibiotics are similar.

Obviously, reaction (1) differs from reaction (2) in that way that the concentration of the more energetic acylating agent (undissociated acid) is low at neutral or alkaline pH, the equilibrium permitting appreciable acylation only in sufficiently acid solution. The proteins from *E. coli* and *B. megaterium*, selected for hydrolysis at alkaline pH, have low activity and low stability under these conditions.

Equilibrium (1) can be shifted to the right by adding miscible solvents and thus reducing the activity of water [15] or by using biphasic solvent systems [16]. The low solubility of certain cephalosporin esters and their precipitation from the reaction system has been exploited to shift reactions of type (3) in the desired direction [17].

Specificity of the Enzymes

Conclusions based on work with whole cells or crude enzyme preparations always involve the risk that more than one enzyme is responsible for the outcome of the experiments.* Investigations dealing with pure or purified proteins [12, 13, 20–28] are therefore particularly valuable in this context.

Obviously, the term "penicillin acylase" reflects the investigator's interest in the application rather than a characteristic of the protein which was designed by nature for a different purpose than the hydrolysis or synthesis of β-lactam antibiotics. Thus any relations between the proteins derive from a common selection principle or from some relevant feature which the different reactions used in screening have in common, e.g. the mechanism of the reactions.

It is perhaps not unfair to say that the misleading name "*penicillin* acylase" or "*penicillin* amidohydrolase" tended to obscure the significance of such observations as the induction of the *E. coli* and *B. megaterium* enzymes, and also their competitive inhibition [21, 22] by *phenylacetic acid*. A more appropriate name for all the well characterized proteins, first used by Takahashi et al. [17, 27, 28] in conjuction with the enzymes from *A. trubidans* and *X. citri*, is α-aminoacid esterase. All these enzymes are esterases as well as more or less efficient hydrolyses of amides of α-aminoacids like phenylglycine; in addition, the *E. coli* and *B. megaterium* proteins also recognize derivatives of phenylacetic acid as substrates. It is probably a result of the selection that the *X. citri* protein [27] is much more efficient as an esterase than an amidohydrolase** and has a pH-activity profile in the acid region; these properties are advantageous in the acylation of 6-APA-type compounds, the very purpose for which the protein was selected.

Reaction Mechanism

The characteristic feature of the proteins is that they catalyze the transfer of the acyl group (Ac) from esters (AcOMe), amides (AcNHR) or acids (AcOH) to water, certain amines (RNH_2) or alcohols (MeOH) as receptors***. Well characterized proteases like chymotrypsin display the same catalytic properties. It is therefore plausible to assume that penicillin-acylse type enzymes react by the same mechanism, namely the formation of the acyl-enzyme intermediate (EAc) from the enzyme (EH) and the acyl donors.

The reaction scheme in Fig. 1 accounts in a simple and compact manner for all the reactions of the proteins, for the kinetics of penicillin G hydrolysis by the *B. megaterium* and *E. coli* enzymes [21, 22], namely, competitive inhibition by phenyl acetic acid and non-competitive inhibition by 6-APA, and also the

---

* Like Okachi and coworkers [18, 19] other Japanese investigators encountered problems with β-lactamases and had to devote considerable effort to the isolation of lactamase-deficient mutants (personal communication)

** In the context of the investigated related reactions; in general, the ratio of amidolytic/estereolytic activities for a given acyl group depends also on the nature of NHR group. Thus the B. megaterium enzyme hydrolyzes efficiently the esters of cyanoacetic, phenyl acetic, α-amino-phenylacetic acids, penicillin G, phenyl acetamide, cephalexin (and [α-amino-phenylacetamido]-cephalosporin) but not a cyanacetamidocephalosporin [13, 14].

*** Acylation of RNH_2 and MeOH by AcOH has not been reported for most of the enzymes. However, no experimental proof of this path is necessary, because it is a thermodynamic consequence of the observed reverse reactions

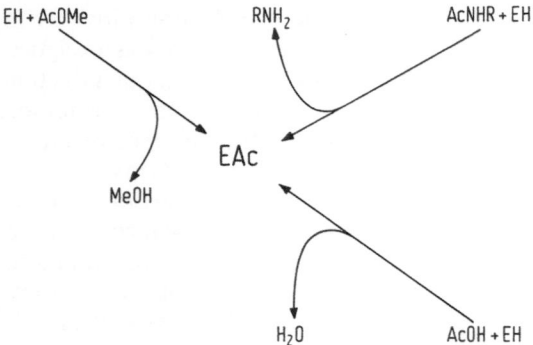

Figure 1. Schematic representation of the coupled reactions involving the ester, amide and acid as enzyme substrates. All reactions are reversible and all equilibria involving AcOH and RNH_2 are pH-dependent, the magnitude of the pH effects depending on the pK of the compounds

kinetics of acyl transfer from esters to amine receptors like 6-APA and 7-amino-deacetoxycephalosporanic acid [9] and deacetyl-7-aminocephalosporanic acid [13, 14] in dilute solutions. This difference in the modes of inhibition is a consequence of the two-step mechanism; the acyl donors compete for the enzyme while the acyl receptors compete for the acyl-enzyme intermediate [9, 11, 13, 14].

In the hydrolysis of penicillin G there is only one acyl receptor (H_2O) and the products are thermodynamically stable. In a reaction like the acylation of 6-APA by an ester, there are two competing acyl receptors, namely 6-APA and water, and the amide produced is thermodynamically unstable with respect to its hydrolysis products, phenyl acetic acid and 6-APA. Being also a substrate, it competes with the ester for the enzyme and undergoes hydrolysis in the process. Ionic strength effects the yield maximum [35]. Thus apart from any thermodynamic considerations, reactions (2) and (3) require an excess of the acyl donors for high yields on account of competing side reactions. Their consequence is that yields of the amide reach a maximum and then decline more [13, 29] or less [13] steeply, depending on the excess of the ester used and also on the relative efficiency of the employed enzyme as an esterase and amidohydrolyse.

A weakness of the proposed mechanism is that the initial rates of ester hydrolysis do not decrease to zero but to a finite limiting value [9, 13, 14] with increasing concentrations of RNH_2. This point deserves further investigation. Recent studies of the reaction mechanisms of proteases [30] and other considerations [14] may be relevant in this context.

Concluding Remarks

In the industrial production 6-APA chemical processes were superceded by enzymatic transformations.

The development of the enzymatic acylations goes back to the early seventies or late sixties, when the initiative was taken by Japanese industrial laboratories.

The incentives were particularly the complexity of the chemical routes to some of the novel antibiotics which require protection of various reactive groups prior to the coupling of the nucleus and the acyl side chain and, after coupling, removal of these protecting groups [31]. The specificity of enzymes as catalysts eliminates these costly complications.

Among other compounds cephalexin, ampicillin and amoxicillin [31–34] were synthesized enzymatically in the laboratory, and some of the antibiotics are now produced industrially in Japan with whole cells as catalysts. However, the interesting acyl side chains derive from costly, optically active acids. Thus the need to use a large excess of the esters to achieve 70–90% yield reduces, no doubt, the economic advantage of the processes over the chemical alternatives. Which of the routes is more economical is a question of the specific compound, the cost and properties of the available enzyme and also, of course, of the effort, ingenuity and experience investigated into the development of the chemical or enzymatic alternative.

References

1. Rolinson GN, Batchelor FR, Butterworth D, Cameron-Wood J, Cole M, Eustace GC, Hart MV, Richards M, Chain EB (1960) Nature 187: 236
2. Kaufmann W, Bauer K (1960) Naturwissenschaften 40: 474
3. Claridge CA, Gourevitch A, Lein J (1960) Nature 187: 237
4. Matsumoto K (1980) Hakko to Kogyo 38: 216
5. Okachi R (1979) Nippon Nogei Kagaku Kaishi 53: R169
6. Vandamme EJ (1977) Adv Appl Microbiol 21: 89
7. Vandamme EJ, Voets JP (1974) Adv Appl Microbiol 17: 311
8. Vandamme EJ (1981) J Chem Tech Biotechnol 31: 637
9. Kato K (1980) Agric Biol Chem 44: 1083
10. Svedas VK, Margolin AL, Berezin IV (1980) Enz Microbiol Technol 2: 138
11. Konecny J (1981) Biotechnol Letters 3: 107
12. Svedas VK, Margolin AL, Borisov EL, Berezin IV (1980) Enz Microbiol Technol 2: 313
13. Konecny J, Sieber M, Schneider A (1981) Biotechnol Letters 3: 507
14. Konecny J, Sieber M, Schneider A (in press) Biotechnol Bioeng
15. McDougal B, Dunnill P, Lilly MD (1982) Enz Microb Technol 4: 114
16. Semenov AM, Martinek K, Shvyadas VYu, Margolin AL, Berezin IV (1981) Dokl Akad Nauk SSSR 258: 1124
17. Takahashi T, Kato K, Yamazaki Y, Isono M (1977) J Antibiotics 30 Suppl, 230
18. Takasawa S, Okachi R, Kawamoto I, Yamamoto M, Nara T (1972) Agr Biol Chem 36: 1701
19. Okachi R, Kawamoto I, Yamamoto M, Takasawa S, Nara T (1973) Agr Biol Chem 37: 335
20. Bondareva NS, Levitov MM, Rabinovitsch MS (1969) Biokhimia (Engl transl) 34: 378
21. Chiang C, Bennett RE (1967) J Bacteriol 93: 302
22. Balasingham K, Warburton D, Dunnill P, Lilly MD (1972) Biochim Biophys Acta 276: 250
23. Kutzbach C, Rauenbusch E (1974) Z Physiol Chem 354: 45
24. Margolin AL, Svedas VK, Berezin IV (1980) Biochim Biophys Acta 616: 283
25. Okachi R, Nara T (1973) Agr Biol Chem 37: 2797
26. Shimizu M, Okachi R, Kimura K, Nara T (1975) Agr Biol Chem 39: 1655
27. Kato K, Kawahara K, Takahashi T, Kakinuma A (1980) Agric Biol Chem 44: 1075
28. Takahashi T, Yamazaki Y, Kato K (1974) Biochem J 137: 497

29. Marconi W, Bartoli F, Cecere F, Galli G, Morisi F (1975) Agric Biol Chem 39: 277
30. Antonov VK (1981) in "Proteinases and their Inhibitors" (Turk V, Vitale L eds), p 141, Pergamon Press, Oxford
31. Fujii T, Matsumoto K, Watanabe T (1976) Process Biochem 11[8]: 21
32. Okachi R, Kato F, Miyamura Y, Nara T (1973) Agr Biol Chem 37: 1953
33. Shimizu M, Masuike T, Fujita H, Kimura K, Okachi R, Nara T (1975) Agr Biol Chem 39: 1225
34. Kato K, Kawahara K, Takahashi T, Igarazi S (1980) Agr Biol Chem 44: 821
35. Kasche V, Galunsky B (1982) Biochem Biophys Res Comm 104: 1215

Biochemistry of Differentiation and Morphogenesis

33. Colloquium 25.–27. März 1982

Editor: L. Jaenicke
1982. 158 figures. XI, 301 pages
(Colloquium der Gesellschaft für Biologische
Chemie in Mosbach, Band 33)
ISBN 3-540-12010-6

Light Reaction Path of Photosynthesis

Editor: F. K. Fong
1982. 118 figures. XI, 342 pages
(Molecular Biology, Biochemistry and
Biophysics, Volume 35)
ISBN 3-540-11379-7

Membrane Proteins

A Laboratory Manual

Editors: A. Azzi, U. Brodbeck, P. Zahler
With contributions by numerous experts
1981. 76 figures. XII, 256 pages
ISBN 3-540-10749-5

Metabolic Interconversion of Enzymes 1980

International Titisee Conference
October 1–5, 1980

Organized by E. J. M. Helmreich, H. Holzer,
H. Schroeder, O. H. Wieland
Editor: H. Holzer
1981. 212 figures. XII, 397 pages
(Proceedings in Life Sciences)
ISBN 3-540-10979-X

Sialic Acids

Chemistry Metabolism, and Function

Editor: R. Schauer
1982. 66 figures. XIX, 344 pages
(Cell Biology Monographs, Volume 10)
ISBN 3-211-81707-7

R. K. Scopes
Protein Purification

Principles and Practice
1982. 145 figures. XIII, 282 pages
(Springer Advanced Texts in Chemistry)
ISBN 3-540-90726-2

Structural and Functional Aspects of Enzyme Catalysis

32. Colloquium 23.–25. April 1981
Editors: H. Eggerer, R. Huber
1981. 116 figures. IX, 216 pages
(Colloquium der Gesellschaft für Biologische
Chemie in Mosbach, Band 32)
ISBN 3-540-11110-7

Trace Element Metabolism in Man and Animals

Proceedings of the Fourth International
Symposium on Trace Element Metabolism in Man
and Animals (TEMA-4), held in Perth,
Western Australia, May 11–15, 1981
Editors: J. Gawthorne, J. McC. Howell, C. L. White
1982. XV, 715 pages
ISBN 3-540-11058-5

Y. A. Vinnikov
Evolution of Receptor Cells

Cytological, Membranous and Molecular Levels
1982. 36 figures. XII, 142 pages
(Molecular Biology, Biochemistry and Biophysics,
Volume 34)
ISBN 3-540-11083-6

Springer-Verlag
Berlin
Heidelberg
New York
Tokyo

Advances in Biochemical Engineering

Managing Editor: A. Fiechter

Volume 17:
Products from Various Feedstocks

1980. 57 figures. VII, 172 pages
ISBN 3-540-09955-7

Volume 18:
Plant Cell Cultures II

1980. 90 figures. VII, 193 pages
ISBN 3-540-09936-0

Volume 19:
Reactors and Reactions

1981. 142 figures, 37 tables. VII, 269 pages
ISBN 3-540-10464-X

Volume 20:
Bioenergy

1981. 58 figures, 56 tables. X, 209 pages
ISBN 3-540-11018-6

Volume 21:
Microbes and Engineering Aspects

1982. 28 figures, 31 tables. VIII, 230 pages
ISBN 3-540-11019-4

Volume 22:
Space and Terrestrial Biotechnology

1982. 136 figures, 29 tables. VII, 230 pages
ISBN 3-540-11464-5

Volume 23:
Microbial Reactions

1982. 63 figures, 47 tables. VII, 194 pages
ISBN 3-540-11698-2

Volume 24:
Reaction Engineering

With contributions by H. Binder, K. Buchholz,
W. D. Deckwer, H. Hustedt, K. H. Kroner,
M. R. Kula, G. Quicker, A. Schumpe, U. Wiesmann
1982. 99 figures, 44 tables. VII, 178 pages
ISBN 3-540-11699-0

Volume 25:
Chromatography

1982. 41 figures, 32 tables. VIII, 145 pages
ISBN 3-540-11829-2

Volume 26:
Downstream Processing

1983. 72 figures, 18 tables. VIII, 209 pages
ISBN 3-540-12096-3

Volume 27:
Pentoses and Lignin

1983. 46 figures, 46 tables. VII, 186 pages
ISBN 3-540-12182-X

Springer-Verlag
Berlin
Heidelberg
New York
Tokyo